Fundamentals of Biology

On the cover: The diagram of the human circulatory system is a fitting reminder that cardiovascular disease is still the number one cause of death worldwide.

Cover design by Danielle Anzelone

Building Foundations in Science Series

Fundamentals of Biology
Michael J. Anzelone, BS, MS

Editor-in-Chief, Randal Saad, BA, MS
Editor, Erin Conklin-Kimaoui, BA

Acknowledgements:
    I am grateful to the following colleagues who have reviewed the manuscript and those who have guided the direction of topics in this text in several focus groups. Any errors are the sole responsibility of the author.

    Paul C. Maccaro, MD, FACC, FACP, Director, Arrhythmia Services, Huntington Hospital, North Shore-LIJ Health System. Assistant Professor of Cardiology Hofstra University, North Shore-LIJ School of Medicine, Huntington, NY.

    Michelle Smedley, BA, MS, Mathematics Instructor, West Islip Public Schools, West Islip, NY.

    Gerald Surprise, BS, MS, RN, St. Francis Hospital, Manhasset, NY.

    Fred L. Wehran, BSCE, MSCE, Consulting Engineer, Mahwah, NJ.

Digital composition and line art by Michael J. Anzelone

Dedications:
    This book is gratefully dedicated to my wife Rita and daughters Danielle and Michelle for their contributions and help in its production.

Published by Angelus 1, Inc.
PO Box 64
Williston Park, NY   11596-0064

Additional copies may be obtained at a discount directly from:
CreateSpace eStore:
https://www.createspace.com/4489770

INTRODUCTION – PLEASE DO NOT SKIP THIS INTRODUCTION.

To the Student:

Fundamentals of Biology will provide students with an essential foundation in science. Successful science students acquired a sound foundation of basic facts and principles as they progressed through school. A sound foundation helps students make *connections between facts* and *helps assimilate* new knowledge easier than students with poor foundations. I have often heard students say "some students get good grades in science because they are smart." This statement is a myth. "Smart" students have a solid foundation.

My goal is to give students a good foundation in science. Fundamentals of Biology builds a good foundation by explaining complex concepts in *plain English* while it omits confusing **jargon**.[1] Another foundation building technique is the use of *repetition*. Repetition aids retention of information. Repetition may be said to be the mother of learning, but *don't rely solely on memorization* to learn science. Reliance on memory alone is a fatal mistake some students make. Science is not a collection of memorized facts; no more than a listing of telephone numbers is a novel. *Reason along with memory helps students gain a true understanding of science.* Another foundation building tool is the inclusion of a *brief history* of science. Science has a fascinating history that helps connect seemingly different ideas. History also clarifies the origins of scientific thought and makes it easier to remember facts.

The pursuit of scientific discovery takes place in different cultural and political environments through the ages. History and culture comprise the **matrix**[2] of science. The naming element 105, discovered in 1968, is a good example of how cultural and political conflict intertwine. The democratic U. S. and the former Communist Soviet Union became hostile towards each other after World War II. Because of this mutual hostility, it took twenty-nine years to agree on a name for element 105! Dubnium was finally chosen in 1997, after the Russian town where it was discovered.

Helpful tips: Pay attention to **headings** and **subheadings** *as you read.* A good study technique is to make a list of headings and **boldfaced words** while reading. Pay attention to *graphics*; they clarify information in the text. Read the *legends* that appear with **figures**[3] **and tables**. Legends explain tables and figures. A unique feature of this text is the inclusion of *over 500 explanatory annotations* for ease and efficiency of reading. Most students do not use dictionaries to look up unfamiliar terms. *Vocabulary is a crucial tool* for all students, especially those who have been out of school for a while and for whom English is a second language. *Fundamentals of Biology* bridges both gaps.

Legge Felicitas[4]

Michael J. Anzelone

---

1 Words and expressions that are specific to a field of study and generally not understood by the general public.

2 Something that encloses something else.

3 A figure may be a graph, photo or diagram. A table consists of columns and rows of data.

4 Read happily.

Table of Contents

Chapter 1 A Short History of Science
Why Present a History of Science?
This brief presentation of the **history**[1] of **science**[2] lists significant discoveries of humankind. Read it for appreciation of science and how scientific principles evolved. Knowledge of the history of science enables students learn science effectively.

Biology provides a foundation to better understand many specialized life sciences, **pure**[3] and **applied**.[4] An *anatomist* is interested in the *structure* of the human body. A *physiologist* is interested in how the body *functions*. Both are pure sciences. Nurses use knowledge of anatomy and physiology in a practical way to heal the sick, but the distinction between pure and applied science often blur. The work of the English nurse **Florence Nightingale**[5] illustrates how pure and applied sciences can overlap. Noted for her *practical nursing skills* and dedication to healing, she also pioneered *graphical representation of sanitary practices* in military and civilian life.

The Process of Science
Science is a process consisting of a series of steps that will yield new information about how nature works. Science studies the natural world by means of **verifiable**[6] **experiments**.[7] *Every experiment yields a new fact*.[8] Each new fact fits in a framework called a *theory* or a *law*.

Theories explain **phenomena**.[9] *Theories are general statements that try to explain how events happen.* Evolutionary theory is one example, "How did life evolve on Earth?" The Theory of Evolution is a combination of fossil evidence and explanations that connect fossils to each other.

Laws describe phenomena in a mathematical way. *Laws are narrow in scope compared to theories.* Isaac Newton's **Laws of Motion** express his experimental findings mathematically. Theories and laws are based on experimentally tested hypotheses. Both are true until proven false.

Do not confuse a theory with a hypothesis. A **hypothesis** is a *proposed solution to one question* that must be experimentally verified. A theory is a general explanation of how things happen in nature.

---

1   Western history is divided into three periods: classical antiquity (700 B.C. to about 500 A.D.), the Middle Ages (500 to 1400) and the modern era (1500 to the present).
2   From the Latin, *scientia* means wisdom.
3   Study for gaining knowledge only for the sake of knowing.
4   Concerned with the practical application of knowledge.
5   Florence Nightingale Founded of modern nursing.
6   Provable or able to be tested.
7   From the Latin, *experīmentum* meaning "test," "proof" or "trial."
8   From the Latin, *factum* meaning something that actually happened.
9   An event. Something that occurred.

The Roots of Science

Science exists because of humankind's curiosity about the natural world. The ancient **Greek**[10] philosophers *invented science*. The ancient Greeks asked questions about *causes* of phenomena. However, their concept of science was based on **philosophical**[11] methods. Greek philosophers **reasoned**[12] thusly: "If this is true, then what follows is true." The Greek philosopher Democritus (460-370 B.C.) hypothesized that matter is composed of small particles. He reasoned that if a cube of gold is cut in half and the resulting halves are repeatedly cut in half, then an infinitely small part of gold will result that cannot be cut in half any further. It is "indivisible." *Átomos* is the ancient Greek word for *indivisible*.

Democritus' atomic theory is very close to modern atomic theory. He used deductive reasoning to create an atomic theory over 2500 years ago! Deductive reasoning can lead to the discovery of truth about the natural world *if* the **premise**[13] is correct, but if the premise is wrong then what follows will be wrong. Democritus reached a conclusion about the atom by using a premise that was, luckily, correct.

The Roots of Modern Science

The ancient Greek method of discovery is different from the *modern scientific method* founded by **Galileo Galilei** (1564-1642). Galileo was an Italian **Renaissance** mathematician, astronomer, physicist and philosopher. He believed truth about nature can be obtained only through experiment, and every experiment must be verifiable by others.

Modern, experimentally verified atomic theory states an atom: (1) is a particle that cannot be broken down further by ordinary means and still be an atom, (2) has the same chemical and physical properties as a visible amount of the element and (3) has a positively charged nucleus surrounded by infinitely small negatively charged electrons.

Science and the Process of Learning

A scholarly journal published an article dealing with people's ability to learn under different conditions. Essentially, the results were that *the more people know, the better they will learn*. This is no surprise! Some students start off on the path to "knowing" at an early age, others, do not.

To know more can only be accomplished by spending more time reading, thinking and making connections between facts.

---

10   Ancient Greek, civilization existed c. 2000 B.C. to about 30 B.C. when all Greek territory came under Roman rule.
11   From the Greek, *philo-* meaning love and *sophos-* meaning wisdom.
12   To reason is to arrive at a logical conclusion to determine the cause of an effect through the use of intellect. Using evidence to make an inference.
13   A statement about something that is assumed to be true.

Scientists, other scholars and students study a great deal. *A good definition of "study" is reading and thinking about what is being read in order to make* **connections** *between known facts.* Study enables "new knowledge" to fit between the gaps of "old knowledge." This is the process of **assimilation** of knowledge. It is exactly like constructing a building. A strong, properly constructed foundation will support a tall **edifice**.[14] In science, a foundation is made of verified facts that fits into theories and laws. Theories and laws support the framework of science.

The Role of History in the Learning Process
History is the place to start learning about any subject. If a student has a poor science background, it is **crucial**[15] to strengthen it. As students learn about events that shaped science, they are learning more than just events. Students learn that *science began with the ancient Greeks.* As students read about ancient Greek science, they learn about *Thales' belief in cause and effect* circa (**c**.)[16] 650 B.C., *Democritus' atomic theory* c. 480 B.C. and *Aristotle's biological studies dealing with classification and embryology* c. 380 B.C. History teaches *who, what, where, when and how* of a discovery. Even a **cursory**[17] reading of history shows that what is known today is built on discoveries of those who came before us.

History anchors knowledge in one's mind. The history of science is placed in the first chapter because it should be read first. Read it for enjoyment. The historical presentation is presented in the first chapter also because it is easy to go back to for *reference* as you read the text.

Over the years, many students have told me they felt as if they were on the surface of an alien planet when they read their first science textbook. If the *serious* student reads this history as they move from chapter to chapter, science will not be so alien. A poor background is one of the stumbling blocks to good academic performance in science. It is never too late to acquire a background. *Study works if the student works.*

Knowledge of the history of science will be very helpful to the student that has been away from school for any length of time, whether it be one year or twenty years. Many historical ideas presented below are acquired by science students over the course of years of study. Knowledge of significant scientific achievements provides a solid foundation for further learning. Students should not have a narrow "splinter-like" view of science by just memorizing facts, nor should students race through their course work. ***Do not be in such a hurry that you miss your education!***

---

14   A building that is imposing or impressive.
15   Crucial means something that is essential in order to resolve a problem.
16   Lower case "**c**." an abbreviation for *circa*, Latin *for* - **about** or **approximately**.
17   Rapid, sketchy or hasty.

How to Present the History of Science

History meant obtaining knowledge of the past through *persuasive arguments*, *oral history*, *written records* and **inquiry**[18] to the ancient Greeks. Persuasive argument is the use of reason and evidence to prove a point. Persuasive argumentation is *not reliable* because false arguments can prevail. Oral histories may be biased and can lead to misinterpretation. Different people can tell different stories. Both are poor choices.

In the Roman world, history was *res gestae*, written accounts of *"things done."* I have tried to use the Roman concept of *res gestae*. If students browse tables 1.1 through 1.14, the list of past scientific achievements will show how modern science was created. A *short introduction* precedes each table of *scientists and their discoveries*.

Ancient peoples and people of the middle ages sought to explain natural events with supernatural causes. This kind of thinking began to fade during the **Renaissance**.[19] The Renaissance was a *rebirth of learning that brought forth a modern view of the world*. A call for reason is first heard in pre-Renaissance Italian society. Educated people began to search for *natural* causes for events. For the first time since the ancient Greeks, science is about the *"how,"* not about the *"why."*

Yet, **vestiges**[20] of **superstition**[21] still persist. For example, the disease commonly called the "flu" is short for *influenza*, an Italian word that means "influence." The "why" of influenza was believed to be caused by the influence of an unlucky star. Today, the "how" is known. Influenza is a **contagious**[22] disease caused by a RNA (ribonucleic acid) virus that is caught directly from infected persons or indirectly by touching **inanimate**[23] objects called *fomites*. Pencils, forks and doorknobs are fomites.

Science and Technology

*Science is the search for the **cause** of an **effect**. **Pure science** tries to understand how nature works just for the sake of knowing. **Applied science** or technology uses scientific discoveries for practical ends. Many of the most significant discoveries were technological, not scientific.

Ancient humans used *fire* to keep warm, to cook food and scare off predators. Ancient humans did not understand the process of *how* a fuel combines with oxygen to produce an *exothermic chemical reaction* that gives off heat and light. Ancients made bread, beer and wine. Modern humans know the process of *how* yeast on the skin of grapes uses sugar in grape juice to make alcohol and carbon dioxide. Ancients did not explore the "how" because they believed in supernatural causes for events.

---

18   Inquiry is questioning and reasoning.
19   *Renaissance* is a French term coined during the 1800s.
20   A trace of something that is left over from long ago.
21   A cause of an event resulting from a power beyond the natural world.
22   From the Italian *contagiare,* "to touch" or "to contaminate."
23   Inanimate means nonliving.

The historical presentation in this chapter will bring students through a series of major scientific discoveries. I call them *"touchstones."* Each touchstone will put modern science into a meaningful **context**.[24]

Significant scientific discoveries often **coincide**[25] with, result from or initiate notable transitions in human history. *The Italian Renaissance began in Italy the late 1200s,* setting the stage for *the scientific revolution.* The scientific revolution was an *irreversible leap forward* in human cultural evolution. Knowledge of the natural world, for the most part, passed from superstition to the accepted belief in *cause and effect.*

The history of science demonstrates the common inventiveness of humankind. Science is not built upon **"eureka"**[26] moments, although such moments do occur. Science developed from methodical investigations of nature and shared ideas among peoples. New discoveries were based upon previous discoveries. The most important developments in human history began with ancient man in Africa.

Science and Technology in Ancient Civilizations
Africa (200,000 B.C.-60,000 B.C.)

Fossil[27] evidence indicates the earliest humans, *Homo*[28] *sapiens,*[29] originated in Africa about **200,000**[30] years ago. The most important discoveries made by these ancient humans are the use of *tools, fire* and *numbers.* Recent evidence indicates humans left Africa around 60,000 B.C., migrating to the Middle East, India, Europe, and Asia. More recent findings suggest that "migrations" were not one way, but several migrations back and forth, probably following ample supplies of food.

China (6000 B.C.-1279 A.D.)

China is home to the oldest continuous culture in the world. Development of culture, literature and philosophy is dated from about 1045-256 B.C. The Chinese people made four great inventions: *papermaking, the compass, gunpowder* and *printing.* They also invented deep well drilling, cast iron and other metallurgical processes. The Chinese were first to publish a written Canon of Medicine, and they devised an amazingly modern system of diagnostic methods that relied on: *interrogation,* **inspection, palpation,**[31] **auscultation,**[32] and **olfaction.**[33]

---

24   A setting, circumstances or events that surround some other event.
25   To happen at the same time.
26   From the Greek, "I have found it!"
27   A preserved part or an imprint of a living thing in stone or some other medium.
28   From Latin, meaning man.
29   From Latin, meaning, wise or wisdom.
30   The universe is 13.8 billion years old. The Earth is about 4.54 billion years old. *Homo* appeared about 2 million years ago, and *Homo sapiens* about 200,000 years ago.
31   To feel.
32   To listen to sounds of the body.
33   The sense of smell.

Mesopotamia[34] (c. 5300 B.C.-600 A.D.)

Peoples of Mesopotamia invented writing (3500-3000 B.C.), made *tools*, had *libraries* and are responsible for some of the oldest significant *astronomical text* (1200 B.C.). Records reveal peoples of Mesopotamia had a concept of **physical examination**,[35] **diagnosis**,[36] **prognosis**[37] and **prescription**.[38] However, if illness could not be cured, **exorcism**[39] was employed. Babylonians (Southern Iraq) had a *true place-value system* where digits written in the left column represented larger values like our base-ten system.

Egypt (3150 B.C. - c. 500 B.C.)

The ancient Egyptians were the first to develop highly sophisticated bronze surgical instruments and the first medical text. The text contained case studies that included the condition, diagnosis and **treatment** for patients. The ancient Egyptians used medicines for killing pain, ridding the body of worms and other infections.

India (3000 B.C.-1190 A.D.)

Peoples of ancient India developed irrigation, **reservoirs**,[40] cultivation of cotton, sugarcane, *indoor bathrooms* connected to *sewers*, *horse-drawn plows*, *cataract surgery* and the *concept of zero*. They had a form of *inoculation against smallpox* (c. 700 B.C.). Indian mathematicians invented what is known today as *"Arabic numerals"* (0,1,2,3,4,5...). The Hindu-Arabic number system eventually made its way to Europe.

Greece (750 -500 B.C.-146 B.C.)

The ancient Greeks invented the concept of science. Some Greek thinkers abandoned superstition and **believed there were natural causes for events**. Their's was the beginning of a scientific way of thinking, but it was not experimental, nor was it widespread. This approach is different from consulting the gods or the **entrails**[41] of animals. The Greeks used philosophical methodology to arrive at truth. The Greek seeds of science were revived, grew, bloomed and were perfected in Italy between the later Middle Ages and the Italian Renaissance. See table 1.1.

---

34   Mesopotamia in Greek means land between rivers. Mesopotamia is between the Tigris and Euphrates rivers; present day Iraq, parts of Syria, Turkey and Iran.

35   Performed after a patient's history is taken. During the examination the patient is examined for signs of disease. Signs are *objective* such as temperature or a growth on the body. Signs are responsible for the symptoms a patient may feel. Symptoms are non-measurable *subjective* feelings noticed by a patient such as malaise, chills or nausea.

36   Determining the cause of disease from a patient's history, signs and symptoms.

37   Determining the possible outcome of a disease.

38   Written or oral directions that direct a plan of care for a patient.

39   A ritual that will drive an evil spirit from an ill individual.

40   From the French, meaning "storehouse."

41   Entrails are the contents of the abdominal cavity of an animal.

Table 1.1. Ancient Greek thinkers build the foundation for Western science.			
Philosopher	Achievement(s)	Philosopher	Achievement(s)
Thales (c. 640-546 B.C.)	Thales was the first of several ancient Greek thinkers to believe in *cause and effect*.	Plato (c. 427- c. 347 B.C.)	Founded *The Academy* in Athens. The Academy was the *first institution of higher learning in the Western world.*
Alcmaeon of Croton* (6th century B.C.) *A town in Southern Italy.	He was *the first to dissect humans for research.* He described internal structures, nerves of the eye (anatomy), brain as the center of intelligence, development of chick in an egg (embryology).	Hippocrates (460 B.C.- 375 B.C.)	Hippocrates is considered to be the *"Father of Medicine."* He was a major influence on medical treatment, even to this day.
Democritus (460-370 B.C.)	Democritus was the first to document a proposal of an *atomic theory.* Democritus' theory resembles modern atomic theory.	Aristotle (384-322 B.C.)	Aristotle is the *First Biologist.* He looked for causes of phenomena, first to make distinctions between different disciplines of biology, organic diversity, classification of invertebrates.
Socrates (469-399 B.C.)	Socrates is a *founder of Western philosophy.* He helped lay the foundation of Western philosophy and science along with Plato and Aristotle.	Archimedes of Syracuse* (287-212 B.C.) *A city in Sicily, Italy.	He would recognized the lever and screw pump today. He used experiment to test hypotheses. His model of physics such as *buoyancy* is equal to − d (density) = m (mass)/V (volume).

Ancient Italy during the Roman Period (c. 700 B.C.- c. 500 A.D.)

The Romans were Italic peoples living on a peninsula that juts into the Mediterranean Sea. Roman society accepted many different peoples into their Republic. Romans dominated, brought their culture to and administered an area that stretched from Germany in the North to the Sahara Desert in the South, from England in the West to present day Mesopotamia in the East for more than 1,000 years. See figure 1.1.

Rome, London (Londinium), Cologne (Colonia Agrippina), Paris (Lutetia Parisiorum), Voorburg (Forum Hadriani) in the Netherlands and Timgad in Algeria are among thousands of cities founded by the Romans.

Figure 1.1. The Roman world c. 117 AD.

To be a Roman citizen required service in the army, paying taxes and *speaking, reading and writing Latin*. The long period of "***Romanization***," made Latin the universal language of commerce, communication and science in the known world. Physicians and anatomists gave Latin names to body structures that are still used.

England was a Roman province for over 400 years, thereby infusing Latin into the English language. *Half of the words in the English language has Latin roots. For this reason, scientists in France would be able to recognize many words in a scientific article published in English.*

The Romans gave the world the concept of a republic. The Romans made great advances in engineering, technology, law and medicine. See table 1.2. Their invention of concrete enabled Roman engineers to build great public works such as the Pantheon, the largest unreinforced dome in the world to this day, Trajan's Market in Rome, the first shopping mall and the Colosseum, the first superdome.

Aqueducts, indoor plumbing and sewer systems greatly improved sanitation; resulting in improved health of its citizens. The Romans invented many different kinds of surgical instruments. *The established the first public medical service.* Roman physicians used anesthetics such as opium poppies (morphine) and henbane seeds (**scopolamine**).[42]

The central Roman administration gradually gave up its control over public administration, the army, and maintenance of public roads c. 500 AD. The classical period was slowly ending. ***The brilliant light of the classical period was dimming. Europe was descending into darkness.***

---

42   A preanesthetic used with narcotic painkillers such as morphine.

Table 1.2. Roman science, medicine and technology.			
Philosopher	Achievement(s)	Philosopher	Achievement(s)
Lucretius (c. 98-33 B.C.)	Wrote *De rerum natura (On the Nature of Things)*.	Claudius Ptolemy (c. AD 100 - 170)	Ptolemy devised a *"geocentric theory of the universe."*
Cicero (106 B.C.- 43 B.C.)	Cicero wrote in his treatise "On Duties." In this treatise he states *"all people should have justice because they are humans."*	Galen or Aelius Galenus (A.D. 131-200). Ancient physician	Galen *influenced the development of: pathology, anatomy, physiology, neurology and pharmacology.* His studies were based on monkeys and pigs.
Aulus Cornelius Celsus (c. 1 A.D.)	Celsus *listed four major signs of inflammation:* heat (*calor*), pain (*dolor*), redness (*rubor*) and swelling (*tumor*). Used antiseptic substances on wounds.	Anonymous	Concrete, the arch, aqueducts, water mill for grinding grain, hydraulic mining and cranes.
Pliny the Elder (A.D. 23-79)	His encyclopedic work *Naturalis Historia* became the model for all encyclopedias.	Anonymous	Public health system, clean water systems and sewage systems.
Dioscorides (A.D. 40 - 90)	*Founder of pharmacology.* Described medicinal properties of many plants.		

Medieval Europe

The Medieval period or **Dark Ages**[43] (500 A.D.-1300s A.D.) saw a general decline of health, education and living standards for Europeans after 500 A.D. The decline was due primarily because of the lack of Roman administration. Absence of supervision and organization resulted in the collapse of infrastructures such as aqueducts for clean water, sewer systems to carry away human waste and roads for transportation.

---

43   The term "Dark Ages" was coined by the Italian scholar Francesco Petrarca (Petrarch) in the early 1300s. He referred to going from the intellectual light of Rome to the intellectual darkness of the Middle Ages.

Islam Preserves Science of the Ancients

Islam has been in contact with the West for ages. Commerce with the Romans brought the Islamic world in contact with Greek and Roman culture, literature and science. During this period, Muslim scholars translated the ancient Latin texts *into* Arabic for their use and thus unwittingly preserved Western knowledge from the ravages of the Middle Ages in Europe. Although many ancient texts were lost during the Middle Ages, many were housed in European monasteries. Europeans began to hunger for ancient texts in the late Middle Ages through the 1500s. Scholars obtained ancient texts from monasteries, directly from the Islamic world or **repositories**[44] in areas dominated by Islam, such as Spain. European scholars were now translating ancient knowledge *from* Arabic back into Latin in European centers of learning. See table 1.3.

Table 1.3. Establishment of medical schools and universities in Europe.			
Salerno, Italy (c. 900)	*The first university* established in Europe. It was organized around a *medical school.* Located near the library at Monte Cassino that housed Greek, Roman and Muslim knowledge.	Montpelier, France (c. 1160)	Medical school and university established.
Bologna, Italy (1088)	*The second university* established in Europe. It was organized around a *law school*.	Oxford, England (c. 1000s)	University
Paris, France (1150)	University	Salamanca, Spain (1218)	University
Cambridge, England (1209)	University	Padua, Italy (1222)	University

Islam and Islamic Science

Contact with Islam also occurred during the **Crusades.**[45] The Crusades were religious wars that took place from 1095 to 1291 when Catholic Europeans battled Muslims for control of the Holy Land. The Crusades also brought Europeans into contact with the West's lost classical works, Muslim science, medicine and mathematics.

Muslim science and medicine flourished during their own "scientific revolution" (600-1400), helped in part by knowledge acquired previously from the West. See table 1.4.

44   A place to store things.
45   Nine Christian military expeditions took place from 1095 to 1291.

Table 1.4. Selected Arab contributions to science and medicine.

Philosopher	Achievement(s)	Philosopher	Achievement(s)
Al-Jahiz (781 - c. 868) Mesopotamia	Described *evolutionary ideas* and the concept of a *food chain*.	Al-Hazen (965-1038) born in Basra, what is now Iraq	*Modern theory of optics*. Light emanates from the object seen.
Shapur ibn Sahl (d. 869) Persian (modern Iran) Christian physician	Authored the first known book dealing with *antidotes to toxic substances*.	Ibn Sahl Persian (c. 940-1000)	Discovered the *law of refraction of lenses* and understandings of curved mirrors.
Abu Ali Sina (890-1037) Persian scholar (Latinized to Avicenna)	Wrote a *Canon of Medicine* that was used in European medical schools until 1700s.	Ibn-al-Nafis (1210-1288) A Syrian physician	The *septum of the heart did not have pores* to allow passage of blood.

Pre-Renaissance Italy (Late Middle Ages)

The European Renaissance began in Florence, Italy during the late 1200s. Italian artists, engineers and philosophers of the late 1200s and early 1300s began to express a modern view of the world termed. Their view was termed the *Rinascimento*, meaning "rebirth" in Italian.

The artistry of **Giotto di Bondone**[46] (1266-1337), writings of **Dante Alighieri**[47] (1265-1321), **Francesco Petrarca** (1304-1374) and the revival of human dissection provided the **impetus**[48] for a humanistic cultural change in Italy. Humanism is a concern for *human values* and an interest in the *classical world*. The Renaissance paved the way for the scientific revolution. See table 1.5.

Figure 1.2. An example of an illuminated (decorated) text produced in a *scriptorium*.

An interest in scholarship had its beginnings in the Italian **Benedictine abbey of Monte Cassino**,[49] established in 529. Ancient Greek, Latin and Arabic texts were copied in the *scriptorium*,[50] exquisitely illustrated and housed in the Abbey's library. See figure 1.2.

46   Giotto was the first of many great Italian painters of the Renaissance.
47   Wrote *The Divine Comedy*, a world masterpiece. Written in Italian, not Latin.
48   Encouragement or stimulus.
49   The Abbey is midway between Rome and the city of Salerno, south of Naples.
50   Means a place for writing in Latin.

Europe's first medical school was established at **Salerno**,[51] Italy in the 800s. the medical school was not far from the library housed at the Benedictine Abby of Monte Cassino. This is a very important point. Access to this information made the medical school at Salerno the most important source of medical knowledge for centuries.

Table 1.5. Selected Pre-Renaissance thinkers.

Philosopher	Achievement(s)	Philosopher	Achievement(s)
Gerard of Cremona (1114-1187) Italian scholar	*Translated Arabic texts of Greek scholars into Latin.*	Roger Bacon (c.1220-1294) English priest and philosopher.	*Stressed using experiment* in the pursuit of knowledge.
Leonardo Fibonacci (1170-1240) Italian mathematician	His writings *introduced Arabic numerals into European mathematics.* Number theory.	Thomas Acquinas (1225-1274) Italian scholar and Catholic priest	*Science and religion are not mutually exclusive.*
Albertus Magnus (1193-1280) German Catholic scholar	Using rediscovered Greek and Arab knowledge he laid the *foundation of medieval science.*	William of Ockham (1288-1348) English scholar and Roman Catholic priest	*"Ockham's Razor"* (the *principle of parsimony*). The best answer is the simplest.

What Was the Renaissance?

The *Rinascimento*, French pronunciation, *Renaissance*, represented a significant change in human values. It was truly a rebirth of social, political and economic life in Europe. Large numbers of people began to shift from a "God" - centered or *theocentric* way of thinking to a "man"- centered or *anthropomorphic* way of thinking. The Renaissance was *humanistic movement* that occurred in the late 1200s in Italy and it spread throughout Europe by the mid-1600s. *Without humanism, there would be no Renaissance; without the Renaissance there would not have been a scientific revolution.*

There are several proposed reasons as to why the Renaissance began in Italy: (1) Italian cities were surrounded by the ruins of ancient Rome, (2) the plague ravaged Italy first and may have caused people to shift away from spirituality and became more concerned with the present. (3) inherited wealth by the survivors of the plague made more money available for commerce, the arts and science.

---

51   *Schola Medica Salernitana* is still a functioning medical school.

Pre- and Early Renaissance and the Rise of Modern Science in Italy

Mondino de' Luzzi[52] (1275-1326), an Italian physician and professor of surgery, broke with tradition and personally dissected cadavers to instruct his students. He wrote the first textbook devoted solely to anatomy. Two-hundred years later Andreas Vasalius (1514-1564) followed Modino's approach to medical education.

Art and science were very much interrelated during this period. Many artists did dissections on cadavers in order to produce a more accurate representation of the human form. In fact, most artists probably understood anatomy better than most trained in medicine during this period. See table 1.6.

Man is part of nature, and nature began to be re-evaluated as seen in the works of **Marsilio Ficino**[53] (1433-1499) and **Pietro Pomponazzi**[54] (1462-1525). Pomponazzi wrote in his *On the Causes of Natural Effects or On Incantations,*[55] *"It is possible to justify any experience by natural causes and natural causes only."* This is a very modern thought. He criticized miracles, magic, demonic possession and **oracles**.[56]

By the 1500s, this new view of the world had spread beyond Italy. The modern period of Western history had begun. Jacob **Bronowski**[57] writes in his book *Magic, Science and Civilization* (1978), *"Something had happened in Italy that made a great inroad in established, authoritarian and traditional views of life."* Renaissance thinking displays a greater concern for the individual. Italian art, literature, architecture and science of the period reflects a "humanist" movement. Philosophers, artists, architects and engineers no longer automatically deferred to authority. Renaissance Italians looked at nature from a *human* **perspective**,[58] not a perspective they were told to have. During this period, the **emergence**[59] of some of the greatest artists/anatomists in the world would take place.

---

52   Dissection of human cadavers ceased after 200 A.D. because of legal and religious prohibitions. When the prohibitions ended, Mondino became the first to carry out a public dissection in Bologna in 1315 for medical students and interested spectators.

53   Marcilio Ficino was an Italian philosopher and theologian skilled in Latin and Greek. He authored several books and translated many others from the Greek into Latin.

54   "When I sing a song to the Sun it is not because I expect the Sun to change its course but [because] I expect to put *myself* into *a different cast of mind* in relation to the Sun." Italics mine. From Bronowski, *Magic, Science and Civilization*, 1978.

55   Published in 1520.

56   A person that could predict the future.

57   Polish-Jewish British mathematician, historian of science and biologist.

58   To see things in their true relationships.

59   Coming into view.

Table 1.6. Selected Pre-Renaissance and Early Renaissance artists/philosopher

Philosopher	Achievement(s)	Philosopher	Achievement(s)
Giotto (c. 1266-1337) Italian painter and architect	The first to **paint the natural world realistically**.	Johann Gutenberg (1388-1468) German goldsmith	Invented the **printing press**.
Emmanuel Chrysoloras (c. 1355-1415) Turkish scholar	A major force that **re-introduced Greek literature to Western Europe**.	Pico della Mirandola (1463-1494) Italian philosopher	Wrote **Oration on the Dignity of Man**.
Phillippus Aureolus Paracelsus Swiss physician	The body is a mass of chemical reactions and as thus chemical medicines should be used to cure illness	Luca Signorelli (c. 1445-1523) Italian painter	**Had an unusual mastery of the human form**. Obtained bodies for dissection from cemeteries?

The Age of Discovery and Exploration Propels Science Further

Marco Polo's explorations and the voyages of Christopher Columbus were two of the most important factors that produced a surge in exploration and discovery in the 1400s through the 1500s. In 1271, 17-year-old Marco Polo, an Italian merchant from Venice, traveled for 25 years throughout Asia. Upon his return to Italy, he authored the book in 1300 that came to be known as *The Adventures of Marco Polo*. It told of the wondrous sights and adventures he encountered in Asia. Marco Polo described new lands and new species of **plants**[60] and **animals**[61] unknown in Europe. His writings widened the horizons of Europeans, especially, Renaissance scientists. See table 1.7.

Christopher Columbus (It. Cristoforo Colombo), an Italian navigator who sailed for Spain, discovered the New World in 1492. Columbus inspired waves of explorations to the Americas. Like Marco Polo, he brought an awareness of new plants and animals to Europeans. Columbus completed four round trip voyages across the Atlantic Ocean in an effort to reach the East Indies. He believed if he sailed West he would find landfall in India, China or Japan. If true, he would be able to enter the lucrative spice trade. Columbus differed from other seafarers because he believed the Earth was round and *hypothesized* that sailing West would bring him to the East. *Columbus' experiment was to sail west.*

60   Multicellular, chlorophyll bearing eukaryotic organisms.
61   Multicellular, eukaryotic heterotrophs.

Table 1.7. Explorers that widened the scope of living things through their discovery of new lands.			
Explorer	Achievement(s)	Explorer	Achievement(s)
Marco Polo (c. 1254-1324) Italian merchant and explorer	*Traveled for 25 years through Asia and China.* Gave a detailed view of Asia.	Henry Hudson (1560/70-1611?) English explorer	*Explored the Hudson River* in New York.
Christopher Columbus (1451-1506) Italian mariner Sailed for Spain.	Sailed for Spain. *Discovered the "New World"* in 1492.	Vasco da Gama (c. 1460 or 1469-1524) Portuguese explorer	First to sail from Europe directly to India.
John Cabot (Giovanni Caboto) (1450-1499) Italian explorer that sailed for England	*Discovered North American continent* (Newfoundland) in 1497.	Ferdinand Magellan (1480-1521) Portuguese explorer	*First to circumnavigate the globe* (1519-1522).
Amerigo Vespucci (1454-1612) Italian mariner that sailed for Portugal. The name of America is derived from Vespucci's first name.	*Explored the East coast of South America.* Determined that North America and South America are separate continents.	Giovanni da Verazzano (1485-1528) Italian mariner that sailed for France	*Explored the Coast of South Carolina to Newfoundland and New York harbor* in 1524.

The Mid-Renaissance

The Mid-Renaissance is dated to about the 1400s. This date is arguable, but it is a good approximation. We see at this time the prominence of the Medici banking family. The Medici's made fortunes in banking and supported the arts and artists in Florence as well as the rest of Italy. *Filippo Brunelleschi* (1377-1446), *Leonardo da Vinci* (1452-1519), and *Michelangelo Buonarroti* (1475-1564) were patrons of this family. See table 1.8.

The greatest genius of all time, Leonardo made major contributions to painting, sculpting, architecture, music, mathematics, engineering, anatomy, geology, cartography, physiology, mathematics, **engineering**,[62] philosophy, botany, **physics**,[63] **hydraulics**[64] and **classical mechanics**.[65]

62 Engineering is the application of scientific knowledge to the construction of buildings and other structures.
63 Physics is the study of the relationship between matter and energy.
64 Hydraulics is the study of the behavior of a fluid (liquids and gases).
65 Studies of the motion of large bodies when forces are applied to them.

Michelangelo was an Italian sculptor, painter, architect, poet and engineer. Many of his works are considered to be the best examples of Western art to this day. Some of his works are the *Pietà*, *David*, the fresco of *The Last Supper* and *Genesis* on the ceiling of the Sistine Chapel in Rome. He studied anatomy using corpses, even though human dissection was forbidden at this time, except for physicians.

Table 1.8. Scientists, philosophers and artists of the Mid-to Late Renaissance.

Scientist/Artist	Achievement(s)	Scientist/Artist	Achievement(s)
Nicolaus Copernicus (1473-1543) Polish astronomer	*Heliocentric theory of the universe.*	Matteo Colombo (Renaldus Columbus) (1516-1559) Italian professor of anatomy and surgeon	*Discovered pulmonary circulation of the blood.* William Harvey built on his findings.
Girolamo Fracastoro (1478-1553) Italian physician/scholar	Epidemic diseases spread by tiny particles that transmit infection. *Coined the word "fomite."*	Johannes Kepler (1594-1600) German astronomer	Kepler devised *Laws of planetary motion.* Planets travel in elliptical orbits.
Andreas Vesalius (1514-1564) Dutch anatomist	**"Father of Modern Human Anatomy."** Wrote *"On the Fabric of the Human Body."* (1543)	Gabriele Falloppio (1523-1562) Italian physician	Described the tube that leads from the ovary to the uterus - *Fallopian tube.*

Founding of European Scientific Societies
Scientific societies *communicate scientific discoveries* and promote *understanding* of science. The first societies were formed in Italy, soon followed by other European nations. See table 1.9.

Table 1.9. Early European scientific societies.

Italy	Germany	France	England
1560s Naples. *Academia Secretorum Naturae*	1622 Rostock. *Societas Ereunetica*	1666 Paris. *Académie des Sciences.*	1660 London. The Royal Society of London.
1615 Rome. *Acedemia del Lincei*	1672 Altdorf bei Nürnberg. *Collegium Curiosum sive Experimentale*		1765 Birmingham. The Lunar Society.
1657 Florence. *Academia Cimento*	1700 Prussia. Berlin Academy.		

Late Renaissance Leads to the Age of Enlightenment

Galileo Galilei, the "Father of Modern Science," was the most prominent scientist of the Late Renaissance and, perhaps of all time. See table 1.10. He was the first to *set standards for measurement* in order to have experiments *reproduced accurately*. He also was *willing to change his views if the evidence demanded*. Like Galileo, modern scientists use standards of measurements and change their views if evidence dictates.

Galileo founded an entirely new science, *physics*, the study relationships between *matter* and *energy*. He formulated the laws of falling bodies and projectile motion. Galileo's experiments proved that bodies of *lesser* **mass**[66] fall to Earth at the same rate of acceleration as those with *greater mass*, if both are dropped from the same height at the same time.

Galileo also made improvements to the telescope and the military compass. He **debunked**[67] the idea that the moon was **translucent**.[68] He was the first to observe and describe sunspots, the **phases**[69] of **Venus**[70] and the rings of **Saturn**.[71] He discovered the **Milky Way**[72] to be a collection of stars in 1610, not just a faint cloud-like band in the sky. His work, *Two New Sciences* formed the basis for the science of kinetics.

Galileo's astronomical observations supported the Polish astronomer *Nicolaus Copernicus'* **heliocentric**[73] view of the universe. The Church leaders believed in Ptolemy's **geocentric**[74] theory of the universe and opposed heliocentrism.

In 1632, Galileo published his *Dialogue on the Two Great World Systems, The Ptolemaic and Copernican*. Gallileo believed the Church should not attach itself to scientific theories because he knew *theories change*. Because of his beliefs, Galileo's was tried by the Inquisition in 1633 for his views, found guilty and placed under house arrest. The Church's authority was not be challenged without penalty!

Contrary to popular belief, the Church was not against science. Many scientists were Catholic priests. A few were: William of Ockham, Lazzaro Spallanzani, Gabriele Falloppio, Gregor Mendel, Claude Bernard, Roger Bacon and **Georges Lemaitre**.[75]

---

66   Mass is the amount of matter in an object, and is constant from place to place.
67   To expose as false.
68   Light can pass through an object.
69   How light falls and varies in appearance on the surface of the planet or moon.
70   The second planet from the Sun. Venus has been known since ancient times.
71   Known since ancient times. Saturn's rings were first seen by Galileo in 1610.
72   The galaxy that the Earth is in. From the Greek, *galaxias* meaning milky.
73   A model of the universe with the Sun at its center.
74   Claudius Ptolemy (c. 100 - c. 170) proposed a model of the universe with the Earth at its center.
75   Proposed an expanding universe, later to be known as the "Big Bang Theory."

Table 1.10. Selected Scientists and artists of the Late Renaissance.

Scientist	Achievement(s)	Scientist	Achievement(s)
Bartolomeo Eustachi (c.1500-1574) Italian anatomist	Described the tube leading from the pharynx to the middle ear, *Eustacian tube*.	Robert Boyle (1627-1691) English scientist	*Gas Laws*. Mathematically described the behavior of gases under different conditions.
Francis Bacon (1561-1626) English philosopher	*Suggested rules for the study of nature*, but made no discovery.	Marcello Malpighi (1628-1694) Italian physician	Malpighi *discovered capillaries*, the link between arteries and veins.
Galileo Galilei (1564-1642) Italian mathematician, physicist, astronomer	**"Father of Modern Science."** **"Father of Modern Physics."**	Issac Newton (1643-1726) English physicist	*Formulated the three Laws of Motion*. Built on Galileo's work.
Johann Baptista van Helmont (1579-1644) Flemish chemist	Coined the term "gas." Studied combustion and fermentation.	Robert Hooke (1635-1703) English scientist	Called rectangular boxes in Cork. He called them them *"cells."*
Francesco Redi (1628-1697) Italian physician	Proved *flies do not come from rotting meat*.	Antonie van Leeuwenhoek (1632-1723) Dutch naturalist	**"Father of Microbiology."** He was the first to publish observations of microbes.

The Age of Enlightenment

The early 1700s marks the beginning of the "The Age of Enlightenment" also known as "The Age of Reason." Reason is considered to be the only authority during this period. Rational thinking was applied to all aspects of thought, including the study of nature. See table 1.11. The individuals that made up the cultural and scientific communities of Europe and the English colonies in North America supported adherence to reason as the primary authority for explanations of natural events. The intellectuals of this period advanced the welfare of society and science through reason alone. The American **"Declaration of Independence"**[76] and the French **"Declaration of the Rights of Man and of the Citizen"**[77] are both a result of this cultural and intellectual movement.

---

76   Approved on July 4, 1776 and signed on August 2, 1776.
77   Adopted on 23 August 1789 during the French Revolution (1789-1799).

Table 1.11. Scientists of the 1700s in Europe and some of their achievements.

Scientist	Achievement(s)	Scientist	Achievement(s)
Laura Bassi (1711-1778) Italian scientist	Taught Newtonian physics in Italy at the University of Bologna.	Luigi Galvani (1737-1798) Italian physician	Muscles of a *dead frog leg twitched when electricity was applied*.
Benjamin Franklin (1705-06?-1790) American scientist, politician and inventor	*First to recognize positive and negative electrical charges*, proved *lightning is electricity*.	Edward Jenner (1749-1823) English surgeon	**"Father of Vaccination."**
Carl von Linne (1707-1778) Swedish biologist and physician	He is considered the **"Father of Modern Taxonomy."**	Lavoisier (1743-1794) French chemist	**"Father of Modern Chemistry."** First to state the Law of Conservation of Mass
Lazzaro Spallanzani (1729-1799) Italian priest and biologist	Proved *standing broth does not make microbes*. First experimental proof against abiogenesis.	Jean-Baptiste Lamarck (1744-1829) French naturalist	The theory of *inheritance of acquired characteristics*. Later proven false.
Carl Scheele (1742-1786) Swedish chemist	*Discovered oxygen* about 1772. (His work was not published until 1777).	John Dalton (1766-1844) English chemist and physicist	Developed *modern atomic theory*.
Alessandro Volta (1745-1827) Italian physicist	*Invented the battery* (voltaic pile). He sent electric current over great distances.	Joseph Priestley (1733-1804) English scientist	Discovered *oxygen* in 1774. He is given credit because he published first.

Technology and Invention in the 1800s and 1900s

Rapid scientific and technological advances took place during the 1800s. Many of the inventions of the 1800s and 1900s are part of everyday life today. New means of travel and communication were born and developed rapidly during this period. Among the major scientific and technological advances were railways, steamships, automobiles and airplanes. Below are some inventions of the 1800s and 1900s that are recognizable today. See tables 1.12 and 1.13.

Table 1.12. A summary of inventions of the 1800s and early 1900s.	
Year	Inventor and Nationality.
1799	James Watt wins in court and is credited as the inventor of the steam engine. Scottish. First century Greek, Hero of Alexandria demonstrated a steam "engine." In 1712, Thomas Newcomen made first working, safe steam engine. It was used to pump water out of mines. English.
1880	Thomas Edison patented a system for electricity distribution. American.
1808	Pellegrino Turri invents the typewriter and carbon paper. Italian.
1813	Peter Durand invents the tin can. A tin can enabled foods to be kept over long periods of time and shipped over long distances. French.
1816	Rene Laennec invents the stethoscope. French.
1829	Louis Braille invents braille printing for the blind. French.
1834	Cyrus McCormick invents the first successful reaper for cutting and gathering crops. American.
1835	Samuel Morse builds the first American telegraph. American.
1853	Dr. Francis Rynd invents the hollow needle that became the hypodermic needle. Irish.
1855	Henry Bessemer invents process for mass producing steel. English.
1867	Alfred Nobel invents dynamite. Swedish.
1869	Thomas Edison invents the electric voice recorder. American.
1872	George Westinghouse invents air brakes. Greatly improved passenger and freight train safety. American.
1876	Disputed: Innocenzo Manzetti (It.), Antonio Meucci (It.), Alexander Graham Bell (Am.) and Thomas Edison (Am.) all claimed to invent the telephone. Bell (Am.) was awarded the first American patent in the U.S.
1878	Thomas Edison invents the phonograph and perfects electric light bulb. American.
1885	Gottlieb Daimler builds the world's first four-wheeled gasoline-powered motor vehicle. German.
1886	John Dunlop patents a pneumatic tire. American.
1886	George Westinghouse builds alternating current (A.C.) generator and distribution system. The same system is in use today. American.
1902	Willis Carrier invents the air conditioner. American.
1903	Wright brothers. First sustained gasoline-powered flight. American.
1903	Willem Einthoven invents electrocardiogram. Dutch.
1908	Henry Ford devises the assembly line for the production of gasoline-powered automobiles. American.

Table 1.13. Selected Scientists of the 1800s and early to mid-1900s in Europe, the United States, Canada and some of their major achievements.

Scientist	Achievement(s)	Scientist	Achievement(s)
Matthias Schleiden (1804-1881) German botanist	Concluded after many investigations that *plants are composed of cells.*	Martinus Beijerinck (1851-1931) Dutch microbiologist	The **"Father of Virology."** Discovered the tobacco mosaic virus.
Charles Darwin (1809-1882) English naturalist	*Published "On the Origin of Species by Means of Natural Selection."*	Henri Becquerel (1852-1908) French physicist and Madame Marie Curie (1867-1934) Polish physicist	Discoverer of *radioactivity* along with Madame curie (Manya Skłodowska) and Pierre Curie.
Theodor Schwann (1810-1882) German physiologist	*"All animals are made of cells."* Discovered neurolemmocytes (Schwann cells) and pepsin. He coined the term "metabolism."	Thomas Hunt Morgan (1866-1945) American geneticist	*Genes are located on chromosomes.*
Claude Bernard (1813-1878) French physiologist	Coined term *"milieu interieur,"* – the modern concept of *homeostasis*.	Karl Landsteiner (1868-1943) Austrian American biologist and physician	Discovered A, B, O *blood group system* and the *Rh factor. Discovered the polio virus.*
Ignaz Semmelweis (1818-1865) Hungarian physician	*Hand washing prevents puerperal fever* (childbed fever), a Staphylococcal disease.	Jules Bordet (1870-1961) Belgian physician	*Isolated Bordetella pertussis* the cause of whooping cough in 1906.
Rudolph Virchow (1821-1902) German physician	*Omnis cellula e cellula* ("every cell originates from another existing cell like it").	George Washington Carver (1865?-1943) American scientist, educator, humanitarian and former slave	Carver developed hundreds of products from peanuts, sweet potatoes, pecans and soybeans.

Table 1.13. Selected Scientists of the 1800s and early to mid-1900s in Europe, the United States, Canada and some of their major achievements.

Scientist	Achievement(s)	Scientist	Achievement(s)
Florence Nightingale (1820-1910) **"The Lady With the Lamp**." English nurse, writer and statistician	**Founder of modern nursing.** Founded first nursing school. Used pie chart to illustrate sources of patient mortality in military and civilian life.	Max Planck (1858-1947) German physicist	Founder of *quantum theory*.
Louis Pasteur (1822-1895) French chemist	*Dealt the final blow to the theory of spontaneous generation*, formulated the *Germ Theory of Disease*, developed *vaccination against rabies* and *anthrax*. He invented *pasteurization* (heating of beverages to prevent spoilage).	Otto Loewi (1873-1961) German pharmacologist	*Transmission of nerve impulses are chemical* in nature.
August Weismann (1834-1914) German biologist	Traits are passed on from parent to offspring only through egg and sperm.	Robert Goddard (1882-1945) American physicist and inventor	**"Father of Modern Rocketry."** Showed a liquid fuel rocket could work in a vacuum of space.
Gregor Mendel (1822-1884) Austrian monk and scientist	*Discovered laws of heredity in pea plants*. Published his findings in 1866.	Alexis Carrel (1873-1944) French physician	Developed new techniques for *suturing blood vessels.* He devised an *artificial heart* with Charles Lindbergh to keep organs alive outside the body.

Table 1.13. Selected Scientists of the 1800s and early to mid-1900s in Europe, the United States, Canada and some of their major achievements.

Scientist	Achievement(s)	Scientist	Achievement(s)
Joseph Lister (1827-1912) English surgeon	Developed *antiseptic surgery*.	Guglielmo Marconi (1874-1937) Italian inventor	*Wireless telegraph*.
August Kekule (1829-1896) German chemist	**"Father of Organic Chemistry."** Explained the role of carbon atoms in organic reactions.	Niels Bohr (1885-1962) Danish physicist	**Bohr Model of the Atom.** A small, positively charged nucleus surrounded by orbiting electrons.
Robert Koch (1843-1910) German physician	*Specific microbes cause specific diseases.*	Selman Waksman (1888-1973) Ukrainian American biochemist	*Coined the term "antibiotic."* His team discovered the antibiotics: *actinomycin, streptomycin and neomycin*.
Camillo Golgi (1843-1926) Italian physician	*Discovered a structure that processes and secretes proteins*, the Golgi body.	Frederick Banting (1891-1941) and Charles Best (1899-1978) Canadian physicians	Banting and Best Were the *first to extract insulin from pancreatic tissue* in 1922.
William Roentgen 1845-1923) German physicist	*Discovered X-rays* (1895).	Raymond Dart (1893-1988) Australian anatomist	Dart *unearthed first human ancestor in Africa in 1924*.
Ilya Mechnikov (1845-1916) Russian biologist	*Discovered phagocytosis*.	Alexander Fleming (1881-1955) Scottish biologist	Fleming *discovered penicillin* in 1928.
Hugo deVries (1848-1935) Dutch, Carl Correns (1864-1933), German and Erich von Tachermak (1871-1962) Austrian. All botanists	*Rediscovered laws of heredity* previously proposed and published by Gregor Mendel in 1866.	Paul Domagk (1895-1964) German pathologist	*Discovered first commercially available antibiotic* (Sulfonamidochrysoidine). Effective against streptococcus, a Gram-positive bacterium.

Table 1.13. Selected Scientists of the 1800s and early to mid-1900s in Europe, the United States, Canada and some of their major achievements.

Scientist	Achievement(s)	Scientist	Achievement(s)
Ivan Pavlov (1849-1936) Russian physiologist	*Classical conditioning.*	Rita Levi-Montalcini (1909 - 2012) Italian neurologist	*Discovered growth factors.*
Emil von Behring (1854-1917) German physiologist	*Discovered diphtheria antitoxin.*	Ernest Rutherford (1871-1937) British chemist and physicist	*Discovered and named the proton.* Concept of half-life.
Joseph Thomson (1856-1940) British physicist	*Discovered the electron.*	Hans Krebs (1900-1981) German physician	Identification of the *citric acid cycle*.
Albert Szent-Györgyi (1893-1986) Hungarian physiologist	*Actin, a short filament* reacts with *myosin* and *adenosine triphosphate* (ATP) to *cause muscle fiber contraction*.	Peyton Rous (1879 -1970) American physician	Discovered that some *viruses can transmit certain types of cancer*.
George Whipple (1878-1976) American physician	*Biermer's anemia* (Addison's anemia) *can be reversed* if affected dogs are fed liver. Applied successfully to humans.	Enrico Fermi (1901-1954) Italian-American theoretical physicist	*Developed first nuclear reactor*, made major contributions to quantum theory, nuclear and particle physics and the first atomic bomb.
Albert Einstein (1879-1955) German American theoretical physicist	*Theory of relativity*, predicted deflection of light by gravity and the relationship between matter and energy ($E=mc^2$).	Oswald Avery, (1877-1955) Maclyn McCarty (1911-2005) and Colin McLeod (1909-1972) American physicians and microbiologists	Published seminal paper in 1944 proving *deoxyribonucleic acid (DNA) is genetic material* in bacterial transformation.

Table 1.13. Selected Scientists of the 1800s and early to mid-1900s in Europe, the United States, Canada and some of their major achievements.

Scientist	Achievement(s)	Scientist	Achievement(s)
Luther Burbank (1849-1926) American plant breeder	*Developed many plant strains.* Bred blight resistant Russet Burbank Potato (Idaho potato).	Luis W. Alvarez (1911-1988) American physicist	Proposed theory that giant *impact on Earth's surface by an asteroid caused the extinction of the dinosaurs.*
Vincent du Vigneaud (1901-1978) American biochemist	Discovered the *structure of oxytocin* and vasopressin	Alfred Hershey (1908-1997) and Martha Chase, (1927-2003) American geneticists	*Confirmed DNA is genetic material* that was first proven by Avery, McCarty and McLeod in 1942.
Rennato Dulbecco (1914-2012) Italian physician and virologist	*Demonstrated that viral genes can be incorporated into a genome of the host.*	Alfred Hershey (1908-1997) Salvador Luria (1912-1991) Max Delbrück 1906-1981)	American geneticists, virologists. *Phage viruses can exchange bacterial genes* between bacterial cells
George W. Beadle (1903- 1989) and Edward L. Tatum (1901-1975) American geneticists	Demonstrated *how genes make enzymes.* ("one gene, one enzyme" hypothesis).	Arthur Kornberg (1918-2007) American biochemist	*Isolated DNA polymerase*, an enzyme that connects the monomers ATGC to make the polymer DNA
Barbara McClintock (1902-1992) American cytogeneticist	Mobile genetic elements: *"jumping genes."*	Frederick Sanger (1918-) English biochemist	Determined *amino acid sequence for insulin.*
Carl Anderson (1905-1991) American physicist	Discovered the *positron* (anti-electron). Proof of anti-matter.	Paul Boyer (1918-) John E. Walker (1941-) American and Canadian chemists respectively	*Discovered mechanism for the biosynthesis of ATP.*

Table 1.13. Selected Scientists of the 1800s and early to mid-1900s in Europe, the United States, Canada and some of their major achievements.

Scientist	Achievement(s)	Scientist	Achievement(s)
Paul Ehrlich (1854-1915) German physician	**"Father of Modern Chemotherapy."**	Howard Florey (1898-1968) Australian pharmacologist Ernst Chain (1906-1979) British chemist	*Developed penicillin into a practical medicine* (1939).
Nikola Tesla (1856-1943)	Work formed basis for alternating current and radar.	Linus Pauling. (1901-1994) American biochemist	Determined *the nature of the chemical bond*: how atoms link up to form molecules.
Robert Whittaker (1920-1980) American plant ecologist	Devised a *five-kingdom system of classification..*	**Charles Drew** (1904-1950) American medical researcher	*Blood banking.* Found that plasma will keep longer than whole blood.

The Second World War Spurs on Scientific Achievement

World War II began on September 1, 1939 in Europe with Germany's attack on Poland. Soon after this unprovoked attack, France and England came to Poland's aid. In 1939, Germany, **Italy**[78] and Japan had agreements that developed into military alliances. The United States entered World War II after Japan's surprise attack on the U.S. Naval base at Pearl Harbor, on December 7, 1941. The major powers of the world lined up. Fascist Germany, Italy and Japan as the axis powers and the United States, England, Russia and France as the allied powers. These events set the stage for the greatest **cataclysm**[79] the world had ever known.

The Second World War was the most tragic period of modern times. World War II caused over 60-80 million deaths and untold suffering. At the same time, the conflict produced a great number of advances in science and technology, accompanied by a **commensurate**[80] advance in moral and humanitarian concerns by *some* nations, but not others. Some scientific discoveries occurred prior to World War II and were developed further during the war. Others were invented in response to needs of war, and some were developed near the end of the war in 1945. *Penicillin, plastics, synthetic rubber, radar, television, electron microscopes, nuclear science, ballistic missiles, jet aircraft, sonar* and *computers* are just a few innovations of this period. See table 1.14.

Major scientific figures fled the fascist countries and came to the U.S. before the war started – Albert Einstein from Germany and Enrico Fermi from Italy. Both men were instrumental in the construction of the atomic bomb used to end the war in the Pacific.

World War II was a major turning point for how the world's people viewed politics, culture and technology. It was a period that marked the end of colonialism and the rise of nationalism in many European and Asian overseas possessions.

In the scientific community, teams of people routinely work on a single problem. Although Enrico Fermi and J. Robert Oppenheimer are considered "Fathers of the Atomic Bomb," thousands of people made contributions to the project. The team of Avery, McCarty and McLeod led to the **seminal**[81] 1944 publication of the discovery of DNA as the "stuff of genes." The electron cloud model of the atom is another interesting example. So many scientists contributed to this concept that a "team of scientists" is often credited with its formulation. See table 1.14.

78 Italy signed an armistice with the Allies in 1943 and joined the Allies in the fight against Nazi Germany.

79 A disaster.

80 Having an equal amount.

81 An event that has the potential to give rise to future discovery.

Table 1.14. Major scientific figures and their achievements of the mid-1900s to the present in Europe, the United States and Canada.

Scientist	Major Achievement(s)	Scientist	Major Achievement(s)
Dmitri Mendeleev (1894-1907) Russian chemist	*Devised a Periodic Table of the elements.*	Jonas Salk (1914-1995) American medical researcher	Developed *first polio vaccine.*
Harold C. Urey (1893-1981) and Stanley L. Miller (1930-2007) American scientists	Proved *simple organic compounds can form under a primitive Earth's conditions* in vitro (1953).	Eric Kandel (1929-) American neuro-scientist	*Memory is stored in neurons.*
Robert Edwards (1925- 2013) British biologist	*Developed in vitro fertilization.*	Paul Berg (1926-) American bio-chemist	*Pioneer in genetic engineering.*
Francois Jacob (1920- 2013) along with Jacques Monod (1910-1976) French biologists	Demonstrated *mechanism of gene repression.*	Carl Woese (1928- 2012) American micro-biologist	*Proved archaea are genetically different from bacteria.* Devised a three domain system of classification.
Joshua Leder-berg (1925-2008) American molecu-lar biologist	Discovered that *bacteria can mate and exchange genes.*	Gerald Edelman (1929- 2014) American biolo-gist	*Discovered antibody structure.*
Rachel Carson (1907-1964) American marine biologist	*Silent Spring*, published in 1962, brought an *awareness of the dangers of pesticides*.	Luc Montagnier (1937-) French virologist	*Discovered the Human immunodeficiency virus* (HIV).
Alan Turing (1912-1954)	**"Father of the Modern Computer"**	Edward Roberts (1941-2010) American engi-neer	*Introduced personal computer as a kit* in 1975.

Table 1. 14. Major figures and their achievements of the mid-1900s to the present in Europe, the United States and Canada.

Rosalind Franklin (1929-1958) British biophysicist	*Produced X-ray diffraction images that elucidated the structure of DNA* in the 1950s.	Buzz Aldrin (1930-) and Neil Armstrong (1930-2012) American astronauts	*Apollo astronauts that land Ed on the Moon.* (1969). Armstrong was first to walk on the moon.
James Watson (1928-) American Francis Crick English (1918-2004)	*Discovered the structure of DNA in 1953.*	Howard Temin (1934-1994), David Baltimore(1938-), Rennato Delbecco (1914-2012) All Americans	Discovered reverse transcriptase in 1970.
Clair Patterson (1922-1995) American geologist	First to determine in 1953 that *the Earth is 4.6 billion years old.*	Centers for Disease Control (CDC) 1981	AIDS officially recognized to be caused by a virus.
Yuri Gagarin (1934-1968) Russian astronaut	*First person in space* (1961).	Stanley Prusiner (1942-) American neurologist	*Discovered prions* (infective proteins) in 1982.
George Palade (1908-2012) Romanian-American cell biologist	Great contributions to electron microscopy and a founder of modern cell biology.	Kary Mullis (1944-) American biochemist	*Invented polymerase chain reaction* (PCR) in 1983.
Murray Gell-Mann (1929-) American physicist.	In 1960, he found protons and neutrons are made up of smaller particles, *quarks.*	Carolyn Greider (1961-) American molecular biologist	*Discovered telomerase* in 1984. Telomeres protect the ends of chromosomes.
John Glenn (1921-) American astronaut	*First person to orbit the Earth* (1962).	Tim White American (1950-) Berhane Asfaw, Ethiopian and Gen Suwa, Japanese. All are paleontologists.	In 1993, *parts of a hominid skull, jaw and arm bones* found in Ethiopia. The bones were dated to 4.4 million years ago.

Chapter 2 What is Life?

To be alive is a simple concept to grasp, however, an exact definition has never been achieved. The ancient Greeks believed everything is made up of four materials: earth, air, fire and water. Aristotle (384 B.C. - 322 B.C.) believed every living thing is made up of matter (material) and a soul (nonmaterial). Another ancient belief is *vitalism*. Vitalism explains life in terms of a nonmaterial energy called a "vital force."

In 1828, Friedrich Wohler (1800-1882), a German chemist, **synthesized**[82] urea from two nonliving chemicals, silver cyanate and ammonium chloride. Urea is a nitrogen containing waste product of metabolism. His experiment began erode scientists' belief in the theory of a "vital force."

Life Characteristics

A *characteristic* is a defining quality. However, it is still a difficult to define life because "life" is a collection of processes that are carried out by nonliving substances. For example, energy rich molecule, **adenosine-5-triphosphate**[83] **(ATP)** is a nonliving chemical substance that is necessary to keep **organisms**[84] alive. ATP allows life processes to continue. It is easier to described life in terms of self-sustaining processes or *life characteristics*. Life characteristics are reproduction, locomotion, responsiveness, organization, circulation, immunity, respiration, metabolism, nutrition, digestion, excretion, growth, homeostasis and adaptation.

1. Reproduction

When an organism makes more of itself, it has *reproduced*. Reproduction the **sine qua non**[85] of life. It is a basic belief that if something reproduces, it is alive. The *two methods of reproduction* are *asexual* and *sexual*. Binary fission is a form of asexual reproduction.

One parent is needed for asexual reproduction. Binary fission is division of an organism's nucleus (mitosis) followed by division of the organism's cytoplasm (cytokinesis). When a nucleus divides, identical chromosomes are distributed to each cell. Each is a *clone* of the other. **Archaea**,[86] **bacteria**[87]**(eubacteria)**[88] and **protists**[89] reproduce this way.

---

82   To make. To put together.

83   An energy bearing molecule that is necessary to keep life activities going. Energy is defined as the ability to do work.

84   An entire living thing. One celled (unicellular) or many-celled (multicellular).

85   Without equal, cannot exist or "be" without this feature.

86   Archaea are ancient bacteria. The oldest life forms on Earth.

87   Bacteria are genetically different from archaea even though both look alike.

88   Eubacteria (*eu-*) meaning true. Sometimes used to distinguish more modern bacteria from ancient bacteria (archaea).

89   Many protists have animal-like, plant-like or fungus-like characteristics. Pro-

Archaea, bacteria and protists look identical because all have identical *genomes* (an organism's total number of genes). Other forms of asexual reproduction are **budding,**[90] **sporulation**[91] and **vegetative propagation.**[92]

Sexual reproduction requires *two* parents. Each parent contributes half of its genome to the new organism. Since offspring are a combination of two *different* genomes, the offspring inherit a *variety* of traits.

When gametes unite at fertilization, the result is a fertilized egg termed a *zygote*. The zygote undergoes repeated cell divisions to develop into a multicellular organism. **Plants** and **animals**[93] reproduce this way.

Most eukaryotes reproduce sexually, except for one group of *unicellular* eukaryotes, the **protists**. Protists reproduce mainly by binary fission. Examples of one celled, eukaryotic protists are *Amoeba proteus*, *Paramecium caudatum* and *Euglena gracilis*. See below.

2. Locomotion

Locomotion (L. *locus* - place and *motu* - motion) is an obvious sign of life. Locomotion is moving from one place to another. Movement can be locomotion in animals or **tropisms**[94] in plants. Protozoa are extremely small organisms that can only be seen with the aid of a microscope.

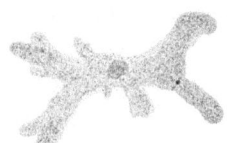 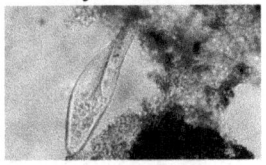 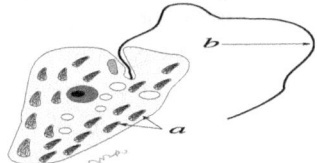

Figure 2.1a. *Amoeba proteus.*

Figure 2.1b. *Paramecium caudatum.*

Figure 2.1c. *Euglena gracilis.* a. Chloroplasts and b. flagellum.

Protozoa are animal-like because of how they move. Amoeba (figure 2.1 a) move by means of **pseudopods**[95] or "false feet." Amoeba "crawls" through its watery environment by *amoeboid motion* to find food. Paramecia (figure 2b) use *cilia*, tiny hair-like projections that propel it through water like oars of a boat. This is *ciliary motion*. Euglena (figure 2c) move by *flagellar motion*. A *flagellum* is a whip-like structure works like a propeller in reverse, pulling euglena through the water.

---

tists have two things in common: they are *eukaryotic* and most are *one celled*.

90   A smaller organism grows out of a larger cell or organism.

91   Functions in survival in bacteria and reproduction in fungi.

92   A cutting from one plant is grafted to another, or placed in soil to grow.

93   Plants and animal are multicellular organisms by definition.

94   A turning toward or away from a stimulus.

95   Cytoplasmic projections from the cell body of an amoeba that surround and engulf food particles and produce movement. From Greek, *pseudo-* false and *poda-* feet.

The sponge, an invertebrate, is the simplest animal. This simple animal lacks a backbone and organs of locomotion. Its **larval**[96] stage is motile. A mature sponge floats on water currents and eventually becomes **sessile**.[97] Vertebrates such as sharks, fish, amphibians, reptiles, birds and mammals use a skeleto-muscular system for locomotion.

Plant movements are called ***tropisms***. Tropisms are movements towards or away from stimuli. Examples of tropisms are ***phototropism***, ***geotropism***, ***hydrotropism*** and ***thigmotropism***. A plant's growth towards light is a positive phototropism. A positive geotropism occurs when roots grow towards the center of the Earth. A plant growing towards water is a positive hydrotropism. Thigmotropism is a response to touch. A vine climbs a wall because the plant growth hormone called ***auxin*** is released one side of the plant stem, causing cells to grow faster on that side. As a result, the plant bends and surrounds whatever it has touched.

3. Responsiveness

Responsiveness (irritability) is an obvious life process. Many animals show a startle reflex by jumping in response to a loud noise. One-celled organisms also respond to stimuli. *Paramecium caudatum* moves toward light and increased **oxygen tension**.[98] Some unicellular organisms move away from light and oxygen.

Observation of representative animals, from simple to complex, reveals increasingly more complex systems for detecting and responding to environmental stimuli. For example, a sponge does not have **neurons**,[99] but it does use chemicals to communicate between its cells. A network of neurons is first observed in jellyfish, but it is *not* a nervous system. A primitive nervous system is first seen in flatworms, such as *Planaria maculata*.

4. Organization

To *organize* is to arrange things in order. For example, ***Atoms*** combine to form molecules– ***molecules*** unite to make **organelles**[100] – organelles make up cells– ***cells*** make up tissues– ***tissues*** make up organs– ***organs***– make up organ systems and ***organ systems*** make up ***multicellular organisms*** like humans. This is a ***hierarchical*** arrangement, an ordered arrangement of one thing above or below another. See figure 2.2 and table 2.1.

---

96   Immature stage, does not resemble the parent and has to undergo change.
97   Permanently attached.
98   Oxygen tension is the same as the partial pressure of oxygen (PPO). The atmosphere is 79% nitrogen and about 20% oxygen. Oxygen tension refers to how much pressure oxygen would exert if it were by itself.
99   A nerve cell is the structural and functional unit of a true nervous system.
100   Organelles are subcellular structures that carry out specific cell functions.

Organization of Living Things Based on Cell Number: Unicellular vs. Multicellular.

The cell is the unit of structure and function of living things. Organisms may be ***unicellular*** (see figure 2.2a and table 2.1), *consisting of one cell* or an organism may be ***multicellular***, *consisting of many cells.* See figure 2.2b and table 2.1. Examples of unicellular organisms are *archaea* (ancient bacteria), *bacteria (eubacteria)* and most *protists*. ***Protists*** are classified as ***animal-like*** (amoeba), ***plant-like*** (algae) or ***fungus-like*** (slime molds).

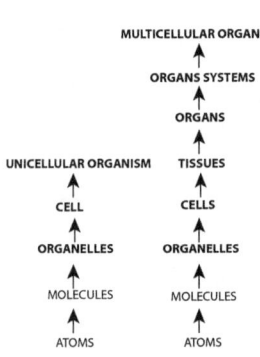

Animals and plants are ***defined*** as multicellular and eukaryotic organisms. Therefore, if an organism is ***eukaryotic*** and ***multicellular***, it must be, by definition, an animal or a plant.

Sponges, worms, fish, frogs, reptiles, birds and humans are ***multicellular animals***. ***Multicellular plants*** are mosses, flowers and trees. Fungi are molds and mushrooms, both are multicellular. Yeast is an exception, is a one celled fungus.

Figure 2.2. A simple hierarchy of life in ascending order.

Organization Based on the Structure of a Cell's Nucleus: Prokaryote vs. Eukaryote

Prokaryotic cells have a ***nucleoid***. A nucleoid is not a nucleus because its DNA is not surrounded by a membrane. Nucleoids float freely in the cytoplasm. See figure 2.3. Archaea and bacteria are prokaryotes.

Eukaryotes have a ***nucleus***. Their chromosomes are surrounded by a double-layered membrane. See figure 2.4.

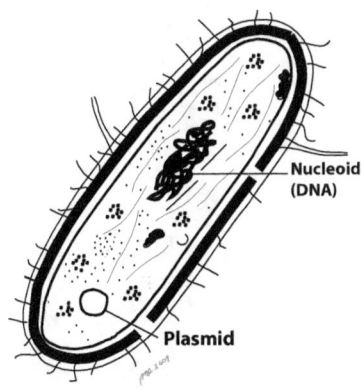

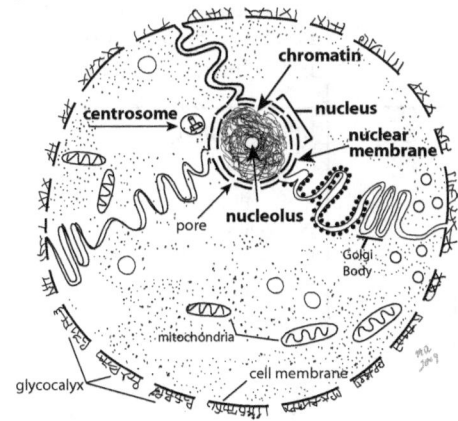

Figure 2.3. A typical prokaryotic (bacterial) cell. The nucleoid is freely-floating DNA in the cytoplasm.

Figure 2.4. A typical eukaryotic cell. Note the DNA is surrounded by double-layered nuclear membrane.

A Hierarchy of Living Things

A *hierarchy* is the placement of things above or below another. See table 2.1. *Atoms and molecules are nonliving,* but are included for completeness in the hierarchical presentation below.

Table 2.1. Hierarchy of life.		
Multicellular Organisms		Unicellular Eukaryotic Organisms
Structure	Component	Component
Atoms	C, H, O, N, S and P	C, H, O, N, S and P
Molecule	Carbohydrates, Fats, Proteins and Nucleic Acids	Carbohydrates, Fats, Proteins and Nucleic Acids
Organelle	Mitochondria Vacuoles Endoplasmic reticulum Golgi apparatus Nucleus Chloroplasts	Mitochondria Vacuoles Endoplasmic reticulum Golgi apparatus Nucleus Chloroplasts
Cells	There are about 200 types of specialized cells in humans. Epithelial, Adipose, Muscle, Nerve, etc.	Cell
Tissue	Epithelium Connective Muscular Nervous	
Organ	Skin, Heart, Muscles, Bones, Brain, Kidney, Lungs, Liver, Pancreas, Small intestine, Large intestine and Spleen	
Organ system	Integumentary Muscular Skeletal Nervous Endocrine Circulatory Digestive Excretory Reproductive Immune Respiratory	
Organism	Man	Amoeba, Paramecium, Euglena

5. Circulation

Circulation is the movement of materials within an organism to deliver oxygen and nutrients *to cells* and remove waste products and carbon dioxide *from cells*. Circulation is not readily observed with the unaided eye, but it can be observed indirectly in *Amoeba proteus* by observing granules floating in the amoeba's cytoplasm with a microscope. Circulation can also be seen indirectly in the **aquatic**[101] plant cell elodea by observing **chloroplasts**[102] move in a circular manner around the cell, also with a microscope. This movement is called *cyclosis*.

Mammals have a four chambered heart and blood vessels. The heart pushes blood through arteries, veins and capillaries to bring blood to every cell of the body. On average, cells are about 40 to 200 micrometer (μm) away from a capillary. This distance is critical to maintaining a diffusion gradient. A human hair is about 100 μm in diameter.

6. Immunity

Immunity is the ability to resist **infection**.[103] Vertebrates have *three lines of defense* for protection against **pathogens**.[104] The *first line* of defense consists of *nonspecific mechanical barriers*, the skin and mucus membranes. The *second line* of defense consists of *nonspecific cellular responses*. The *third line* of defense consists of *antibodies*, a specific humoral response.

The first line of defense, the skin and mucus membranes are *nonspecific* mechanical barriers that physically prevent foreign pathogens from entering an organism. Intact skin and mucus membranes keep everything out. Mucus membranes of the eye contain *lysozyme*, an enzyme that brakes down bacterial cell walls. Cilia in the bronchial tubes sweep back and forth to keep bacteria and debris out of the lungs.

If a pathogen gets past the first line of defense, then the nonspecific, second line of defense, white blood cells or **phagocytes**[105] goes into action. Phagocytes ingest *any* pathogens to prevent them from reproducing freely in the blood. The first and second lines of defense make up the *innate immune system*. If a pathogen gets past the cellular response, then the third line of defense goes into action, the *antibody-mediated system* also called the *specific*, humoral or adaptive immune system. The antibody-mediated system remembers pathogens and rapidly mobilize specific antibodies against specific pathogens. Only vertebrates have an antibody-mediated system in addition to an innate immune system.

101   Aqueous, pertaining to water or consisting of water.
102   A chloroplast is a plant cell organelle. Photosynthesis takes in chloroplasts.
103   The reproduction of any organism in a host organism that results in disease.
104   A living thing that has the ability to cause disease in another living thing.
105   Neutrophils, monocytes, macrophages, dendritic cells and mast cells.

7. Respiration

Getting Oxygen to Cells

Mammalian respiration occurs on three levels. The *first level* is **external respiration**, commonly called **breathing** or ventilation of the lungs. Breathing is the exchange of gases between air in the environment and the lungs. Breathing consists of **inspiration**[106] and **expiration**.[107] The *second level* of respiration is *internal respiration*. Here oxygen diffuses into hollow air sacs called **alveoli**. The oxygen then diffuses into the blood of the capillary beds that surrounding each alveolus. The *third level* of respiration is aerobic cellular respiration.

Aerobic cellular respiration takes place inside of cells. ***Oxygen*** and *food* molecules *enter* cells and release ***carbon dioxide***, ***water*** and biologically useful ***energy*** (ATP). *Mitochondrial* membranes of eukaryotes contain enzymes for aerobic cell respiration. Some prokaryotes have these enzymes too, but the enzymes are attached to their *cell membranes*.

Metabolic reactions extract energy stored in chemical bonds of organic *food* molecules ($C_6H_{12}O_6$) and transfers this energy to *ADP* molecules to form *ATP*. See figure 2.6. ***Toxic free electrons*** (e^-) and toxic ***hydrogen ions*** (H^+) result from this process. When oxygen enters a cell, it serves as a *final electron acceptor*. Oxygen combines with free electrons and hydrogen ions in mitochondria, to produce harmless water. Molecular oxygen neutralizes these toxic products. Animals, plants, protists, some fungi and some bacteria use aerobic respiration. See figure 2.5.

$$C_6H_{12}O_6 + 6O_2 \longrightarrow 6CO_2 + 6H_2O + 36\ ATP$$

(1) Glucose + (6) Oxygen $\xrightarrow[\text{RESPIRATION}]{\text{ENZYMES OF CELL}}$ (6) Carbon Dioxide + (6) Water + (36) ATP

Figure 2.5. A balanced chemical equation describing the overall chemical reactions of aerobic cellular respiration.

Anaerobic respiration uses oxygen combined with sulfur (SO_4^{2-}) and nitrogen combined with oxygen (NO_3^{1-}) as *final electron acceptors*. Many archaea and bacteria use anaerobic respiration because they live in low oxygen environments. Bacteria and archaea recycle nitrogen, carbon and sulfur in nature using anaerobic reactions.

8. Metabolism

Metabolism consists of all of the chemical reactions that occur in an organism. Metabolic reactions may be *anabolic* or *catabolic*. **Anabolic** reactions synthesize large molecules out of small molecules. **Catabolic** reactions make small molecules out of large molecules.

---

106   Taking air into the lungs.
107   Expelling air from the lungs.

9. Nutrition

Nutrition is the process of ingesting, digesting and using food in order to maintain life. Foods supply substances needed for *growth*, *main-

ADP + P~ ⟶ ATP

~~Denotes a high energy phosphate that will bond to ADP to create ATP

Figure 2.6. The conversion of ADP to ATP.

tenance and *repair* of an organism. Food molecules are digested and absorbed into an organism's cells. Energy stored in the chemical bonds of foods is extracted and transferred to molecules of *ADP* to form biologically usable energy, *ATP*. See figure 2.6. ATP enables *metabolic pathways* to work. Metabolic pathways provide energy for life processes.

Autotrophic Nutrition

Autotrophs make their food. Plants, algae and photosynthetic bacteria, the **cyanobacteria**,[108] are *photosynthetic autotrophs*. Autotrophs use *chlorophyll*, an organic **catalyst**,[109] to transfer *energy* of the sun to the chemical bonds of glucose molecules. Autotrophs combine *carbon dioxide* from the atmosphere with *water* from the soil to make *simple sugars* and release *oxygen* to the atmosphere. See figure 2.7.

$$6\,CO_2 \ + \ 6\,H_2O \ \longrightarrow \ C_6H_{12}O_6 + 6\,O_2$$

(6) Carbon Dioxide + (6) Water $\xrightarrow[\text{CHLOROPHYLL}]{\text{ENERGY OF SUNLIGHT}}$ (1) Glucose + (6) Oxygen

Figure 2.7. The overall chemical equation for photosynthesis.

Heterotrophic Nutrition

Heterotrophs get their carbon source (food) from the environment. Heterotrophs chemically extract energy from chemical bonds that hold carbon atoms together. There are several kinds of heterotrophic nutrition. Some bacteria and fungi are heterotrophs called *saprobes*. Saprobes secrete digestive enzymes into the environment where food is digested **extracellularly**.[110] Digested food molecules then diffuse into the cell.

Amoeba surround a food source with pseudopods, a process called *phagocytosis*. Phagocytosis is a form of heterotrophy called **holo-trophic**[111] *nutrition*. In this process, a food vacuole forms around an ingested particle of food. The food vacuole fuses with an enzyme filled vacuole. The food particle is *digested*, *absorbed* and **assimilated**[112] into the amoeba's structure. Humans and other multicellular animals ingest food through a mouth opening and digest food in a digestive tract.

108   Cyanobacteria are aquatic bacteria that probably produced the earth's present oxygen levels. They are formerly known as blue-green bacteria or blue-green algae.
109   Changes the rate of a chemical reaction, but is not used up in the reaction.
110   Outside of the cell.
111   A method of feeding in which large particles are taken in.
112   To make a part of something. In biology to make part of the organism.

10. Digestion

Digestion is a ***catabolic*** process that ***breaks down*** large water ***insoluble*** molecules into small, water ***soluble*** molecules. Insoluble proteins, starches, **lipids**[113] and nucleic acids are polymers made up of water soluble monomers such as, amino acids, simple sugars, fatty acids and nucleotides. These monomers that are soluble in water that can diffuse from the small intestine into the blood for transport to body cells.

Organisms consume two major classes of nutrients: ***macronutrients***, those *s*ubstances that are required in large amounts, and ***micronutrients***, those needed in small amounts. ***Proteins, carbohydrates, lipids*** and ***nucleic acids*** are macronutrients. ***Vitamins*** **and** ***minerals*** are micronutrients. ***Water*** is not a nutrient but it is necessary for life.

11. Excretion

Excretion is the elimination of toxic waste products from the body. In vertebrates, ***kidneys*** filter blood to remove urea, a toxic, nitrogen containing waste. Urea becomes mixed with water to form ***urine***. The ***large intestine or colon*** removes water and salts from feces (solid wastes) and returns them to the body. The ***rectum*** temporarily stores and then eliminates feces by the process of **defecation.**[114] The *lungs* rid the blood of gaseous carbon dioxide. The *skin* also functions in excretion by secreting a small amount of urea and dissolved salts in sweat.

12. Development

Mammalian development and growth are genetically controlled. *Development* of **viviparous**[115] vertebrates passes through distinct stages. ***Fertilization*** occurs in the **ampulla**[116] of the uterine tube first. For a period of eight weeks *after* fertilization, a single celled embryo develops into a multicellular ***embryo***. The formation of the multicellular embryo is the *embryonic stage of development* called ***embryogenesis***. Embryogenesis begins at fertilization. Development of the embryo consists of four distinct stages: **morula,**[117] *blastula, gastrula* and **organogenesis.**[118] The morula is a solid ball of cells that develops into a blastula. The blastula, a hollow ball of cells, folds to form a gastrula. The gastrula has three primary germ layers: ***ectoderm***, ***mesoderm*** and ***endoderm***. *After* the eighth week, but *before* birth, the embryo is termed a *fetus*. The length of time from fertilization to birth of the fetus is the ***prenatal period***.

---

113   Fats, waxes, fat-soluble vitamins (A, D, E and K), monoglycerides, diglycerides, triglycerides and phospholipids.
114   About half of eliminated solid waste consists of bacteria.
115   Organisms that bear their young alive. Viviparous animals do not lay eggs.
116   The second part of the uterine tube. It curves around the ovary.
117   Morula means mulberry in Latin.
118   Germ layers gives rise to specific organs in the process of organogenesis.

13. Homeostasis

Homeostasis is the process of keeping an organism's physiological mechanisms in *balance*. Maintenance of homeostatic balance is accomplished with the help of the brain, endocrine system, liver, kidneys and other organs. See Table 2.2. If homeostatic mechanisms fail to keep physiological processes within normal ranges, disease or death can result. Common *homeostatic imbalances* are diabetes and high cholesterol levels. Normal *Blood glucose* levels after **fasting** 8 to 12 hours should be about 70-110 mg/dL. Fasting blood glucose levels higher than 120 mg/dL may be pre-diabetes. Pre-diabetes can lead to diabetes. *Cholesterol* levels increase with age, but the levels should be about 200 mg/dL.

Table 2.2 Partial List of Organs that Function in Homeostasis in Mammals	
Organ	Function
Hypothalamus and Pituitary	Body temperature, hunger, thirst and circadian cycles
Lungs	Oxygen, carbon dioxide and pH balance
Skin	Thermoregulation and excretion of uric acid
Kidneys	Water, salt and pH regulation of the blood
Liver	Blood glucose level regulation and detoxification
Pancreas	Insulin decreases blood glucose levels. Glucagon increases blood glucose levels.

14. Organic Evolution

Organic evolution is a process whereby a **population** changes over generations due to selective processes. A population is all the same **species**[119] of organism living in the same geographic location. When an organism acquires a beneficial **trait**,[120] the trait enables the organism to be successful and reproduce faster than those without the trait. A beneficial trait favors survival and is **conserved**[121] in a population. A trait that does not favor an organism's survival is usually lost in the population. Organic "evolution" is not a new idea. Evolutionary thought is seen in the writings of the ancient Greeks. **Charles Darwin**,[122] an English naturalist, *had a new view of the natural history of life forms* and unlike his predecessors he presented *evidence* and gave an explanation for his theory of evolution.

119   Species is the smallest unit of biological classification and taxonomy. Singular for species is "species" abbreviated sp. Pleural is "species" and is abbreviated ssp.
120   An observable characteristic. Hair, skin or eye color are traits. Traits can be inherited behavior patterns called instincts. Nest building by birds is an instinct. Traits can also be biochemical such as the sickle cell trait or lactose intolerance.
121·   Retained or kept.
122   An English naturalist. He proposed a theory that explained that all life forms evolved slowly overtime from a common ancestor.

A Classic Example of Organic Evolution

The color of moths' wings in a changing environment illustrates the selection of a favorable and elimination of an unfavorable trait. Moth's wings in the population occurred in two varieties: light and dark. Buildings and trees were *not* covered with soot before the **industrial revolution**.[123] Moths with light-colored wings blended in with the environment. White winged moths survived and reproduced in greater numbers than the dark-winged moths. White wings were not as visible to birds. White wings are a *favorable trait*. Dark-winged moths on light colored objects were easy prey for birds; therefore, dark wings are an *unfavorable trait*.

Soot from burning coal began to cover trees and buildings. White-winged moths were now clearly visible to birds, and the dark-winged moths blended in with the dark environment. The dark-winged moths survived and reproduced in greater numbers than white-winged moths. Dark wings are now a *beneficial trait*. This is the process of *natural selection*. Nature selects for favorable traits. *Natural selection is the mechanism of evolution,* and it explains evolutionary theory.

Theories of the Origin of Life Forms

Humans have always speculated on what life is and how new living things come into being. Prior to the late 1800s, most people believed living things came from nonliving things. It was commonly believed that frogs came from mud, flies came from spoiled meat and microorganisms came from broth. These beliefs belong to the *discarded* theory of **spontaneous**[124] **generation** (*abiogenesis*).

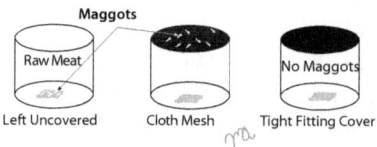

Figure 2.8. Diagram of Francesco Redi's experiment proving maggots do not originate from spoiled meat.

The theory of abiogenesis was challenged by a number of scientists, most notably Francesco Redi (1628-1697), Lazzaro Spallanzani (1729-1799) and Louis Pasteur (1822-1895).

Redi was the first to hypothesize that cells come from pre-existing cells in his **dictum**[125] "*Omne vivum ex ovo,*"[126] living things come from living things. Redi proved spoiled *meat does not produce flies*. He placed meat in three jars. Each jar had different conditions. The first jar was left open, the second was covered with gauze and the third was tightly sealed. See figure 2.8. The uncovered jars had **maggots**[127] on the meat. The jar

---

123   When coal burning steam engines replaced manual labor during the 1700s.
124   To happen suddenly, to appear at once.
125   A statement considered important and authoritative.
126   Although Redi was the first to state this idea, William Harvey (1578-1657) and Rudolph Virchow (1821-1902) are often credited for it.
127   The stages of a fly's development are: egg, larva or maggot, pupa, adult fly.

covered with gauze had fly eggs and maggots on the gauze, but not on the meat. The tightly sealed jar had no maggots at all, not on its cover nor on the meat. This proved that flies lay eggs on the meat, and eggs hatch into **larvae**,[128] which in turn gives rise to flies. Right? Well, remember the idea of a "vital life force?" Vitalists said the tight lid kept the vital force away from the meat, preventing the production of flies from the meat.

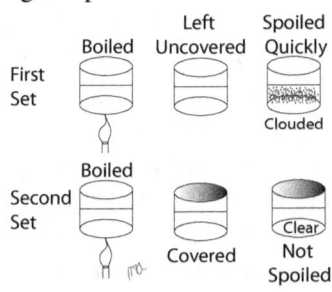

Spallanzani, building on Redi's work, *disproved the idea that microbes come from broth.* He placed freshly made boiled broth in two sets of containers. The first set of jars containing broth were left *uncovered*. The second set of jars containing broth was *immediately covered*. The freshly made broth left uncovered spoiled rapidly. It was teeming with microbes. No microbes were seen in the second set of jars containing broth that was immediately made and covered. Vitalists objected on the grounds that heat destroyed the vital force. See figure 2.9.

Figure 2.9. Diagram of Lazzaro Spallanzani's experiment.

There seemed to be no way around the vitalists' objections until Louis Pasteur's ingenious, simple and elegant experiment proved that microbes are in the air. Pasteur believed microbes cause food to spoil. He believed food does not cause microbes to appear because of some invisible vital force. Pasteur constructed a flask with a narrow swan neck shape. This shape allowed dust and microbes to settle at the lowest end of the neck because the bacteria were unable to make its way up the tube and into the broth.

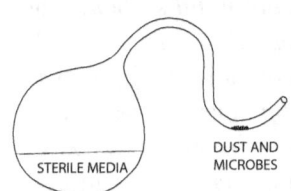

Figure 2.10. Pasteur's swan necked flask.

This arrangement allowed oxygen and any "special" force to come in contact with the cool broth, but microbes could not. The broth did not spoil. The vitalists could not object to this experiment because the "vital force," if it existed, had free access to the broth. This experiment essentially ended belief in the theory of spontaneous generation. Elements of this experiment were incorporated into Pasteur's germ theory of disease. See figure 2.10.

The germ theory of disease was not original with Louis Pasteur, but Pasteur was the first to *experimentally proved microbes were in the air* using his swan necked flask apparatus.

---

128   A worm-like, crawling, feeding life form that hatches from an insect egg.

Chapter 3 An Introduction to Taxonomy and Classification

Humans like to put things into groups based on similarities and differences. Biological *taxonomy* is the science of *identifying and naming* living things based on an organism's characteristics. Biological *classification*, on the other hand, is *placing living things in an ordered hierarchy* based on the taxonomist's descriptions.

Biological classification can trace its heritage to the ancient Greek philosopher, Aristotle. Aristotle classified animals in groups that are similar to modern groupings of amphibians, reptiles, birds and mammals over 2000 years ago. He classified animals based on observable characteristics such as **morphology**,[129] modes of reproduction, feeding habits and **habitat**.[130] Aristotle was truly the first biologist.

The first *modern system of biological classification* was devised by the Swedish naturalist **Carl von Linné** (1707-1778), the "Father of Modern Taxonomy," best known by his Latinized name, *Carolus Linnaeus*. *Linnaeus* classified living things into two kingdoms: *Animale* and *Vegetabile (Plantae)*. He published his **Systema Naturae**, in 1735, a pioneering text that dealt with classification of living things. *Linnaeus was the first to place humans with the primates*. His *"Systema"* became the foundation of modern biological classification.

The remarkable difference between Linnaeus' method of classification and previous systems is that he replaced common names with *two Latin names* for each living thing. It was the first **binomial system**. The first name is the **genus**; the last name is the **species**. Genus is a general category consisting of many different, but related organisms. Species is the most specific category. A species is a *unique* organism. Species is a group of organisms that can interbreed and produce fertile offspring. *Linnaeus' system is the first formal, uniform system of classification*. His binomial system became the universal system of classification.

DOMAIN
KINGDOM
PHYLUM
CLASS
ORDER
FAMILY
GENUS
SPECIES

Modern classification is based on the observable characteristics of organisms, but with advances in technology, cell structure, biochemical pathways, DNA and RNA information is used. Modern taxonomic ranks have expanded to eight hierarchical levels. Domains are now the largest category of living things, of which there are three: Archaea, Bacteria and Eukarya. Domains are subdivided into kingdoms. Each kingdom divides its members into smaller related groups – phyla. Phyla are divided into classes. See figure 3.1 and table 3.1

Figure 3.1.
Modern taxonomic ranks.

---

129   Appearance of a living thing; its shape and structure.
130   A habitat is an area where a species of animal, plant or other organism lives.

General Principles of Animal Classification

Since *animals, plants,* **fungi** [131] and *protists* have nuclei, this structure qualifies them to be placed in the **domain**[132] Eukarya. Classification

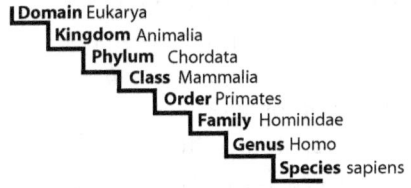

of modern humans begins with *domain eukarya, kingdom animalia.*

Humans are eukaryotic *animals*. Clearly, humans are not one celled protists, filamentous fungi or chlorophyll bearing plants.

Figure 3.2. Classification of the human combining the Domain and Linnaean system.

Humans are animals placed in the *phylum chordata* because they have a dorsal (back) nerve cord. Humans are placed in the *subphylum vertebrata* because they have a backbone made of vertebrae that protects a dorsal nerve cord. Fish, frogs, snakes and birds are vertebrates too. Clearly, humans are none of these. Humans are placed in the *class mammalia* because humans have **mammary glands,**[133] but bats and whales are also mammals. Clearly humans are not bats or whales. Humans are placed in the *order* **primate**[134] because primates walk upright, have backbones composed of vertebrae, forward-facing eyes, large brains, limbs with five digits and opposable thumbs. Apes, chimpanzees and monkeys are primates that have the same characteristics. Humans are not apes or monkeys. Humans have a highly-developed brain and high level thought processes of *abstract reasoning, language* and *complex problem solving*. Human hands are free to manipulate objects more than any other species and they make tools. Chimpanzee, monkeys and apes do not have these characteristics. Therefore, humans are placed in the *genus Homo (man)*, distinct from other primates. There are several species of *Homo*. All are **extinct**[135] except for the *species Homo sapiens*. *Homo habilis*, the toolmaker, lived 2.3 to 1.4 million years ago (ma). *Homo erectus*, (upright man), lived about 2 million years ago. *Homo sapiens neanderthalensis*, (**Neanderthal man**[136]), lived about 130,000 to 30,000 years ago. *Homo sapiens* appeared about 200,000 years ago. Humans now inhabit almost every corner of Earth. Having a first and last name allows a taxonomist to be specific as to which "man" is being discussed. See figure 3.2.

---

131   A microscopic yeast cell, a mold or a mushroom. All lack chlorophyll.
132   A large taxonomic group equal to a superkingdom.
133   Females of species produce milk to feed their young. Non-functional in males.
134   Animals that have hands, different kinds of abilities to move from place to place, exhibited complex behaviors and social and cultural interactions.
135   No living examples exist.
136   First discovered in the Neander Valley in Germany.

General Principles of Plant Classification

Plants are placed in the domain Eukarya along with *animals, fungi* and *protists* because all have nuclei. The plant kingdom includes all *multicellular*, *eukaryotic* living things that possess the green pigment *chlorophyll*. Plants are photosynthetic. Simply put, if it is *green* due to the presence of chlorophyll, *photosynthetic, multicellular and eukaryotic,* then it belongs in the domain Eukarya, kingdom Plantae.

There are two distinct groups of eukaryotic, multicellular green plants. Some are **non-vascular**[137] green plants such as low growing **mosses**[138] and some are **vascular**[139] green plants from ferns to **giant conifers**.[140] Vascular plants have vascular[141] tissue called *xylem* and *phloem*.

The oak tree is in the genus *Quercus*, but there are hundreds of living

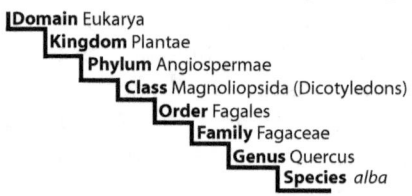

species of oak trees. *Quercus alba* is the white oak. The white oak grows in Eastern North America. See figure 3.3. *Quercus nigra*, the water oak, grows in the Southeastern region of North America.

Figure 3.3. Classification of the white oak combining the Domain and Linnaen systems.

Having a first and last name allows a botanist be specific as to which oak tree is being referred to. *Linnaean* taxonomy uses the *ladder of biological classification*. The ladder arrangement consists of groups within groups. A narrowing of criteria allows an organism to belong to each smaller group until the smallest group the species, is reached.

The Tree of Life

Using taxonomic descriptions, a "tree of life" can be constructed that demonstrates *patterns of relationships* among **taxa**.[142] The tree of life highlights the diversity of living things and reveals an evolutionary history among the groups of living things. See figure 3.4.

Systematics is the study of *relationships among taxa*. The main branches of the tree show common evolutionary connections. Data obtained from chemistry and DNA technologies allows for better understanding of these relationships. In addition, DNA information combined with the fossil record give indications how species evolved.

---

137   Non-vascular refers to lacking xylem (carries water) and phloem (carries food).
138   Mosses do not have rigid tissue that would support its structure.
139   Tube-like structures: xylem transports water and phloem transports food.
140   Conifers retain their needlelike leaves year round.
141   Tubes that conduct water (xylem) and tubes that conduct food (phloem).
142   Taxa is the plural of a taxon, a group of one kind of organism.

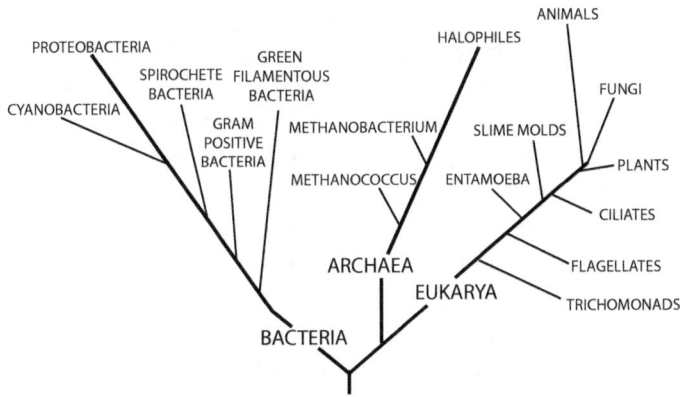

Figure 3.4. The phylogenetic tree of life depicting the three main branches or domains of living things: Archaea, Bacteria and Eukarya (Eukaryota).

Schemes of Classification

Linnaeus' system consisted of *two kingdoms* of *living things,* the kingdom *Animalia* and kingdom *Plantae.* The two kingdom system of Linnaeus was replaced by the *five kingdom system* of Robert Whittaker in 1969. Whittaker proposed the kingdoms we are most familiar with: *Animalia, Plantae, Fungi, Protista* and *Monera. The Monera contained the domains Archaea and Bacteria.* Imagine, Linnaeus' two kingdom system for the classification of living things lasted from the mid-1700s to 1969! The five kingdom system is a result of ecological, physiological and morphological information collected over a period of 300 years. The Carl Woese's *three domain system* replaced the five kingdom scheme. It is based on ribosomal RNA similarities among organisms. See table 3.1. The domain replaced the kingdom as the largest group of classification.

Table 3.1. Comparison of the five kingdom system of nomenclature of living things to the three domain system.

Five Kingdoms	Examples	Three Domains	Examples
1. Animals	Sponge to Man	1. Eukarya	Animals, Plants, Fungi and Protists
2. Plants	Mosses to Trees	2. Archaea	Archaeal species.
3. Fungi	Molds and Yeast	3. Bacteria	Bacterial species.
4. Protists	One celled organisms: Amoeba, Paramecium and Euglena, Slime molds, Water molds	Prions and viruses are additional entities that are not living, but that interact with living things.	
5. Monera	Bacteria and Archaea		

Modern Methods of Grouping Living Things

In 1977, Carl Woese devised a three *domain* scheme for classifying living things. Analysis of **ribosomal**[143] RNA (rRNA) from many organisms, revealed three distinct groups (domains): **archaea,**[144] **bacteria**[145] and **eukarya.**[146] *Each domain is based on the genetic make up of an organism's rRNA.* See figure 3.5.

Table 3.1 lists the domains archaea and bacteria. The rRNA of archaea is different from the rRNA of bacteria. The rRNA of archaea and bacteria is different from rRNA of eukarya (animals, plants, fungi and protists). Eukarya rRNA are related to each other, but not to archaea or bacteria.

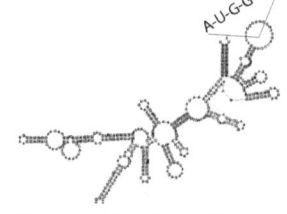

Figure 3.5. A small subunit of ribosomal RNA.

Eukaryotes have nuclei. Archaea and bacteria lack nuclei. Archaea and bacteria have areas called *nucleoids* because their chromosomes float freely in the cell's cytoplasm. *The kingdom Monera no longer exists.* The archaea and bacteria were once classified in the moneran kingdom.

1. Domain Archaea

Archaea have existed on Earth for almost three billion years. **Archaeal morphology**[147] is very much like bacterial morphology. Archaea live in varied environments. It was *once* thought that archaea only lived in extreme environments such as deep hot sulfur springs. However, archaea are found living in mild conditions such as marshes, oceans, the human gut and sewage. *No parasites are found among the archaea.*

2. Domain Bacteria

Gram staining allows bacteria to be placed into one of the two groups: *Gram-positive* or *Gram-negative*. *Gram staining* takes advantage of *differences* in bacterial cell wall structure. Cell wall structure allows a purple dye to be retained or not be retained under specific staining conditions. The Danish bacteriologist, Hans Gram (1853-1938), developed this technique in order to distinguish *Streptococcus pneumonia* (Gram-positive), from *Klebsiella pneumoniae* (Gram-negative).

Some bacteria are *acid fast*. Mycobacteria (see page 59 and 137) are acid fast. Mycobacteria retain a pink dye after washing with an acid/alcohol mixture. *Mycobacterium tuberculosis* is an example.

143   Ribosomes are the cell's protein factories.
144   Do not have nuclei. Archaea are prokaryotes.
145   Do not have nuclei. Bacteria are prokaryotes.
146   Have nuclei making them eukaryotes, (animals, plants, fungi and protists).
147   Some archaea have novel forms: flagellated cocci, lobed cocci, flat and brick-like. Bacteria have three main shapes: cocci, bacilli and curved.

3. Domain Eukarya

Eukaryotes have nuclei and *enzymes* for **transcription** *and* **translation**. Transcription is making a RNA copy (messenger RNA) of the DNA code. Translation is using the RNA copy as a blueprint to build proteins. Archaea have the same enzymes. Bacteria do not have these enzymes. Having the enzymes for translation and transcription make Archaea more closely related to the Eukarya than to the members of the domain bacteria. Archaea also differ from bacteria because archaea have **glycoprotein,**[148] **pseudopeptidoglycan**[149] or **polysaccharides**[150] in their cell walls. Bacteria have mainly **peptidoglycan**[151] in their cell walls.

Viruses and Prions: Nonliving Things that Cause Disease

Viruses[152] *and prions are nonliving* **entities**[153] *that affect living things.* Viruses are *not* placed in any domain of living things because viruses are non-living. The terms **"biological entities"**[154] for viruses and "infective proteins" for **prions**[155] seems most fitting. The phrase "biological entity" implies living and nonliving, an apt description since viruses were always considered a question mark. Reproduction is an essential characteristic of living things. If an organism reproduces on its own, it is alive; if it does not reproduce on its own, it is not a life form. Viruses reproduce only in a living cell. Viruses cannot reproduce on their own. Viruses can be crystallized. Crystals are **nonliving.**[156]

Viruses

Viruses are **obligate** [157] **intracellular**[158] parasites. Viruses cannot reproduce on their own. Viruses reproduce *only in living cells* by using the cell's biochemical and genetic machinery. Viruses are about 0.01th the size of a bacterial cell and can only be seen with an electron microscope. Viral classification is not covered here except in a general sense. A brief overview of structure, shape, method of replication and viral diseases are presented.

148   A protein that contains chains of sugar molecules.
149   Peptidoglycans are a combination of sugars and amino acids that make up the cell wall in bacteria. Archaea's cell walls are made up of pseudopeptidoglycan.
150   Long chains of polysaccharide molecules.
151   Peptidoglycan consists primarily of sugars and amino acids.
152   From the Latin, meaning poison. Virion is used interchangeably with virus.
153   Anything that exists by itself, an object. Entity does not imply living.
154   Coined by Salvador Luria (1912-1991). Italian-American microbiologist.
155   Prion is a term coined by Stanley Prusiner (1942-), an American neurologist and biochemist. The word prion is composed of protein and infectious.
156   Demonstrated by Wendell Stanley in 1935.
157   Indispensable. An organism that can survive only under certain conditions. Obligate aerobes must have oxygen. Obligate or strict anaerobes must avoid oxygen.
158   *Intra-* within, referring to the inside of a cell.

Viruses were first discovered by the Russian botanist Dmitri Iva-novsky (1864-1920) in 1892. Martinus Beijerinck (1851-1931), a Dutch microbiologist, discovered the tobacco mosaic virus. Both men are considered the founders of the science of *virology*.

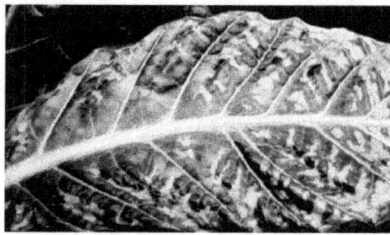

Figure 3.6. Photograph of a tobacco leaf infected with TMV. The leaf has mottled patterns of light and dark coloration.

Viruses are known to infect all living things. Wendell Stanley crystallized the *tobacco mosaic* **virus**[159] (TMV) in 1935. Crystals are nonliving substances. Stanley placed TMV crystals on a tobacco leaf. The crystals transformed into viruses within the tobacco leaf cells to cause the characteristic appearance of tobacco mosaic disease. See figure 3.6. The tobacco leaf became infected by the *transformed*, nonliving crystals.

Viral structure consists of a core of either DNA or RNA as their genetic material. See figure 3.7a. The core is surrounded by a protein coat called a *capsid*. See figure 3.7b. The capsid is made up of smaller units called *capsomeres*. Viruses **mutate**[160] and reproduce at fast rates.

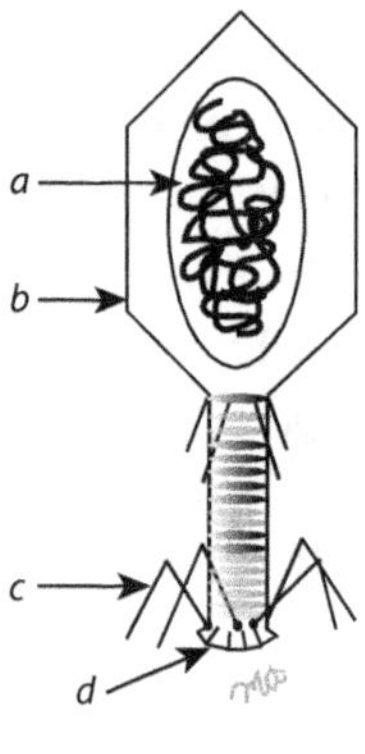

The capsid protects the viral *genome* and delivers it to a living cell where the virus can reproduce. Some viral capsids have an *envelope* that serves as an outer covering. The envelope is composed of lipids. The envelope may be derived from the host cell membrane.

A *bacteriophage* is a virus that infects bacterial cells. The bacteriophage uses tail fibers (figure 3.7c) near its base (figure 3.7d) to attach itself to a bacterial cell. Many viruses attach to cell membranes using receptors located on a viruses' outer coat.

Figure 3.7. The general structure of a bacterio-phage virus.

Viruses stimulate the production of **interferon**[161] in the host cell. Interferon is a protein that triggers the immune system to defend against foreign invaders such as bacteria, parasites and viruses. Interferon may help eliminate tumor cells in the body.

Transmission of viruses from one organism to another takes place in several ways. Insects spread viruses from plant to plant. Viruses that infect animals and humans are spread by insect **vectors**[162] that feed on blood.

159   Martinus Beijerinck, coined the term "virus" from the Latin *virus* (poison).
160   A change in the DNA structure.
161   Viruses, bacteria and other parasites stimulate the production of interferon.
162   An organism that transmits a pathogen to a host, usually a mosquito or a tick.

Viral transmission can occur when *insects* inject viruses into a new host along with blood taken from a previous host. **Yellow fever,**[163] **West Nile fever,**[164] and **dengue fever**[165] are viruses that are spread by insect vectors. Other viruses, such as the *influenza virus*, are spread by *droplets* in the air produced by sneezing, coughing or talking. *Norovirus and rotavirus* spread by *oral-fecal* contact causing severe gastroenteritis. Rotavirus was responsible for over 55,000 hospitalizations and from 20 to 60 infant deaths before the rotavirus vaccine was introduced in 2008.

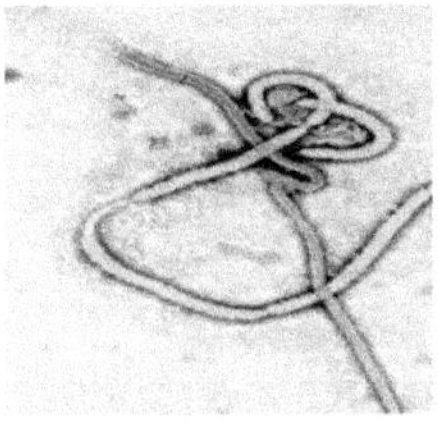

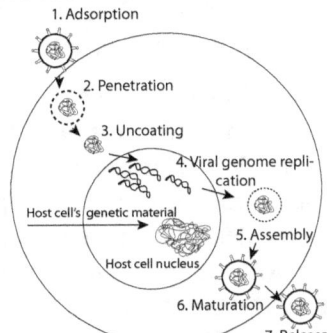

The *Ebola virus* (figure 3.8) causes Ebola hemorrhagic fever (Ebola HF). Its mortality rate is as high as 70%. Ebola HF spreads *to humans* by the *bite of the Egyptian fruit bat, contact with the bat's urine or feces* or from the *body fluids* of an infected *monkey or human*. Human to human infection is exclusively through body fluids. The incubation period of the virus is from 2 to 21 days.

Figure 3.8. Electron micrograph of the Ebola virus, a filovirus type.

In 2014, a severe epidemic of Ebola HF occurred in several West African countries. Sierra Leone, Guinea and Liberia had the most widespread transmission of EHF. Deaths totaled over 9000 as of February 10, 2015 in these countries.

On September 30, 2014, CDC announced that a man in Texas, who left Liberia on September 19, became the first diagnosed case of Ebola in the United States on September 30, 2014. He died on October 8, 2014.

Human immunodeficiency virus (HIV) causes acquired immunodeficiency syndrome (AIDS). AIDS is spread by *direct contact* with bodily fluids through sexual intercourse, sharing needles and syringes.

Viral Replication

Adsorbtion or "sticking" to the host cell membrane occurs first. The virus then *penetrates* the cell. Once inside the cell, the virus *uncoats* by removing its protein coat exposing its genome. The genome is then replicated within the cell. *Assembly* of new DNA or RNA molecules proceeds, followed by *maturation* and *synthesis* of new viral particles. New viruses are *released* from the infected cell. See figure 3.9.

1. Adsorption
2. Penetration
3. Uncoating
4. Viral genome replication
Host cell's genetic material
5. Assembly
Host cell nucleus
6. Maturation
7. Release

Figure 3.9. Major events of viral infection.

---

163   *Aedes aegypti* mosquito transmits virus from monkey to monkey, monkey to human and human to human. Vaccination and mosquito control recommended.

164   Originated in Africa. Many species of *Aedes, Anopheles and Culex* mosquitoes can transmit West Nile virus. Birds are primary reservoir. No vaccine exists presently.

165   Transmission is from person to person via mosquito. There is no vaccine.

Summary of Viral Replication

1. *Adsorption* – Proteins on the surface of the cell called *receptors* allow for viral attachment. If the right receptors are not present, then binding will not occur and no infection will occur.

2. *Penetration* – The viral genome gets into the host cell. Different viruses get their genome into the host cell in different ways.

3. *Uncoating* – Protein coat removal exposes the viral DNA or RNA.

4. *Viral Genome Replication* – The viral genome replicates using the host's cellular machinery.

5. *Assembly* – Parts of the virus and the necessary enzymes are produced and begin to be **assembled**[166] into new viruses.

6. *Maturation* – Viruses develop completely.

7. *Release* – New viruses are released from the host cell. The cell may burst and release viral particles, killing the cell or viral particles can also exit the cell by forming a bud. The virus surrounds itself with the host's cell membrane. This process may not kill the cell, rather it may be responsible for a persistent infection.

Sites of Viral Replication

DNA viruses **replicate**[167] *in the nucleus of cells* using a *DNA-dependent DNA polymerase*. DNA polymerase depends on having DNA present in order to work. DNA polymerase builds new DNA from smaller molecules of adenine, thymine, cytosine and guanine.

RNA viruses *replicate in the cytoplasm* of cells using *RNA-dependent RNA-polymerase* to replicate their RNA. These viruses make RNA from an RNA template. Another strategy viruses use is the enzyme *reverse transcriptase* or *RNA-dependent DNA polymerase*. This enzyme uses an RNA template to make DNA. *Retroviruses* such as human immunodeficiency virus (HIV) use this method.

Viral disease conditions can be can be **chronic**,[168] **latent**,[169] **oncogenic**[170] or **teratogenic**.[171] See table 3.2.

---

166   Put together.
167   To make an exact copy.
168   Chronic means having a long duration; lasts over a long period of time.
169   Something that is present, but not visible, but can become visible at a later time. *Varicella zoster*, the chicken pox virus remains present and can re-emerge as shingles.
170   Oncogenic is something that can cause cancer.
171   Something that causes a deformity of the organism.

Table 3.2. The Nature of several viral diseases.			
Chronic	*Latent*	*Oncogenic*	*Teratogenic*
Hepatitis B and D viruses. Prevented by vaccination.	Human herpes simplex virus (HSV) 1 and 2.	Human papilloma virus 16 and 18 Cervical cancer.	Rubella (German measles). Birth defects or spontaneous abortion. **Reportable**
Herpes viruses Can re-emerge later in life as shingles.	*Varicella zoster* virus (HVZ) Chickenpox and shingles. **Vaccine available. Reportable**	Hepatitis B virus is transmitted by body fluids. **A vaccine is available.** (HBV), liver cancers. **Reportable**	Human cytomegalovirus. The virus may cause birth defects in pregnant women.
Epstein-Barr virus (infectious mononucleosis).	Human immunodeficiency virus *(HIV)*. **Reportable**	Hepatitis C virus -blood to blood transmission. Liver cancers. **No vaccine** is available, but it is treatable and curable. **Reportable**	Varicella Zoster virus. Birth defects in pregnant women.
Human papilloma virus 2 and 7 common warts (*Verruca vulgaris*).	Hepatitis B virus (HBV) and liver cancers.	Epstein-Barr virus. Nasopharyngeal carcinoma and Burkitt's lymphoma. All are rare in the U.S.	Fifth Disease B19 parvovirus (Significant harm to an unborn child).
		Kaposi sarcoma-associated herpes virus (KSHV or HHV8).	

Viral Nomenclature

Several criteria are used to name viruses. Nucleic acid content (see table 3.3), **morphology**,[172] growth characteristics (see table 3.4), the host it infects and the type of disease it causes. See table 3.5.

Most viruses are named after the diseases they cause. See tables 3.3-3.5. Over 2000 species of viruses are placed in six orders: *Caudovirales* (bacteriophages infect bacteria), *Herpesvirales (*infect animals), *Mononegavirales* (infect animals and plants), *Nidovirales* (infect vertebrates), *Picornavirales* (infect primarily plants, insects and animals) and *Tymovirales* (infect plants).

David Baltimore devised an alternate scheme based on the type of nucleic acid content, and whether it is double or single stranded.

---

172   Shape.

Target Organs of Viruses

Blood is the most common **portal of entry**[173] for viruses. Virus are spread by the circulatory system to target organs. Some viruses seek out and attack the central nervous system or peripheral nerves. These are *neuroinvasive* or *neurotropic* viruses. The rabies virus is a **zoonotic**[174] and highly neuroinvasive. **Polio virus**[175] is spread by the circulatory system, but then attacks **motor neurons**[176] of the nervous system. *Herpes simplex*, *Varicella zoster* and **cytomegalovirus**[177] are neurotropic.

Rhinoviruses cause the common cold and will preferentially infect the upper respiratory tract. *This group of viruses are the most common infective agents of humans.* There are over 90 different viruses that cause the common cold. Rhinoviruses are different from each other because molecules on the viral surface are different from each other.

The common cold can be contracted by inhaling droplets from infected individuals, from touching objects that have viruses on them or from direct person-to-person contact. The most effective way to prevent the common cold is frequent handwashing. *Droplet particles containing viruses remain suspended in the air of an unventilated room for up to 36 hours. Changeover of a room's air will help lessen the risk of infection.*

Ebola virus is a RNA virus in the order Mononegavirales, family **filoviridae**[178] (meaning fine filament). See figure 3.7. It causes a bleeding disorder called *hemorrhagic fever*. Ebola virus has a high **mortality**[179] rate. It is so infective that it must be handled at **biosafety level 4**. Level 4 containment uses precautions necessary to keep a dangerous agent isolated. Biosafety levels range from 1 the lowest, to 4 the highest.

The *mumps* virus is a RNA virus that infects the **parotid gland**.[180] Mumps virus is the cause epidemic parotitis (mumps) in children. *Measles* virus attacks the respiratory system and skin. *Rubella* (German measles) is contracted by close contact or droplet infection. Measles produces a rash all over the body. **MMR vaccine**[181] prevents mumps, measles (rubeola) and rubella (German measles). If *varicella* (chickenpox) is included in the MMR vaccine, the vaccine is abbreviated **MMRV**.

173   The part of the body through which a microorganism enters to cause an infection. In most cases there are many different portals of entry to an organism's body.
174   A disease that is transferable between species such as anthrax and rabies.
175   The polio virus is strictly a human pathogen.
176   Neurons within the central nervous system that control a muscle.
177   Known as human CMV (HCMV). Is a human herpesvirus-5 (HHV-5).
178   From the Latin, *filum* meaning a fine filament.
179   Death rate. Morbidity is a relative incidence of a disease in a specific locality.
180   Salivary gland.
181   A vaccine is a preparation of live, dead or a part of an infective agent that produces immunity.

Table 3.3. Selected DNA and RNA viruses and the diseases they cause.			
DNA Viruses	Disease	RNA Viruses	Disease
Human Cytomegalovirus (CMV) Human herpes virus 5	A common virus that infects people of all ages.	Orthomyxoviruses Only A, B and C strains cause influenza:	Influenza Strains a.) 1918: H1N1 (Spanish Flu). b.)1957: H2N2 (Asian Flu). c.) 1968: H3N2 (Hong Kong Flu). d.)1997: H5N1 (Avian Flu).
Herpes Simplex Virus 1 (HSV1)	Cold sores.	Human Immunodeficiency Virus (HIV)	Acquired Immune Deficiency Syndrome (AIDS). Reportable
Herpes Simplex Virus 2 (HSV2)	Genital infections.	West Nile Virus	West Nile fever. Reservoir is birds/vector is a mosquito. Reportable
Human herpes virus 6	Childhood Roseola.	Picornaviruses: 1) Human rhinovirus (75 types) 2) Hepatitis A virus	1) Common Cold. 2) Hepatitis (HAV) or Infectious hepatitis. Reportable
Herpes virus 8 (KSEV 8)	Kaposi's sarcoma.	Hepatitis C virus	Hepatitis (HCV) blood, contaminated needles. Reportable
Varicella zoster (VZV)	Chicken pox and shingles. Reportable	Hantavirus	Hantavirus Hemorrhagic fever with renal syndrome. Reportable
Herpes zoster	Shingles.	Rubella virus	Rubella (German measles). Reportable Birth defects (teratogenic).
Epstein-Barr virus (EBV) Human herpesvirus 4	Infectious mononucleosis, Hodgkin's & Burkitt's lymphoma.	Rabies virus	Rabies. In the saliva of an infected dog, bat or raccoon. Zoonotic. Reportable
Variola	Smallpox Reportable Vaccination is the only prevention.	Measles virus	Measles. (rubeola or morbilli). Rare infection in the U. S. About one out of 1,000 die. Reportable

Table 3.4. Morphological Characteristics Used for Naming Viruses.			
Family Name	Shape	Disease	Virus Name
Rhabdoviridae	Bullet shaped.	Rabies **Reportable**	Rabies virus
Filoviridae	Long, thin.	Ebola hemorrhagic fever **Reportable**	Ebola virus
Togaviridae	Outer covering like a toga.	Tobacco Mosaic Disease	Tobacco Mosaic Virus
Retroviridae genus lenti-virus (Latin *lente* meaning slow)	Slow growing.	Hand, foot and mouth (HFM) disease	Coxsackie A virus
		AIDS **Reportable**	HIV virus

Table 3.5. Some viruses named after the host they infect.			
Organism	Disease	Host	Disease
Blue crab	Blue crab virus.	Human	Human rhinovirus numbers 1 to 100.
Tobacco plant	Tobacco Mosaic. Disease	Human	Human adenoviruses numbers 1 to 47.
Cattle	Bovine diarrhea virus.	Human	Human herpesvirus numbers 1-7.
African green monkey	African green monkey cytomegalovirus.	Human	Human immunodeficiency virus numbers 1-2.
Mouse	Mouse mammary tumor virus.	Human	Human papillomavirus.

Prions

Prions are *transmissible, infectious proteins* with a long incubation period. Prions were discovered by Stanley Prusiner at Columbia University in 1982. Prions cause transmissible spongiform encephalopathies (TSEs). Transmissible spongiform encephalopathy affect the **central nervous system**[182] of many animals. Prions cause neurodegenerative diseases in **ruminants**[183] and humans causing a marked decrease in mental ability, physical coordination and personality change occurs over time. See table 3.6. A common characteristic of a neurodegenerative disorder caused by prions is a sponge-like appearance of brain tissue at autopsy.

---

182   Brain and spinal cord of vertebrate animals.
183   Ruminants include cattle, goats and sheep among others that predigest plant matter in different compartments of their stomachs.

Prion Diseases

Normal protein structure has four levels of folding. Prions are misfolded proteins. Diseases caused by prions can be sporadic (spontaneous), familial (genetic/inherited) or acquired (transmitted by infection). Prions can be transmitted from animal to animal, animal to human or human to human by contaminated meat. *Bovine spongiform encephalopathy* (BSE) or "mad cow disease," *scrapie* in sheep and *feline spongiform encephalopathy* are among the known animal diseases caused by prions.

Table 3.6. Prion diseases of humans and animals.

Human Prion Diseases	Animal Prion Diseases
Creutzfeldt-Jakob disease (CJD)	Bovine spongiform encephalopathy (BSE)
Gerstmann-Straussler-Scheinker Syndrome (GSS)	Scrapie
Kuru	Transmissible mink encephalopathy
Fatal familial insomnia (FFI) Rare disease. Death occurs about 18 months from onset of first symptoms.	Feline spongiform encephalopathy is known to affect domestic cats and cat species held in captivity.

Creutzfeldt – Jakob disease (CJD) is the most common prion disease in humans. CJD is a rare, degenerative, incurable fatal disease. Symptoms include depression loss of memory, inability to think clearly and eventually, coma and death. Presently diagnosis is only by autopsy.

There is evidence of acquiring CJD by **iatrogenic**[184] means. Transplanted corneas and transplanted tissues from cadavers were suspected to cause this infection. There are documented cases of CJD acquired from transplanted corneas of infected donors and of blood transfusions.

Gerstmann-Sträussler-Scheinker syndrome (GSS) is a very rare neurodegenerative inherited form of CJD. The incidence of the disease is as low as 1 in 100 million.

Kuru was acquired by cannibalism among the natives of the Fore tribe of **Papua New Guinea**.[185] The disease has all but disappeared due to the elimination of cannibalism.

Sheep and goats are can be infected with a spongiform encephalitis that leads to a fatal, degenerative disease that affects their nervous systems called *scrapie* because infected animals scrape their bodies against objects in the fields when they graze for food.

---

184   Iatrogenic refers to acquiring a disease from a physician either directly or by a procedure performed by a physician.
185   Located in the southwestern Pacific Ocean.

Chapter 4 Classification and the World of Living Things

Biological classification is the framework that helps organize living things. Classification reveals relationships among groups of living things that may not be evident otherwise.

There are two major themes in Linnaeus' great work, *Systema Naturae*. His efforts at classification of living things were first and foremost to demonstrate God's order based on **immutability**[186] of species. That is, living things came into being and have never undergone change. Second, he assigned *two names* to every living thing known to him at the time; this is known as the *binomial system of classification*. See table 4.1. Linnaeus placed all living things in either a plant or animal kingdom. The idea of kingdoms was not new. What was new and original was Linnaeus' assignment of a *first* and *last* name to all plants and animals. Scientists still use the binomial system.

Table 4.1. Scientific names of some common animals		
Common Name	Genus	Species
Lion	*Felis*	*leo*
Tiger	*Panthera*	*tigris*
Leopard	*Panthera*	*pardus*
Domesticated cat	*Felis*	*catus*
Humankind	*Homo*	*sapiens*

An extraordinarily **diverse**[187] **spectrum**[188] of living things is presented in the following pages. The mechanism of how the array of life came into existence was described by *Charles Darwin* in his *On the Origin of Species by Means of Natural Selection* published in 1859.

Darwin's travels along the East and West coasts of South America revealed a great variety of living things not found in Europe. For example, a small number of birds found on one of the Galapagos Islands located off the West coast of South America, obtains its water and nutrients by pecking at the skin of other birds and drinking their blood. This trait, in the course of evolution, probably arose as a mutation in the population. The larger population depended on fresh water of ponds and streams for their existence. As a result of climate change, the area became dry, making very little fresh water available. The trait of drinking blood became a survival trait. This group, called vampire finches, survived and reproduced in greater numbers than those that depended on scarce fresh water. The mechanism of evolution, natural selection, was at work.

186   Not changeable.
187   Different from each other.
188   A range.

The World of Life Explored
How Life May Have Begun

The Earth is about 4.5 to 5 billion years old. There was no life on Earth for about 1 billion years. About 3.5 - 3.8 billion years ago archaea appeared on Earth.

The primitive Earth had only inorganic materials present. **Precursor**[189] molecules of living things appeared *spontaneously* at least once in the Earth's history. It is probable that self-replicating RNA molecules evolved into simple organisms. These, in turn, gave rise to primitive organisms with DNA as its genetic material.

In 1952, Stanley Miller and Harold Urey offered a possible explanation for the spontaneous appearance of life on Earth. The scientists placed substances thought to be present on the ancient Earth in a closed container and passed electricity through the mixture. See figure 4.1. Organic molecules were produced, proving substances associated with life can be made from nonliving substances. These simple organic molecules *could* have given rise to molecules that evolved into life forms.

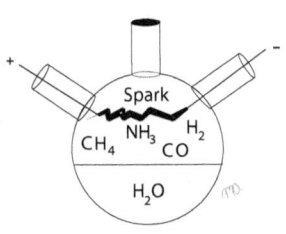

Figure 4.1. Apparatus used in the Miller-Urey experiment.

The First Life Form – the Archaea?

Archaea are the most primitive life forms **extant**[190] on Earth. Archaea are the size of bacteria and look like bacteria, but they are *not* bacteria. Once thought to live only in extreme environments of boiling springs and waters with high salt concentrations, archaea have been found in salt marshes, oceans, lakes, soil and the human colon.

Cell walls of archaea are composed of *glycoprotein, pseudopeptidoglycan* or *polysaccharides*. Archaea, like bacteria, have a *nucleoid* where their chromosomes float freely in the cytoplasm. There are no known **parasites**[191] among archaea, nor do archaea form spores. Most archaea are **commensals**[192] some archaea display **mutualistic**[193] relationships with other organisms. Certain archaea, such as the *cyanobacteria*, were among the first life forms on the primitive Earth that lacked oxygen. Cyanobacteria are aquatic and photosynthetic. These aquatic, photosynthetic archaea probably produced the Earth's present oxygen levels.

---

189   An event or substance that precedes something.
190   In existence.
191   One organism benefits and the other is harmed.
192   One organisms benefits and the other is not affected.
193   A relationship in which both organisms benefit.

Domains of Living Things: *Archaea, Bacteria and Eukarya*

The following is short presentation of biological classification. It is a starting point for the interested student. Included are major groupings in each of the three domains. There are about 5 to 100 million species of living things on Earth. Approximately two million different species have been described to date. *Each is placed in one of three domains:* **Archaea, Bacteria** *or* **Eukarya.**

I. Domain Archaea

General Characteristics of Domain Archaea

Archaea comprise the domain Archaea based on ribosomal RNA studies. Archaea are the oldest living things on Earth. Archaea are **prokaryotic**, **unicellular** *organisms that resemble bacteria, but they are* **not bacteria***. The shapes of archaea may be round, rod, square, flat, or strands. Chromosomes of archaea float in the cell's cytoplasm in an area called the nucleoid.* **Archaea do not have membrane bound organelles or membrane bound DNA. Pseudopeptidoglycan** *is the major chemical in archaeal cell walls. Bacteria have peptidoglycan. Most archaea are anaerobic and use sulfur (S), nitrates (NO^{-2}), ammonia (NH_3), nitrites (NO_2^-) as final electron acceptors.*

Archaea are more closely related to eukarya than they are to bacteria. Once thought to be found only in extreme environments of high temperatures and high salt concentrations. It is now known that archaea exist in less extreme environments such as marshes, oceans, soil, sewage and the human colon. There are no known parasites among the archaea. Of the five archaeal **phyla***.*[194] *Three are discussed below.*

A. Kingdom *Euryarchaeota*
1. Methanogens:

Methanopyrus kandleri is a methane (CH_4) producer. It lives in low to zero oxygen environments. *M. kandleri* is found in marshlands. *M. kandleri* produce methane (CH_4), also known as "marsh gas." *M. kandleri* also produces methane in the intestines of some animals and humans. *Methanobrevibacter smithii* is a commensal that lives in the human gut where it helps digest carbohydrates.

2. Extreme Halophiles:

Extreme halophiles live in environments with salt concentrations five times greater than ocean water **salinity**[195] or about 17.5%. That's 175 grams of salt per liter! *Halobacterium salinarum* lives in extremely salty lakes and on salted fish and salted hides.

3. Extreme Thermophiles:

Extreme thermophiles are "heat loving" archaea that live at an optimum temperature of about 50 °C (122 °F). Some hyperthermophiles live at temperatures of 80 to 105 °C (176 to 221 °F). Extreme thermophiles are found in hot sulfur springs. *Thermus aquaticus* reproduces at temperatures that would kill other organisms. It uses the enzyme *DNA polymerase* replicate its DNA. DNA polymerase is used in DNA amplifying machines to replicate DNA *in vitro*.[196]

194   A smaller taxonomic group below the larger kingdom.
195   The world's oceans on average are 3.5% (35 g/L) salt.
196   From the Latin, meaning "in glass" (in the test tube).

II. *Domain Bacteria*

General Characteristics of Domain Bacteria
The domain Bacteria consists only of bacteria. Bacteria are **prokaryotic, unicellular** organisms that are widely distributed on Earth. **Bacteria do not have membrane bound organelles or membrane bound DNA**. Bacteria range in size from 1 to 10 micrometers (μm)[197] long to 0.2 to 1 μm in width. Bacterial shapes are typically round (cocci), rod shaped (bacilli) and spiral. Cell walls of bacteria are mostly **peptidoglycan**. Bacteria live in a variety of habitats – surface soil, deep layer of the Earth, as well as an animal's skin, mouth and colon.

Bacteria are vital for recycling the Earth's nutrients. Most bacteria are helpful; some are pathogens. Some species make antibiotics. Human insulin and growth hormone can be produced by bacteria using **genetic engineering**.[198]

A. Kingdom Eubacteria
Phylum I *Actinobacteria – Gram-positive, some form branching filaments, lives in water and soil, decomposes organic matter especially cellulose and chitin, some are vital to the* **carbon cycle**,[199] *some cause disease.*

Genus: Corynebacterium – Corynebacteria are Gram-positive, aerobic to facultatively anaerobic, live in soil, water; most are harmless. Some are pathogenic.

1. Species: *C. diphtheriae*   **Disease: Diphtheria**   **Reportable**[200]
Special Remarks: Pathogen, contagious, spread by aerosol, **nosocomial**,[201] produces diphtheria toxin, forms *pseudomembrane* of dead epithelial cells that adheres to the tonsils, affects respiratory system. No diphtheria cases were reported in U.S. between 2004 and 2008. Diphtheria is endemic in developing countries with low vaccination rates.

Genus: Mycobacteria – Mycobacteria are bacilli, **acid fast**[202] and aerobic.

1. Species: *M. tuberculosis*   **Disease: Tuberculosis (TB) Reportable**
Special Remarks: An acid fast, aerobic bacillus (rods). It is an obligate parasite spread by coughs, sneezes, speaking; it can be **latent**[203] in some people. Detection: skin testing and x-rays. *M. tuberculosis* attacks the lungs, but can attack the kidney, spine, bone and the brain. Tuberculosis can be fatal. Affects almost 2 billion people a year.

2. Species: *M. leprae*   **Disease: Leprosy**   **Reportable**
Special Remarks: *M. leprae* is a slow growing, acid fast, aerobic bacillus that causes a chronic infection of the skin and peripheral nerves. In the U.S. 100 cases are reported yearly. Leprosy is *curable* with rifampin, dapsone and clofazimine individually or combined.

197   One μm is equal to 1 millionth of a meter. Thickness of human hair is 100 μm.
198   Making a deliberate change in the genome of an organism.
199   The recycling of the element carbon for reuse by living things.
200   These diseases are reported to the Center for Disease Control when they are first diagnosed. Reportable diseases are those that present a public health danger.
201   Acquired in a hospital or other healthcare facility.
202   A stain cannot be washed out with an acid and alcohol mixture.
203   Latent from the Latin, hidden. A dormant or inactive stage of a disease.

(Domain Bacteria continued)

Genus: Streptomyces – Gram-positive
1. Species: *S. griseus*   **Important use(s):** Produces streptomycin
Special Remarks: *S. griseus* lives in soil, but rarely causes disease.

Genus: Propionibacterium – makes proprionic acid using anaerobic respiration.
1. Species: *P. acnes*   **Disease: Acne**
Special Remarks: Bacilli, lives on healthy skin and in human GI tract. Commensals live in association with sweat glands and sebaceous glands.

Phylum II *Firmicutes – Gram-positive, cocci or bacilli, some produce* **endospores,**[204] *commonly found in soil.*
Genus: Bacillus – Rods, spore forming.
1. Species: *B. anthracis*   **Disease: Anthrax   Reportable**
Special Remarks: Zoonotic. It affects cattle and can be transmitted to humans.

2. Species: *B. cereus*   **Disease: Food poisoning**
Special Remarks: Can be found in cooked cereals left out at room temperature.

3. Species: *B. subtilis*
Special Remarks: *B. subtilis* can be used as a **probiotic.** Probiotics are beneficial because they encourage growth of beneficial bacteria in the human gut.

Genus: Listeria – Non-spore forming rods.
1. Species: *L. monocytogenes*   **Disease: Listeriosis   Reportable**
Special Remarks: Serious food born pathogen. Diarrhea and meningitis will result. *L. monocytogenes* presents a great danger to pregnant women.

Genus: Staphylococcus –   Round bacteria that form *grape-like clusters*, Gram-positive, non-spore forming and **beta hemolytic.** Staphylococci are commensals on the skin, nose and throat. They are **nosocomial.**[205] *S. aureus* resist drying out. Some staphylococci produces a *beta toxin* that lysis sheep blood cell membranes to leave a *clear zone* of *complete hemolysis* around the colony. *Beta lactams (penicillins) are effective against streptococci* by blocking cell wall synthesis, allowing cytoplasm leak out of the cell, thus killing the bacteria.
Diseases: Many - Several are listed below 1-11.

1) folliculitis	5) boils (furuncles)	8) *toxic shock syndrome*
2) carbuncles	6) abscesses	9) *osteomyelitis*
3) *meningitis* (**Reportable**)	7) *endocarditis*	10) *staphylococcal impetigo*
4) *sepsis* (postsurgical wound infection)		11) cellulitis

1. Species: *S. aureus* Variants are **MRSA** (methicillin resistant). **MRSAs** are resistant to beta-lactam antibiotics. Beta-lactam antibiotics penicillins (methicillin, dicloxacillin, oxacillin and cephalosporins) have a beta lactam ring in their structure. Other *S. aureus resistant forms:* **VISA** (Vancomycin-intermediate) and **VRSA** Vancomycin-resistant.   All resistant form are **Reportable**.

---

204   Endospores form on the ***inside*** of a bacterial cell. Endospores have a protective wall around the genome that helps the organism survive adverse conditions.
205   Diseases that can be acquired in hospitals and other health care facilities.

60

Genus: Lactobacillus – Rods, common, benign, metabolizes lactose to lactic acid, a symbiont in the GI tract, vagina. Many species are used in the food industry (butter, yogurt, cheeses and sauerkraut). Important probiotics.

  1. Species: *L. lactis*   **Important use(s):** manufacture of cheese

  2. Species: *Leuconostoc*   **Important use(s):** making of sauerkraut

Genus: Enterococcus – cocci.

  1. Species: *E. faecalis*

Special Remarks: *E. faecalis* is a normal commensal in the human intestine. It is one marker for human contamination of salt water swimming areas. *E. coli* (rods) is a marker of fecal contamination in fresh water swimming areas.

  2. Species: *E. faecium*

Special Remarks: *E. faecium* is a normal commensal in human intestine. It is a marker for human contamination of salt water swimming areas.

Lancefield Groupings:[206]

Group A Streptococci (GAS): Group A streptococci are **beta hemolytic.**[207] The organism is a cause of many major human diseases.

  1. Species: *S. pyogenes*

Diseases(s):	**Reportable**
1) streptococcal **pharyngitis**,[208]	5) localized skin infections
2) scarlet fever (a streptococcal sequela - rare)	6) streptococcal toxic shock syndrome
3) **rheumatic fever**[210] (a streptococcal sequela)	7) **glomerulonephritis**[209]
4) necrotizing fasciitis ("flesh eating disease")	8) *streptococcal impetigo*

Group B Streptococci

  1. Species: *S. algalactiae*[211]   **Disease(s):** septicemia of newborn that can spread to the vagina. Pathogen of cows.

Special Remarks: *S. algalactiae* is found in the normal flora in GI tract.

Group C Streptococci

  1. Species: *S. dysgalactiae*   **Disease(s):** mastitis in cows.

  2. Species: *S. zooepidemicus*   **Disease(s):** wound infection in horses.

Group D Enterococci   **Disease(s):** urinary tract infections, bacterial endocarditis and meningitis

Group E *Streptococci S. mutans*   **Disease(s):** dental caries.

Special Remarks: An opportunist that can enter the bloodstream and colonize mitral and aortic valves. They are normal inhabitants of upper respiratory tract.

---

206   Rebecca Lancefield, an American microbiologist, did pioneering work with group A streptococci and their association with rheumatic fever.

207   Results from an alpha toxin.

208   "Sore throat."

209   May also result after *S. pyogenes* pharyngitis infection.

210   May result after pharyngitis caused by *S. pyogenes* infection.

211   Means no milk.

The following sections (a), and (b) do not have Lancefield Groupings:

(a) Genus: Streptococcus – Gram positive cocci, arranged in long chains, **alpha hemolytic**, has a capsule of polysaccharides that surrounds and protects it.

1. Species: *S. pneumoniae*   **Disease** Pneumococcal meningitis **Reportable**
S. pneumonia **causes many different pneumococcal diseases:**

1) *lobar (bacterial pneumonia)*   4) pneumococcal meningitis   7) *brain abscess*
2) otitis media   5) pericarditis   8) *peritonitis*
3) acute sinusitis   6) *osteomyelitis*   9) cellulitis

Special Remarks: Streptococci are alpha hemolytic. Alpha hemolysis is a greenish hemolysis on blood agar. Many streptococci are serious pathogens. Vaccines are available for 23 capsulated strains (85 to 90 percent of virulent types). *Beta lactams (penicillins) are effective against Streptococci.* Penicillins block cell wall synthesis, the bacterial cytoplasm leaks out and the cell dies. However many strains are resistant to penicillin, cephalosporins and **macrolides.**[212]

(b) Viridans Streptococci: Cocci, arranged in long chains, **alpha hemolytic**, found in the human mouth.

1. Species: *S. sanguinis*   **Disease(s):** Dental caries
2. Species: *S. mitis*
Special Remarks: Normal inhabitant of the human mouth.
3. Species: *S. salivarius*   **Disease(s):** Rare. Opportunistic pathogen.
Special Remarks: *S. salivarius* colonizes the human mouth soon after birth.
4. Species: *S. sanguis*   **Disease(s):** Dental caries
Special Remarks: *S. sanguis* is a normal inhabitant of the human mouth.

Genus: Clostridium – rods, spore formers

1. Species: *C. botulinum*   **Disease(s):** Botulism   **Reportable**
2. Species: *C. difficile*   **Disease(s):** *antibiotic-associated diarrhea*.
Special Remarks: Nosocomial.   Can be life threatening. Contracted in hospitals and nursing homes.
3. Species: *C. perfringens*   **Disease(s):** Gas gangrene
4. Species: *C. tetani*   **Disease(s):** Tetanus   **Reportable**
Special Remarks: Incubation for tetanus is about eight days.

Phylum III *Tenericutes – Do not have cell walls.*
 Genus: Mycoplasma: *Mycoplasma lack cell walls.* Mycoplasma are not affected by **beta lactams,** (penicillins) because beta lactams are effective against bacteria that have cell walls.
1. Species: *M. pneumoniae*   **Disease(s):** "Walking pneumonia." Primary atypical pneumonia (PAP).

---

212   Erythromycin, clarithromycin and azithromycin are examples of macrolides.

Phylum IV *Proteobacteria* – Proteobacteria are Gram-negative rods, possess an outer membrane, some are purple *photoautotrophs*, many are flagellated and many convert gaseous nitrogen to nitrogen compounds. There are many pathogens in this group.

Genus: Brucella
    1. Species: *B. Abortus*   **Disease(s):** Brucellosis   **Reportable**
Special Remarks: Brucellosis is a zoonotic disease that is transmissible from cattle to humans through the consumption of unpasteurized milk and cheeses made with unpasteurized milk. Brucellosis is also called undulant fever because of a wave-like rise and fall in temperature.

  Genus: Escherichia – Gram-negative rods, commensals that live in the gut of warm-blooded animals.
    1. Species: *E. coli*
Important use(s): *E. coli* normally makes vitamin K_2 in human gut. It serves as a marker of fecal contamination in fresh water swimming areas.
Special Remarks: Some strains are serious pathogens. See below.

Escherichia coli serotypes   **Disease(s):**
    Strain: Enterohemorrhagic *E.coli* (**O157:H7**)   **Reportable**
             Can lead to kidney failure and death in young and old.

    Strain: Enteropathogenic *E. coli* (**O127:H6**)   **Reportable**
             Emerging Disease.[213] severe and persistent infant diarrhea.

    Strain: Enteroaggregative *E. coli* (**O128:H2**)   **Reportable**
             Travelers' diarrhea.

  Genus: Pseudomonas – Pseudomonads are aerobic Gram-negative rods that live in soil, water, plant seeds and skin. Pseudomonads do not form spores. Pseudomonads have one or more polar flagella.
    Species: *P. aeruginosa*   **Disease(s):** inflammation and sepsis.
Special Remarks: *P. aeruginosa* is an opportunist that presents great danger to burn victims.

  Genus: Salmonella – rod shaped, non-spore former and motile.
Species:
    1. *S. enterica enterica* (serovar[214]*typhi*) Disease(s): Salmonellosis **Reportable**

  Genus: Shigella
    1. Species: *S. dysenteriae*   **Disease(s):** Shigellosis   **Reportable**

  Genus: Helicobacter
    1. Species: *H. pylori*   **Disease(s):** Gastritis and gastric ulcers

---

213   A new agent. Recognizing a disease that has not been detected previously, or an environmental change that allows organisms to jump to humans.
214   Serotype (serovar) is variant within a species of microorganism based on antigen molecules located on their cell surfaces.

Genus: Neisseria
  1. **Species:** *N. gonorrhoeae*   **Disease(s):** Gonorrhoeae   **Reportable**
  2. **Species:** *N. meningitidis*   **Disease(s):** Meningococcal disease **Report.**

Genus: Francisella
  1. **Species:** *F. tularensis*   **Disease(s):** Tularemia   **Reportable**

Genus: Bordetella
  1. **Species:** *B. pertussis*   **Disease(s):** Pertussis   **Reportable**
Special Remarks: The common name is whooping cough.

Genus: Vibrio
  1. **Species:** *V. cholerae*   **Disease(s):** Cholera   **Reportable**
Special Remarks: Lives in natural and in contaminated fresh waters.

Genus: Legionella
  1. **Species:** *L. pneumophila*   **Disease(s):** Legionellosis **Reportable**

Genus: Serratia
  1. **Species:** *S. marcescens*   **Disease(s):** Catheter-associated bacteremia, nosocomial infections, especially urinary tract and wound infections.

Genus: Klebsiella
  1. **Species:** *K. pneumoniae*   **Disease(s):** Pneumonia
Special Remarks: A normal resident of the human mouth, skin and intestines in humans. *K. pneumoniae* is a nosocomial, opportunistic pathogen.

Genus: Yersinia
  1. **Species:** *Y. pestis*   **Disease(s):** Plague   **Reportable**
Special Remarks: *Y. pestis* it is a true pathogen commonly called the **"Black Death."**[215] It has **bubonic** and **pneumonic** forms. Bubonic plague is so named because of characteristic swelling of lymph notes called "buboes." Buboes have a black coloration. Plague is transmitted to *humans* by a *flea* carried by an infected black *rat* (*Rattus rattus*).

    Pneumonic plague is a form of plague that is spread by droplets coughed up by infected individuals. Plague occurs as an *urban* phenomenon and a *sylvatic* form found in forested areas. The disease is endemic in some parts of the world.

Genus: Proteus
  1. **Species:** *P. vulgaris*   **Disease(s):** wound and urinary tract infections
Special Remarks: *P. vulgaris* is an opportunistic pathogen of humans. Normal resident of GI tract of animals and humans.

215   An epidemic in 1346–1350 that wiped out 30 to 70% of Europe's population.

(Domain Bacteria continued)

Phylum V *Chlamydiae* – Members of this phylum are obligate, intracellular parasites. It lives as an asymptomatic pathogen. Many are smaller than viruses. Chlamydia cannot be grown on bacteriological media.
Genus: Chlamydia
1. Species: *C. trachomatis*   **Disease(s):** Chlamydia   **Reportable**
Special Remarks: *Lymphogranuloma venereum* strain. Obligate intracellular parasite, sexually transmitted. Conditions in females presents as cervical inflammation. Condition in males presents as urethral inflammation. A chlamydia infection can lead to infertility in both male and female.

2. Species: *C. trachomatis* (Trachoma strain) **Disease(s):** Trachoma
Special Remarks: *C. trachomatis* is a leading cause of blindness worldwide. Trachoma is transmitted by direct contact, fomites and flies.

Phylum VI Spirochaetes – Gram-negative, long, helical in shape.
Genus: Brachyspira
1. Species: *B. aalborgi*   **Disease(s):** Appendicitis
Special Remarks: *B. aalborgi* is a normal inhabitant of the human appendix.

Genus: Leptospira
1. Species: *L. interrogans*   **Disease(s):** Leptospirosis **Reportable**

Genus: Borrelia
1. Species: *B. burgdorferi:*   **Disease(s):** Lyme disease **Reportable**
See figure 4.1.
Special Remarks: *B. burgdorferi* is transmitted to humans by the blacklegged tick (*Ixodes scapularis* or deer tick) in the Northeastern, Mid-Atlantic and north-central regions of the United States. It is the **most common tick born infectious disease in North America.** See figure 4.2. *Ixodes pacificus,* the western blacklegged tick, spreads the disease on the Pacific Coast. In general, *B. burgdorferi* cannot be introduced into the human bloodstream unless the tick is attached to the skin for 36-48 hours.

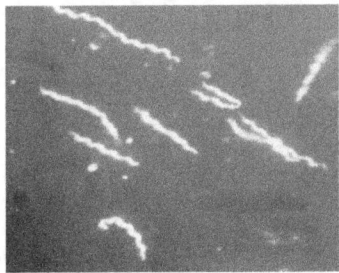

Figure 4.1. Negative staining of *B. burgdorferi*, the agent of Lyme disease.

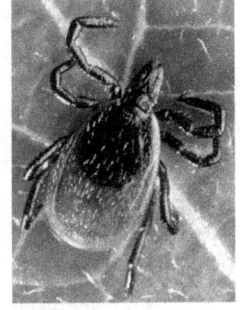

Figure 4.2. The deer tick, *Ixodes scapularis*, the vector of Lyme disease.

Note well: Any tick bite should be reported to a health care provider in a region affected by Lyme disease.

(Domain Bacteria continued)

Genus: Treponema
   1. Species: *T. pallidum pallidum*. **Disease(s):** Syphilis   **Reportable**
Special Remarks: Syphilis is a sexually transmitted disease (STD). New cases have doubled from 2005 (8724) to 2013 (16,663) in the United States. According to the CDC, seventy-two percent of new primary and secondary syphilis cases occurred in men who have sex with men. See figure 4.3. Syphilis presents as four distinct stages: *1. primary* **(chancre sore)**,[216] *2. secondary* (rash on palms or soles of feet), *3. latent* (no symptoms) and *4. tertiary* **(gummas)**.[217]

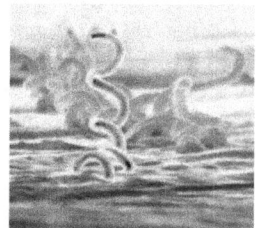

2. Species: *T. pallidum endemicum* **Disease(s):** Bejal

3. Species: *T. pallidum carateum*   **Disease(s):** Pinta
4. Species: *T. pallidum pertenue*   **Disease(s):** Yaws

Figure 4.3. *Treponema palllidum pallidum*. Electron micrograph.

Phylum VII Chrysiogenetes – Uses arsenate as a final electron acceptor, grows in arsenic contaminated environments.
Genus: *Chrysiogen*
   1. Species: *C. arsenatis*   **Important use(s):** Used to get arsenic out of drinking water

Phylum VIII Cyanobacteria[218] – The Cyanobacteria, once called blue-green algae, are photosynthetic, aquatic and **oxygenic**.[219] Cyanobacteria are found in every environment. Cyanobacteria can **"fix"**[220] molecular nitrogen of the atmosphere into the compound ammonia (NH_3), and the anions nitrite (NO^{-2}) and nitrate (NO^{-3}). Plants can make protein from these nitrogen containing substances.

    It is believed that members of the cyanobacteria are responsible for the present day oxygen-rich atmosphere of the Earth. The cyanobacteria are thought to have had an **endosymbiotic**[221] relationship with primitive plant life and evolved into chloroplasts of modern plants.
Genus: Prochlorococcus
   1. Species: *P. marinus*
Important function(s): The dominant photosynthetic organism in the ocean.

Phylum VIX Thermotogae – Chemoorganotrophic, extreme thermophiles. Chemoorganotrophic bacteria use many different sources of organic compounds.
Important Use(s): papermaking.

---

216   A painless ulcer.
217   Non cancerous growth.
218   German taken from the Greek, *kayanos* meaning dark blue.
219   Produce molecular oxygen (O_2) found in air.
220   Conversion of gaseous atmospheric nitrogen into a nitrogen compound.
221   An organism that takes up residence within a larger cell and performs a function helpful to the larger host. The large host provides shelter for the endosymbiont.

III. Domain Eukarya
General Characteristics of Domain Eukarya
Domain Eukarya consists only of organisms that have a true nucleus. It consists of four kingdoms: **Protista, Fungi, Plantae** (**Veridaeplante**)[222] and **Animalia**. Members of the domain Eukarya have a membrane-bound nuclei and organelles. Eukarya range from unicellular to multicellular organisms and have a wide variety of ways to obtain food. Nutrition may be autotrophic, heterotrophic or parasitic. Eukaryotes use mitosis, an orderly division of their nuclei before cell division takes place. Some eukaryotes may live in water for their entire lifetime, part of their lifetime, or some may spend their entire life on land.

A. Kingdom Protista
General Characteristics of Protista
The four major groups in the protist kingdom are: **Protozoa, algae, fungus-like protists** and **apicomplexans.** Members of this kingdom are mostly *eukaryotic* and *unicellular*. Protists do not have specialized tissues. Protist nutrition may be *heterotrophic, autotrophic* or *parasitic*. Protists live in natural waters such as lakes, streams or oceans. Some protists may even live in a thin film of water surrounding a dirt particle. Some may live in the watery environment of an animal's body as parasites. Each group is very different from the other.

Group 1: Protozoa (Animal-Like Protists)
General Characteristics of Protozoans
Protozoa are single-celled, eukaryotic organisms that use pseudopods, flagella or cilia for locomotion. Protozoa were considered to be primitive animals in the past. Protozoa are not animals. Protozoa reproduce asexually by binary fission or sexually by conjugation. Protozoa, such as amoeba and paramecia are heterotrophic. Amoeba and paramecia engulf large particles of food by a *type of feeding* called **phagotrophy**.[223] Protists are traditionally placed in one of three groups: **rhizopoda** (sarcodina), *ciliata* and *mastigophora*. Each group has a characteristic means of locomotion. Cell wall structure varies.

Phylum I Rhizopoda – Rhizopoda demonstrates **amoeboid**[224] locomotion. The most outstanding feature is the formation of **cytoplasmic**[225] extensions called *pseudopods* or "false feet." Pseudopods are formed by flowing of the cytoplasm in response to a stimulus. Amoeba engulf food particles by a process termed phagocytosis, or "cell eating." Phagocytosis is a *type of nutrition* called **phagotrophy**.[226] Amoeba have *no outer covering* to maintain shape. The cell membrane will flow along with the cytoplasm in certain areas to create pseudopods. *Amoeba proteus*[227] is the representative organism for this group. A few are serious pathogens.

222   Also called Chlorobionta, or simply *Plantae*.
223   Phagotropy from the ancient Greek, *phago-* to devour and *troph-* to feed.
224   From the ancient Greek, to change.
225   Liquid portion of a cell, outside the nucleus, that contains that cell organelles.
226   *Phago-* to eat and *troph-* to feed.
227   From the ancient Greek, *protean* meaning able to change or be versatile.

(Domain Eukarya continued, Protists – Protozoa – Animal-like)

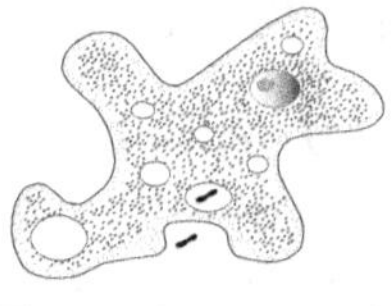

Superclass: Sarcodina
  Genus: Amoeba
    1. Species: *A. proteus* See figure 4.4.
Special Remarks: Lives in fresh water, non-pathogenic.

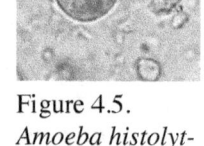

Figure 4.4 *Amoeba proteus.*

    2. Species: *A. histolytica* See figure 4.5.
Disease(s): Amoebiasis (amoebic dysentery)   **Reportable**
Special Remarks: Lives in fresh water and is pathogenic. Does not show typical pseudopod structures.

  Genus: Naegleria
    1. Species: *N. fowleri*
Disease(s): Primary amoebic meningoencephalitis.   **Reportable**
Special Remarks: *N. fowleri* lives in warm, fresh waters. Known as "the brain-eating amoeba," it attacks nerves and the brain. It is a rare infection with a 99% rate of mortality. It can be fatal in days.

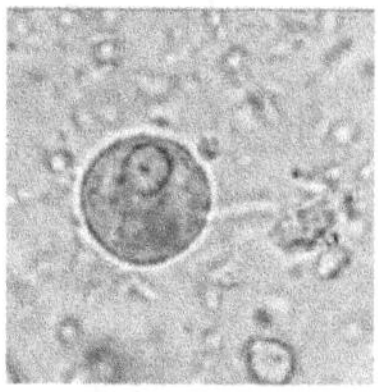

  Genus: Acanthamoeba
    1. Species: *A. encephalitis*
Disease(s): Amoebic keratitis and encephalitis
Special Remarks: Common protozoan in soil.

Phylum II *Ciliophora* – *Ciliates*, such as *Paramecium caudatum*, move by means of **cilia**. Cilia are tiny, hair-like structures that cover the surface of paramecia that propel the organism through water in a tumbling, twisting motion.

Figure 4.5.
Amoeba histolytica. Wet mount.

Paramecia have an outer covering called a **pellicle** that maintains the shape of the organism.
  Genus: Paramecium
    1. Species: *P. caudatum:* freshwater
Special Remarks: Not pathogenic.

  Genus: Balantidium
    1. Species: *B. coli*. The only known pathogen of sheep.

Phylum III Mastigophora (Flagellates) – Moves by means of a whip-like structure called a *flagellum.* A flagellum moves like a propeller in reverse that pulls the organism through the water.

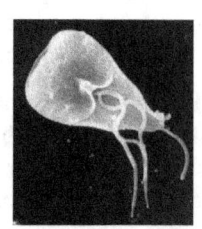

Figure 4.6. *Giardia lamblia.*

  Genus: Giardia. A parasite. See figure 4.6.
    1. Species: *G. lamblia*. Stays in the **lumen**[228] of the small intestine.
Disease(s): Giardiasis   **Reportable**
Special Remarks: Giardiasis is contracted from sewage contaminated water. The parasite is a commonly isolated from diaper changing stations. Use proper handwashing after potential exposure to feces. Giardia is the world's most common pathogenic parasitic infection in humans.

228   The space inside of any tube-like structure such as an artery or a vein.

Group 2: Algal Protists (Plant-Like Protists)
General Characteristics

Algal protists are eukaryotic single-celled, aquatic autotrophs that preceded green plants, the **veridiplantae** (**true plants**), in evolutionary history. Algae are **plant-like** because they have chlorophyll. Algae carry on **photosynthesis** just like true plants, but **algae are not green plants.** Algae do not have specialized tissues like multicellular plants. For this reason algae are placed in the protist kingdom. Algae are primary food producers. Primary producers make organic food substances like glucose ($C_6H_{12}O_6$) from **inorganic**[229] carbon dioxide (CO_2) and water (H_2O). Energy of sunlight is transferred to the chemical bonds of the organic food molecules that algae produce. See figure 4.7.

$$6\,CO_2 + 6\,H_2O \longrightarrow C_6H_{12}O_6 + 6\,O_2$$

(6) Carbon Dioxide + (6) Water $\xrightarrow[\text{CHLOROPHYLL}]{\text{ENERGY OF SUNLIGHT}}$ (1) Glucose + (6) Oxygen

Figure 4.7. The overall chemical equation for photosynthsis.

Algae have cell walls composed of **cellulose**. The largest algal protists are seaweeds. Algal and plant cells have organelles called **plastids** that contain chlorophyll. Chlorophyll gives plants their green color and is a catalyst for photosynthesis. Plastids are found only in algae and true plants. It is in these organelles where photosynthesis takes place and starch is stored.

Phylum I Chrysophyta – Have chlorophyll and carotenoids.
Class: Bacillariophyta: (diatomes) have beautiful cell walls of silica (SO_2).[230] There are about 200 genera (pl.) of diatomes and about 100,000 species. Diatomes are found in marine environments and fresh waters.
 Genus: Ochromonas golden/ algae. Mostly freshwater algae.

Phylum II Chlorophyta (green algae) – Unicellular, flagellated green algae.
 Genus: Chlamydomonas
 1. **Species: *C. reinhardtii*.** *C. reinhardtii* is a microscopic, autotrophic aquatic algal cell that has two flagella.
Special Remarks: *C. reinhardtii* is used as a model organism in biological research.

Phylum III Rhodophyta (red algae) – One of the oldest and largest algal groups. Mostly multicellular and marine and macroscopic.
Special Remarks: Used for food, medicines and fertilizer. Rhodophyta is responsible for production of limestone cliffs.

Phylum IV Phaeophyceae (brown algae) – Macroscopic, marine and multicellular seaweed.

---

229   Not organic. Inorganic substances do not contain carbon atoms, are simple, do not have a biological origin. Carbon dioxide is an inorganic gas that does contain carbon.
230   Silicon dioxide.

Group 3: Fungus-Like Protists
General Characteristics of Fungal-Like Protists

Fungus-like protists are eukaryotic, heterotrophic decomposers that look like fungi, but are not fungi. Fungus-like protists lack chitin, a cell wall component found in true fungi, and they lack specialized tissues.

Fungus-like protists feed on dead organisms or the nonliving products of living things just like fungi. Fungus-like protists feed via heterotrophic nutrition that is **saprobic**.[231] Saprobes release digestive enzymes into the environment and digest food externally. The externally digested food enters the organism by diffusion. This group consists of slime molds (myxomycota) and water molds (Oomycetes).

Phylum I Myxomycota – slime molds[232]
Genus: Physarum
  1. Species: *P. polycephalum*
Special Remarks: There are two forms in the life cycle of *P. polycephalum*. At the first stage it produces spores housed in a spore case, the ***sporangium***. In the second stage, *P. polycephalum* resembles an amoeba. It is widely used in biological research, especially for research into protoplasmic streaming.

Phylum II Oomycota – Filamentous protists.
Disease(s): Some are parasites of fish and amphibians.
  Genus: Phytophthora
  1. Species: *P. infestans*
Special Remarks: A water mold that was responsible for the great **Irish potato blight**.[233] The mold destroys the leaves and the tuber (potato) of the plant.

  Genus: Lagenidium
  1. Species: *L. giganteum*
Special Remarks: Used as a biological control agent. This fungus-like protist feeds on mosquito larvae.

---

231   Heterotrophs carry on extracellular digestion. Saprobes are a form of heterotrophy. Saprobes ingest chemical building blocks they cannot make.
232   Slime molds used to be classified with the fungi. Slime molds live on dead material.
233   The potato crop failed every year in Ireland from 1845 - 49. Cool, moist weather during this period helped *P. infestans* thrive. About 1 million Irish died from starvation.

Group 4: Apicomplexans (formally known as Sporozoa)
General Characteristics of Apicomplexans

The Apicomplexa are a large group of eukaryotic, unicellular, endoparasitic, spore-forming organisms that cause serious human diseases. Apicomplexa do not have specialized tissues, but they do have an *apicoplast* at its apical[234] end. The apicoplast is a specialized structure that is used to get into the host cell. Apicomplexa have a flexible outer cell covering called a *pellicle*.
Diseases: Apicomplexa cause **malaria**,[235] **babesiosis**,[236] **toxoplasmosis, crypto-sporidiosis** and **cyclosporiasis**.　　　　　　All are **Reportable**

Phylum I Apicomplexa – Apicomplexa are parasitic protists that are transmitted from animal to animal by arthropods *vectors*.
 Genus: Plasmodium (malaria parasite) – The *vector* is the *Anopheles* **mosquito**.[237] See page 86. *Members of the genus plasmodium cause malaria.* Only 11 species infect humans out of the 200 species. Infection occurs in two stages:
　　　　1. A *sexually* reproducing form, the *gametocyte*, is found in the *mosquito*. Gametocytes produce asexually reproducing *sporozoites*. *Sporozoites mature in the mosquito's salivary glands.* The mosquito is the *primary host*. The bite of a mosquito's injects sporozoites into the host. The sporozoites then move to *liver cells (hepatic stage)*. From hepatocytes, sporozoites move to red blood cells (*erythrocytic stage*) where they form asexually reproducing *merozoites*.
　　　　2. Asexually reproducing merozoites develop into *gametocytes* in the *secondary (vertebrate) host's red blood cells*, completing the life cycle.
Special Remarks: Malaria kills about 1 million people worldwide annually (2013) and 26 in US (2001-2008). Each year about 250 million cases of malaria are diagnosed worldwide. In the United States, about 1,500 cases of malaria are diagnosed every year, mostly in travelers to warm climates. Synthetic tetracycline, *doxycycline*, is approved by the FDA for **prophylaxis**[238] in affected areas.
　1. Species: *P. falciparum* 　　　**Disease(s):** The most deadly form of *malaria*.
Special Remarks: Vector for the parasite is *P. falciparum* is the *Anopheles*[239] *gambiae* mosquito.

　2. Species: *P. vivax* 　　　　**Disease(s):** Seldom fatal.
Special Remarks: *Anopheles* Found in U.S. and Central America. **Reportable**
　3. Species: *P. malariae* 　　　**Disease(s):** Has mild clinical manifestations.
Special Remarks: Most studied. Vector is *Anopheles* mosquito.

234　　Apical refers to an apex or tip.
235　　From the Italian, "bad air," *mal'aria*.
236　　Tick born infection similar to malaria. Usually infects the white-footed mouse, but has been known to infect humans, mostly in the Midwest and Northeast U.S.
237　　Class *insecta* is in the phylum arthropoda. Arthropods have jointed legs and an exoskeleton. The insecta have six jointed legs. Spiders have eight jointed legs.
238　　An action taken to prevent a disease.
239　　Out of over 460 species of this genus about 30-40 are vectors for plasmodia.

4. Species: *P. ovale*   **Disease(s):** Tertian malaria. **Reportable**
Special Remarks: Only the female *Anopheles* mosquito is the vector.

  Genus: Toxoplasma
    1. Species: *T. Gondii*   **Disease(s):** Toxoplasmosis **Reportable**

Special Remarks: Toxoplasmosis affects humans and animals worldwide. Cats
are the definitive host of the parasite. See figure 4.8. The disease is spread by
careless handling of cat litter, blood transfusions, or by eating undercooked
meat. ***Pregnant women should avoid cleaning cat litter boxes to avoid contact
with cat feces.***

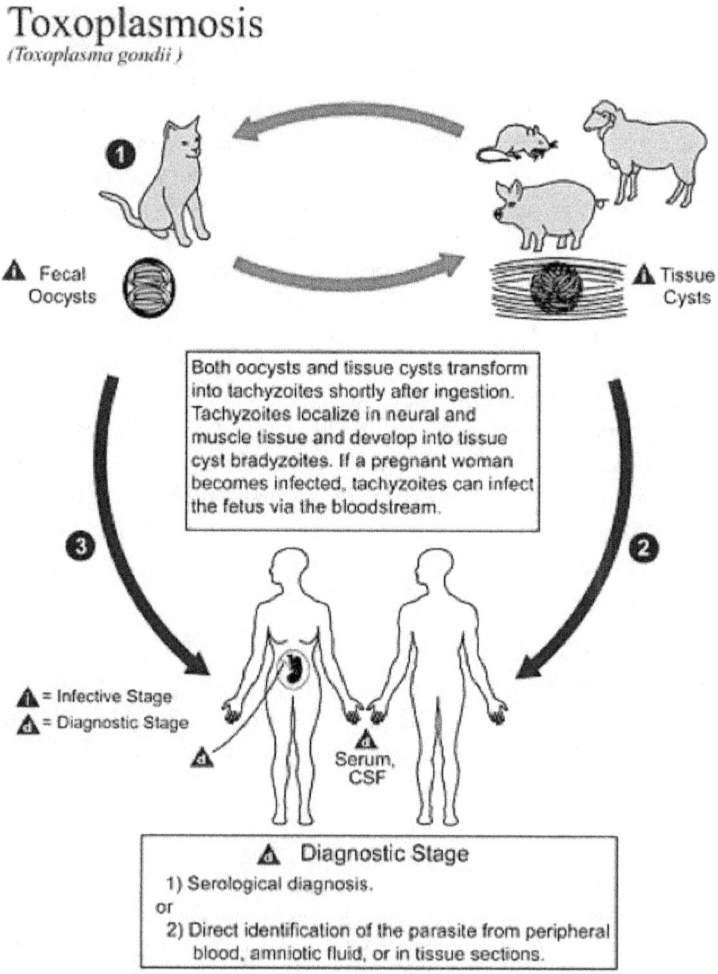

Figure 4.8. Life cycle of *Toxoplasma Gondii*.

B. Kingdom Fungi

General Characteristics of Kingdom Fungi

Mycology is the branch of biology that studies fungi. **Fungi** are *eukaryotic heterotrophs that lack chlorophyll*. Fungal cell walls are composed of **chitin**[240] (plants cell walls are composed of cellulose, bacterial cell walls are made up of peptidoglycan and archaeal cell walls are mainly of pseudopeptidoglycan.). Fungi range in size from microscopic one celled yeast to macroscopic mushrooms. The fungi first appear in the fossil record 1.43 billion years ago. Fungi are classified into four phyla. See page 42.

Some fungi are eaten as **food**[241] or used in the processing of food products. *Some fungi produce antibiotics, while some fungi produce harmful toxins and hallucinogens. Fungal enzymes are used in detergents.*

Only 5% of the 1.5 million species of fungi have been classified. Fungi reproduce asexually by **conidia**[242] produced on a **conidiophore**.[243] Many have a sexual stage. If a fungus does not have a sexual stage, it is placed in the group *fungi imperfecti*. Yeast, a one celled fungus, reproduce asexually by budding. Budding in yeast takes place when a small yeast cell grows out of the parent.

Phylum I Chytridiomycota: Most primitive fungi, are mostly saprobes.
 Genus: Batrachochytrium
 1. Species: *B. dendrobatidis*
Disease(s): May have caused large numbers of amphibian deaths worldwide.

Phylum II: Zygomycota (zygote fungi) bread molds
 Genus: Rhizopus
 1. Species: *R. stolonifer*. Black bread mold
Phylum III Ascomycota (sac fungi)
 Genus: Penicillium
 1. Species: *P. chrysogenum* (old name is *P. notatum*) produces penicillin.
 2. Species: *P. camemberti* used in brie and camembert cheesemaking.

 Genus: Aspergillus This genus has many important pathogens. Some can infect the external ear, cause skin lesions and ulcers called *mycetomas. Madura foot*, first discovered in Madura, India, is a mycetoma. Some aspergillus species have commercial value such as those used to ferment food such as saki wine.
 1. Species: *A. fumigatus*
Special Remarks: Grows on peanuts, corn and water damaged carpets. Causes pulmonary aspergillosis and aspergilloma (mycetoma).
 2. Species: *A. niger*
Special Remarks: A black mold that spoils fruits and vegetables.

240 A polymer derived from glucose found in fungal cell walls and exoskeletons of arthropods (crabs, lobsters, insects).
241 The fruiting body (sporocarp) of mushrooms is consumed as food. The fruiting body commonly recognized as the cap on top of a stem. It contains fungal spores.
242 Spores produced by a fungus. Conidia are asexual reproductive structures.
243 Aerial structures that bear spore cases containing conidia or spores.

3. Species: *A. flavus*
Special Remarks: *A. flavus* produces a toxin called, aflatoxin that can cause liver cancers. *A. flavus* typically grows on nuts, cereal grains and legumes that have been harvested and in the process of being transported,

Genus: Saccharomyces
1. Species: *S. cerevisiae*
Special Remarks: Used in baking and brewing.

Genus: Candida
1. Species: *C. albicans*   **Disease:** Candidiasis or thrush
Special Remarks: *C. albicans* is a commensal of the gut. It can be an ***opportunistic pathogen*** of the mouth and genital tracts. *C. albicans* causes serious infection in immunocompromised individuals, especially AIDS patients. Antibiotics, steroids, birth control pills, and foods high in sugars or starches promote growth of *C. albicans*.

Genus: Claviceps
1. Species: *C. purpurea*
Special Remarks: Produces ergot, an alkaloid hallucinogen.

Genus: Coccidioides
1. Species: *C. immitis*
Disease(s): Coccidioidomycosis is a fungal disease that presents symptoms of acute respiratory illness in about 40% of those infected. About 60% of those infected are asymptomatic. The disease is **endemic**[244] in the Southwestern U.S.
Special Remarks: The disease is caused by inhalation of airborne spores.

Phylum IV Basidiomycota: Club fungi, mushrooms
1. Genus Malassezia
Species: *M. furfur*
Special Remarks: Commensal on skin of humans.

Genus: Amantia
1. Species: *A. muscaria*
Special Remarks: A poisonous mushroom. Deaths are rare.

2. Species: *A. phalloides*
Special Remarks: *A. phalloides* is a poisonous mushroom. It is responsible for most deaths are caused by eating wild mushrooms.

Genus: Boletus
1. Species: *B. edulis*
Special Remarks: Edible as food (penny bun, porcino mushroom).

---

244   Commonly found in a particular region.

(Domain Eukarya continued, the Fungi)

Genus: Ustilago
  1. Species: *U. maydis*
Special Remarks: Attacks corn. Corn is the most produced food in the world.

Genus: Puccinia
  1. Species: *P. triticina*
Special Remarks: Attacks wheat.

Genus: Cryptococcus
  1. Species: *C. neoformans*   **Disease:** Cryptococcosis **Reportable**
Special Remarks: Cryptococcosis is a lung, brain and spinal cord disease that affects mostly AIDS patients. Found in pigeon droppings.

Fungi Imperfecti - (Deuteromycota): Fungi that do not have a sexual cycle are placed in this group.

Genus: Tinea (ringworms)   **Diseases:** See table 4.2.

Table 4.2. Tinea species and disease conditions.	
Species	Area of body affected
T. capitis	Scalp
T. barbae	Beard (barber's itch)
T. pedis	Foot (athlete's foot)
T. cruris	Groin (jock itch)
T. corporis	Anywhere on the body

Lichens: Lichens are a combination of two organisms. Some scientists may place the fungus in phylum Ascomycota. Here, as is mostly done, lichens are without a taxonomic group. A fungal phylum is not assigned to lichens.

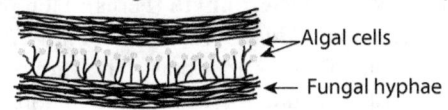

Figure 4.9. Diagram of lichen structure.

Special Remarks: Fungus living with algae in a **mutalistic**[245] **symbiotic**[246] relationship. The fungus provides shelter for the algae, while the algae provides food for the fungus. The photosynthetic partner may be the an algae, *Trebouxia* spp[247] or a cyanobacterium, *Nostoc* spp. See figure 4.9.

---

245   Both organisms benefit from the relationship.
246   Living in close association.
247   The abbreviation ssp. refers to pleural of species, several species. To refer to a single species sp. is used.

C. Kingdom Veridiplantae (True Plants)
General Characteristics of Viridiplantae

True plants are **eukaryotic, green (possess chlorophyll), multicellular** and **photosynthetic**. Viridiplantae have cell walls composed of **cellulose** and include familiar plants: mosses, ferns, conifers and flowering plants. Viridiplantae display **alternation of generations**,[248] *tropisms*, some are *aquatic*, but most are *terrestrial* species. There are about 300-315,000 species of plants. The first land plants appeared on Earth about 500 million years ago. Botanists generally use the term "division" rather than phylum in their classification schemes.

Division[249] I   Bryophyta (mosses) **Hepatophyta** (liverworts): Mosses have no vascular system, thought to cure diseases of the liver in times past. They are about 10 cm (4 in) long, 1-10 cm (0.5 - 4 in) tall, no vascular system, tiny leaves, *no flowers*, no seeds, produces spores, grow in clumps, live in low light and damp environments.
Genus: Sphagnum: Dead and compacted, sphagnum is called **peat** or peat **moss**.[250] Due to high acidity, bogs have preserved human bodies for thousands of years.

Division II   Psilophyta: Primitive tree-like vascular plant, lived 300 million years ago. Example: Whisk fern (*Psilotum nudum*).

Vascular Plants: All species of the divisions below are tracheophytes
Division III   Lycophyta (Club mosses): Lycophyta have no seeds or *flowers*. Lycophyta have tiny *true leaves*. They are the oldest living vascular plant.

Division IV   Sphenophyta (Horsetails): Sphenophyta have *not have flowers* or seeds. They do have tiny leaves and sporangia called sori.

Division V   Pterophyta (Ferns): Pterophyta is the second largest division of the Kingdom *plantae*. Ferns *do not have flowers* or seeds. Ferns do have leaves (fronds), stems and roots. This group is the first to display a more advanced leaf structure. Ferns are used in floral arrangements and have been know to improve contaminated soil.

Division VI   Cycadophyta (Cycads): **Cycadophyta** are **gymnosperms**.[251] Cycadophyta have a thick woody stem, one large seed, a pollen cone, palm-like evergreen leaves, but *no flowers*. One tree may be male and another female. Egg and pollen are not found on the same tree. Cycads can be up to 1,000 years old.

---

248   Parents and offspring have different appearances.
249   Plant taxonomists generally use the term division in place of phylum.
250   Increases soil's ability to hold water. Used in the cultivation of mushrooms.
251   Uncovered seeds. Unfertilized ovules (seeds) are fertilized directly because they are exposed to the air.

(Domain Eukarya continued, the True Plants)

Division VII   **Ginkgophyta** (Maidenhair tree, Ginkgo): Have pairs of seeds, *no flowers*. ***It is a living fossil.***

Division VIII   **Gnetophyta** (Gnetum & Welwitschia): *Does not have flowers.*

Division IX   **Coniferophyta[252] (Pinophyta):** Conifers are cone-bearing trees and shrubs. They *do not have flowers*. They produce seeds on cones that have woody scales. Gymnosperm means uncovered seed. The Great Basin Bristlecone Pine (*Pinus longaeva*) has been dated to 5064 years. This is the oldest tree known tree in the world.
 Family: Pinaceae: 202 known genera and about 2600 species.
  Genus: Picea (spruce)
  1. Species: *P. pungens* (blue spruce)

  Genus: Juniperus (juniper)
  Genus: Cupressus (cypress)
  Genus: Pinus (pine)
  1. Species: *P. sylvestris* (Scots pine)

Division X   Anthophyta: Angiosperms (flowering plants)
Angiosperms are the largest group of plants on Earth. They have flowers with reproductive organs. See figure 4.10. Angiosperms have stamens with pollen sacs. Flowering plants produce a **fruit[253]** containing seeds. ***Seed can be monocots or dicots.*** *Flowering plants consist of two classes, monocots and dicots.*

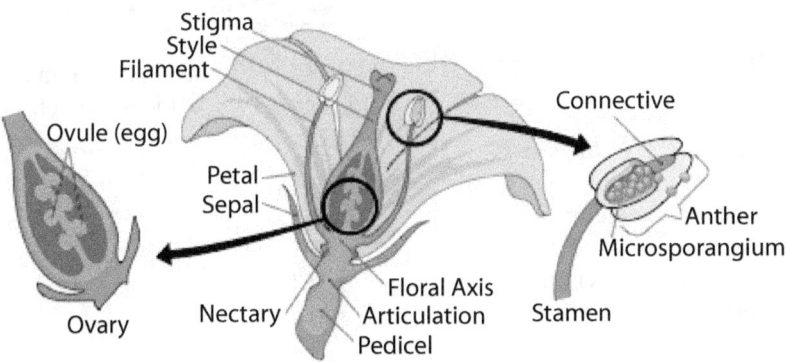

Figure 4.10. Generalized diagram of a mature flower.

Class: *Monocotyledoneae*: **(monocots)** Have one embryonic leaf in its seed. Veins on leaves are parallel to each other and flowers occur in threes or multiples of three.

---

252   Also known as *pinophyta* or coniferae.
253   A fruit is a way flowering plants spread their seeds. Fruits generally develop from an ovary. Many fruits are edible. Examples are apples, oranges and tomatoes.

Family: Liliaceae (Lily family)
Genus Lilium

Genus: Crocus

Family: Orchidaceae (Orchid family)
Special Remarks: 880 genera

Genus: Vani
1. Species: *V. planifolia* Vanilla flavoring derived from this species:

Family: Poaceae (true grasses) Cereals are gotten from these seeds: *corn ,wheat* and *rice*.
Special Remarks: Humans get half their calories from the three grasses above.

Class: *Dicotyledoneae: (Dicots)*
Dicots have two embryonic leaves in its seed (see figure 4.11). Its leaf veins are branched, flower parts occur in 4s or 5s .

Order: Magnoliales (5 families and 3000 species)
  Genus: Magnolia
  1. Species: *M. grandiflora*

Order: Laurales (7 families, 2900 species)
  1. Species: *Persea americana* (avocado)

Order: Urticales (6 families)
  Genus: Ulmus (elm)
  1. Species: *U. americana*

Genus: Cannabis *(*hemp)
  1. Species: *C. sativa*

Order: Fagales
  Genus: Betula (birch family)
  1. Species: *B. papyrifera* (paper birch)

Order: Ericales
  Genus: Camellia
  1. Species: *Camellia sinensis*[254] (tea)

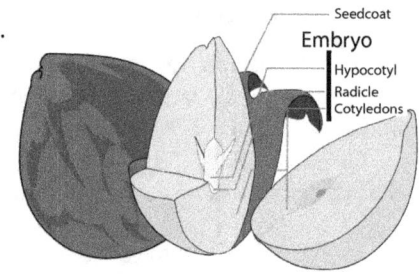

Figure 4.11. Generalized diagram of a dicot seed. Note the two embryonic seed leaves.

---
254   *Sinensis*, from the Latin, meaning from China.

(Domain Eukarya continued, the True Plants)

Order: Violales
  Family: *Begoniaceae* (begonia)
  Family: *Violaceae* (violet and pansies)

Order: Rosales
  Rosaceae (rose family)
  Rose, apples, apricots, plums, cherries, peaches and pears.

Order: Fabales (legumes)
  Phaseolus vulgaris (beans)
  Pisum sativum (pea)
  Glycyrrhiza glabra (licorice)

Order: Rhamnales
  Family: *Vitaceae* (grape family)

Order: Sapindales
Genus Citrus: (maples, mahogany, mangos, frankincense and myrrh)

Genus: Toxicodendron
  1. Species *T. radicans* (Eastern poison ivy) See figure 4.11.
Special Remarks: Note the ***three leaf arrangement***. *T. radicans* has ***reddish coloration*** and are generally ***shiny***. See figure 4.12.

Order: Apiales: (carrots, celery, parsley and ivy)

Order: Solanales: includes nightshade plants (potatoes, eggplants, tomatoes, peppers, tobacco and petunias)

Figure 4.12. Leaves of the poison ivy plant.

Order: Asterales
Genus: Lactuca
  1. Species: *L. sativa*[255] (lettuce)

Genus: *Ambrosia* (there are over 50 species of ragweed) Ambrosia is a North American **weed**.[256] It is responsible for mild to severe ***allergic rhinitis*** reactions.
  1. Species: *A. artemisiifolia* (common ragweed)
  1. Species: *A. trifida* (great ragweed)

---

255   From the Latin, meaning cultivated.
256   An unwanted or undesirable plant.

(Domain Eukarya continued, Animals)

D. Kingdom Animalia

General Characteristics of the Animal Kingdom

Animals are multicellular, **heterotrophic**,[257] *eukaryotic organisms that* **lack chlorophyll and cell walls**. With the exception of sponges and other lower **phyla**, most animals produce a single egg that develops into a multicellular organism. Some undergo **metamorphosis**[258] and *most display locomotion or movement from place to place.* The fossil record indicates that the first animals were marine. The simplest animals appeared 600 million years ago. Most animals appeared in what is termed the **Cambrian**[259] explosion 542 million years ago.

Invertebrates: The term invertebrate is convenient to use, but is not a recognized taxonomic group any longer. It is however, an easy way to group animals into those animals that have a backbone (vertebral column) and those that do not. Invertebrate animals makes up over 90% of all animal species that are still alive on earth, that is, not extinct. Invertebrates include **sponges, starfish, jellyfish, rotifers, nemerteans** (ribbon worms), **roundworms, flatworms, annelids** (earthworms)**, arthropods** (crabs, shrimp, spiders, lobsters and insects).
Subkingdom: Parazoans – not microscopic, have differentiated tissues.

Phylum I Porifera (sponges – bearers of pores). **Sponges have only one germ layer**.[260] Sponges do not have true tissues. They are aquatic filter feeders, with pores in the body wall. The sponge's body is **asymmetric**[261]. The body cavity is called a *spongocoel* and is lined with **choanocytes**.[262] When alive sponges have a gelatin-like *mesohyl* (mesenchyme) that is bounded above and below by a single layer of cells. Mesohyl is made stiff with **spicules**[263] of calcium carbonate. Sponges do not have nervous, digestive or circulatory systems. Sponges are used to filter water supplies, cleaning utensils and padding. It is the only living parazoan. Most sponges are marine, but some live in fresh water.
Subkingdom: Eumetazoa – Eumetazoa have true tissues and the tissues are organized into germ layers. Eumetazoa have an embryonic gastrula stage.
Class: Calcarea – have spicules of calcium carbonate ($CaCO_3$)

Class: Hexactinellida – have spicules of silica or silicon dioxide (SiO_2)

Class: Demospongiae – have spicules of protein and/or silica

---

257   Heterotrophs obtain food from the environment and digest it in a space within body cavity.
258   Change in appearance from the young to the adult form.
259   The first period of Paleozoic era. The fossil ancestors of most of the modern animal phyla known today are found in this era.
260   Gives rise to body tissues.
261   Lacking symmetry.
262   Flagellated cells that line the body cavity of sponges. Flagella keep water moving through pores that filter small food particles from the water.
263   Needle-like or spike-like.

80

Radiata: (unranked).[264] – Symmetry is **radial;**[265] **two germ layers** are present (ectoderm and endoderm).

Phylum II Echinodermata – Marine animals
Class: Asteroids (starfish)
 Genus: Asterias
  1. **Species:** *A. rubens* (common starfish); has five arms

Phylum III Cnidaria (jellyfish) – Cnidarians have stinging cells called **cnido-cytes**[266] and a gelatin-like substance between two one cell layers of epithelial cells. Cnidaria are free swimming, primarily marine and some freshwater. Cnidaria have a network of nerves called a "nerve net" and organs that can detect light. Cnidarians reproduce sexually and undergo three stages of development after fertilization: **planula**[267] larva, **polyps**[268] and **medusae.**[269] Some cnidaria reproduce asexually by budding. The average lifespan of some species is hours to a few months. Cnidaria lack a tube-like digestive system.

Class: Hydrozoa (hydroids): Hydra-like animals, live in fresh water and some live in salt water. They have two thick layers of cells. Each layer is one cell thick.

 Genus: Hydra
  1. **Species:** *H. vulgaris (attenuata)* See figure 4.13.

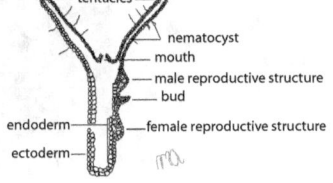

    H. vulgaris lives in fresh water. Its mouth is surrounded with tentacles that has stinging cells called *nematocysts*. The mouth leads to a single gastric cavity. Hydra live attached to a structure by its base. Reproduction can be sexual with egg and sperm, or asexual by budding. Hydra has regenerative properties. If a hydra is cut into many pieces, each piece can regenerate into a new hydra.

Figure 4.13. Cross section of hydra.

Class: Scyphozoa (true jellyfish): Its life phases are: *polyp* and *medusa*. The polyp is motile and the medusa is sessile. The moon jelly *Aurelia aurita* can cause major disruptions in fishing grounds in seas between Japan and China. True jellyfish are also used for food.

Phylum IV Rotifera – Rotifers are almost microscopic. Most rotifers live in freshwater. Some are marine. Rotifers exhibit bilateral symmetry.
 Genus: Rotifer
  1. **Species:** *R. vulgaris*

---

264   Names of groups no longer used in classification but these names act as a guide to where the traditional classification left off.
265   Like a pie. Cutting anywhere through the center will give two identical halves.
266   Used for capturing prey.
267   Flat, ciliated and free-swimming form.
268   Has a sac-like shape with two germ layers, an ectoderm and endoderm.
269   The familiar form of the jellyfish is the medusa stage.

Phylum VI Nematoda (roundworms): This phylum contains about 28,000 species of nematodes. ***Sixteen thousand are parasitic***. Nematodes have a slender body ranging in size from the microscopic to 5 m in length. The mouth may have teeth. Its mouth leads to a pharynx, intestine, rectum and *terminates at an anus*. A cuticle covers single layer of cells that makes up the epidermis. A muscle layer is under the epidermis and the body cavity is below the muscle layer. Nematodes lack blood and a stomach. Four nerves run the length of the body (dorsal, ventral and two lateral nerves). There are separate male and female worms, but some are **hermaphroditic**[270] and some **parthenogenic**.[271]

Class: Dorylaimea: A nematode that has a tooth or hollow spear-like structure protruding from the wall of the mouth.

Genus: Trichuris

1. **Species:** *T. trichiura* (human whipworm) **Disease:** Trichuriasis. See figure 4.14.

Special Notes: Causes an infection of the large intestine, transmission - soil.

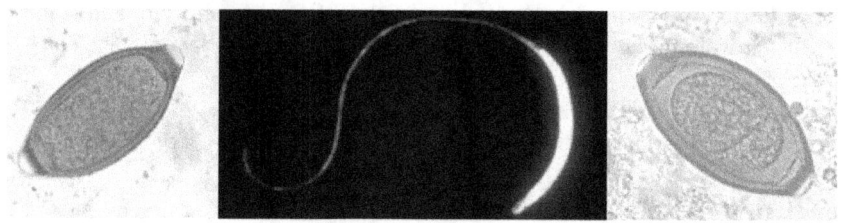

Figure 4.14. Left: Egg of *T. trichiura* in an iodine-stained wet mount. Right: Egg of *T. trichiura* in an unstained wet mount. Center: Micrograph of an adult female Trichuris human whipworm that is approximately 4 cm long.

Class: Secernentea
 Genus: Ancylostoma
  1. **Species:** *A. duodenale*
Special Notes: **"Old World"**[272] human hookworm.

 Genus: Necator
  1.**Species:** *N. Americanus* See figure 4.15.
Special Notes: New World[273] human hookworm.

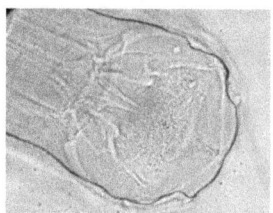

Figure 4.15. Adult *Necator americanus*.

 Genus: Caenorhabditis
  1. **Species:** *C. elegans*
Special Notes: *C. elegans* is a transparent nematode about 1 mm in length that lives in soil. It is a model organism for study of how organisms develop, differentiate and grow.

---

270   Male and female sex organs are on the same organism. Can self fertilize.
271   Development of an embryo from an *unfertilized* egg.
272   Africa, Europe, and Asia are considered to be the "Old World."
273   Referring to North and South America. The term was coined by the Italian navigator and explorer, Amerigo Vespucci (1454-1512) in the 1500s.

(Domain Eukarya continued, Animals)

Order: Ascaridida
 Genus: Ascaris
  1. Species: *A. lumbricoides* See figure 4.16.
Disease: Ascariasis
Special Notes: *A. lumbricoides* infections are one of
the most common worm infections in humans. A few
worms may pass unnoticed. Large numbers of worms
in children can cause intestinal blockage. In adults,
the worms can block the bile duct or they may migrate to the lungs. Awareness of the infection occurs
when the worms are seen in feces or when they try to
exit the body through the nose or mouth.

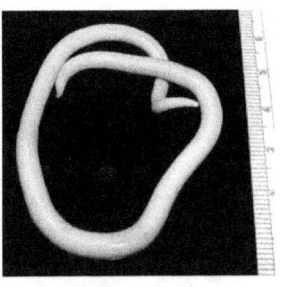

Figure 4.16. The
parasitic roundworm *A.
lumbricoides.*

Order: Trichurida
 Genus: Trichinella
  1. Species: *T. spiralis* See figure 4.17.
Disease: Trichinosis
Special Notes: *T. spiralis* is a parasite of pigs, bears and
humans. Pigs are primary hosts(definitive host). Trichinosis
in humans (intermediate host/secondary host) is gotten from
eating undercooked pork or wild game containing the worm
larvae. Worms can migrate from the intestines to striated
muscle. Invasion of the central nervous system is the most
dangerous.

Figure 4.17. *T.
spiralis.* The
"pork worm."

Phylum VII Platyhelminthes (flatworms)
Has a *ventral* nerve cord. Live in fresh water,
marine and terrestrial.
Class: Cestoda
 Genus: Taenia
  1. Species: *T. saginata* See figure 4.18.
Disease: Taeniasis
Special Notes: Parasitic *flatworms* are called
tapeworms. The cow is the intermediate host.

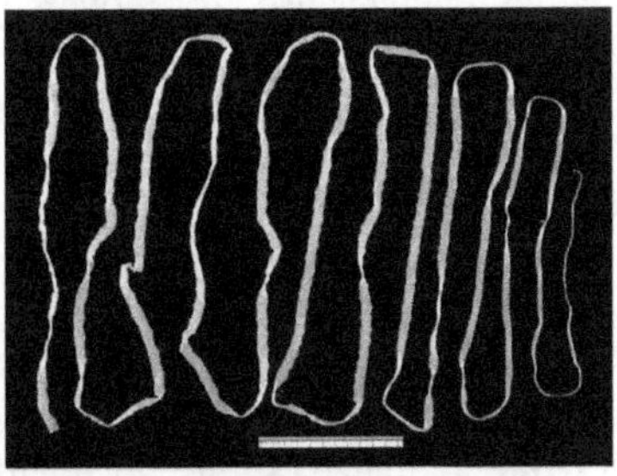

Figure 4.18. *Taenia saginata,*
the beef tapeworm. Notice the
twelve inch ruler at center bottom of this photo.

Class: Trematoda (flukes)
 Genus: Shistosoma
  1. Species: *S. mansoni*
Disease: Schistosomiasis (called bilharzia, snail fever and Katayama fever)

Genus: Planaria
 1. Species: *P. maculata:* Freshwater flatworm used for instruction.

Phylum VIII Nemertea (ribbon worms or proboscis worms) – Marine roundworms that feed on annelids, clams and crustaceans.

Phylum IX Annelida (aquatic and terrestrial segmented worms) – Burrowing worm that comes to the surface to feed. Annelids aerate the soil.
Class: Polychaeta (marine worms)
Class: Oligochaeta (earthworms)
 Genus: Lumbricus
  1. Species: *L. terrestris:* The common earthworm, the "farmer's friend."
Class: Hirudinea (leeches) – Have anterior and posterior suckers. The anterior sucker has a jaw and teeth. The leech injects anticoagulants and anesthetic molecules into the patient. Presently medicinal leeches are used to help improve blood circulation in damaged tissues such as skin grafts.
 Genus: *Hirudo*
  1. Species: *H. medicinalis*

Phylum X Mullusca (clams, oysters, mussels, scallops) – Aquatic animals, very diverse group.
Class: Bivalvia: two-hinged shells (clams, scallops, oysters, mussels)
 Genus: Mercenaria (clams)
  1. Species: *M. mercenaria* (the quahog, a hard edible clam)

 Genus: Ostrea (oyster)
  1. Species: *O. lurida*: Found on North American Pacific coast.

Class: Cephalopoda (squid): Literally head-foot. It has long, flexible organs called tentacles used for movement and grasping prey.

Class: Gastropoda (snails and slugs): Literally stomach-foot. Gastropods are univalves as opposed to clams which are bivalves (hinged shell).

Phylum XI Arthropoda – Have jointed appendages.
Class: Trilobites (extinct)

Class: Arachnida (spiders, ticks, scorpions)
Order: Araneae (8 legs, air breathing, fangs to inject venom)
Family: Theraphosidae (tarantulas)
 Genus: Brachypelma
  1. Species: *B. smithii* (Mexican tarantulas)

Family: Ixodidae (hard ticks): ectoparasites
 Genus: Ixodida
  1. Species: *I. scapularis* (deer tick): It is an arthropod vector for Lyme disease.
 Genus: Dermacentor
  1. Species: *D. variabilis* (dog tick): Can transmit Rocky Mountain spotted fever and tularemia.
Class: Chilopoda (centipedes): One pair of legs per body segment, found in soil, under stones, inside logs and in dead wood. Centipedes can bite humans.

(Domain Eukarya continued, Animals)

Class: Diplopoda (millipedes): two pair of legs per body segment, most are herbivores, poison fangs, may prey on insects and earthworms.

Subphylum: Hexapoda – Hexapods have *three body segments* and *six legs*.
Class: Insecta:[274] About 1 million species exist and have been described. Insecta have an *exoskeleton* composed of chitin, the body divided into head, thorax and abdomen. Insects have *one pair of antenae*, *six-jointed legs* and compound eyes. Some have wings and undergo **metamorphosis.**[275]

Metamorphosis is complete change in appearance. For example, a caterpillar larva is "worm-like" and it is completely different from the adult butterfly.
Order: Hymenoptera: Social insects, live in colonies (bees, wasps and ants).
Order: Blattodea termites
Order: Phthiraptera lice and bed bugs. See figure 4.19.
Order: Orthoptera grasshoppers, crickets and locusts.
Order: Lepidoptera (butterflies and moths)
Suborder Frenatae Includes Most moths.
Special Notes: Moths are important agricultural pests. Larvae of certain moths destroy corn and others destroy forest trees.

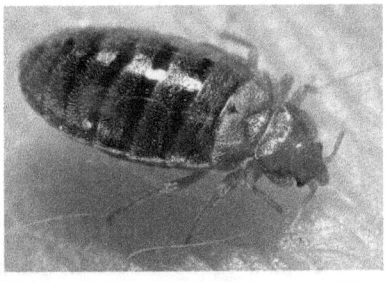

Figure 4.19. *Cimex lectularius* The common bed bug.

Superfamily: Papilionoidea (butterflies)
Order: Diptera: Housefly, mayflies, dragonflies, fireflies, mosquitoes and any other flying insects with "fly" in the name.
Family: Muscidae
 Genus: Musca

 1. **Species:** *Musca domestica* (housefly) See figure 4.20.
Special Notes: Houseflies carries diseases that are easily communicable. Houseflies feed on feces and spread serious human disease such as: *campylobacteriosis, salmonellosis, tuberculosis, Q fever, typhoid, cholera, dysentery* and eggs of parasitic worms.

Campylobacteriosis is a severe gastroenteric human bacterial disease caused by *Campylobacter jejuni*. Millions of people are infected every year in the industrialized world, making it the most common food borne infection.

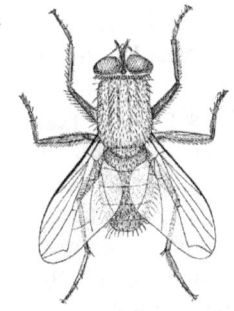

Figure 4.20. Illustration of the common housefly, *Musca domestica.*

Q fever is caused by the bacterium *Coxiella burnetii*. Humans may develop mild fever, pneumonia, hepatitis. Q fever can be fatal.

---

274   From the Latin, *insectum* meaning to cut into sections.
275   A process of developing. Amphibians have a fish-like stage (tadpole) that lives in water. The adult frog looks quite different and lives on land and in water.

(Domain Eukarya continued, Animals)

Family: Culicidae (Mosquitoes)

Mosquitoes suck blood of humans and other vertebrates. Mosquitoes are vectors for many serious human infectious diseases, millions of people are affected every year.

Genus: Aedes

  1. **Species:** *aegypti* See figure 4.21.
Spreads **yellow fever**[276] and dengue fever.

Genus: Anopheles

  2. **Species:** *gambiae*

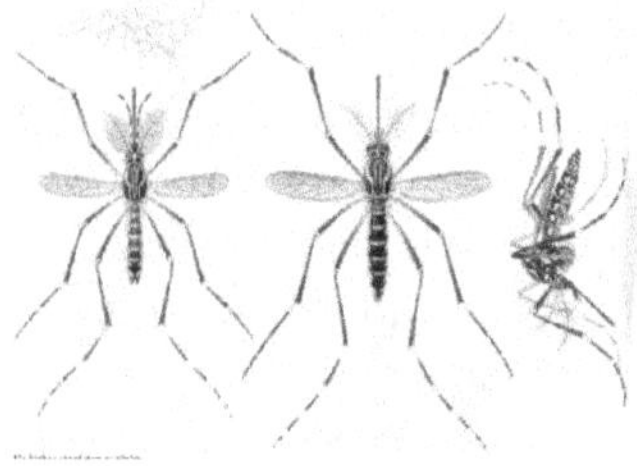

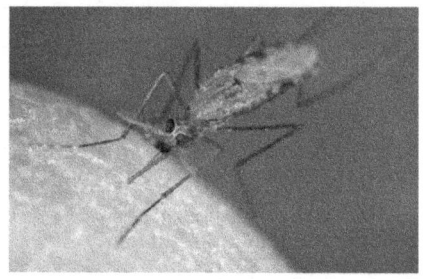

Figure 4.21. *Aedes aegypti.*

See figure 4.22. Of the hundreds of anopheles mosquito species that exist, only 30-40 species carry malarial parasite **plasmodium**. Prevalence is low except in West Africa. Data for 2010 indicate over 200,000 new cases occurred and over 500,000 people died worldwide.

Special Notes: In the United States about 1,500 cases of malaria are reported yearly. Most of these cases are seen in travelers and immigrants from malaria prone areas of the world.

Figure 4.22. *Anopheles gambiae.*

Class: Malacostraca (woodlice)
Order: Isopoda
  Genus: Armadillidium
  1. **Species:** *A. vulgare* (woodlouse)
Special Notes: Rolls up into a ball as a protective mechanism.
Infraorder: Brachyura, king crabs, horseshoe crabs
Species: *Pagurus bernhardus* (hermit crab)

Class: Maxillopoda (barnacles) – Barnacles grows on natural and man made underwater structures. The barnacle cements itself to underwater structures to a depth of about 20 feet. Barnacles are **hermaphroditic**.[277] At one stage in its life cycle the larvae float on the current and at the next stage in its life cycle it attaches to a surface. Barnacles are mostly an annoyance in the maritime industry.
Order: Sessilia (barnacles)
  Genus: Balanus
  1. **Species:** *B. improvisus* (bay barnacle)
  2. **Species:** *B. crenatus* (rock barnacle)

---

276   Walter Reed, an American physician proved the disease is transmitted by *Aedes aegypti*. The disease is a viral hemorrhagic disease called historically "Yellow Jack." Because a yellow flag had to be flown on ships if the disease was on board.
277   Male and female reproductive structures are on the same organism.

Subphylum: Crustacea – A large group of arthropods
Class: Malacostraca (crayfish, crabs, lobsters, prawns and shrimp)
Class: Branchiopoda (brine shrimp)
 Genus: Artemia (a brine shrimp)
  1. Species: *A. salina* lives in salt lakes, sold to children as "Sea Monkeys."
Special Notes: Used to test different chemicals for toxicity

Order: Decapoda, the ten footed, most are scavengers
Family: Nephropidae – Clawed lobsters, marine, lives on the seas floor, five pairs of legs, possess claws, can live to be over 100 years old, come in somewhat rare colors such as blue, red, and recently a yellow/orange one, recently caught and donated to the New England Aquarium, a commercially valuable food
 Genus: Heterocarpus
  1. Species: *H. ensifer* (shrimp)
Genus Homarus
  1. Species: *H. americanus* American lobster (Main lobster)
  2. Species: *H. gammarus* – European lobster

Genus: Metanephrop
  1. Species: *M. japonicus* Found in waters off Japan. See figure 4.23.

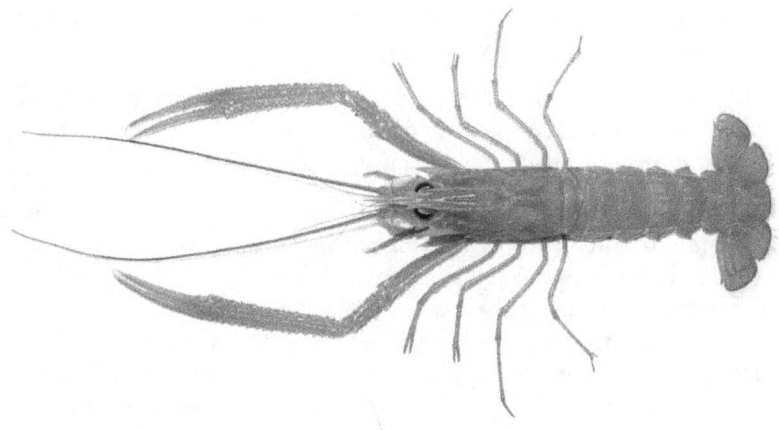

Figure 4.23. *Metanephrops japonicus.*

Phylum XII Chordata: Chordates have a **dorsal**[278] nerve cord as opposed to some invertebrates such as nematodes (roundworms), annelids (ringed worms) and arthropods (insects, arachnids, and crustaceans) that have a ventral nerve chord.

Subphylum: Tunicata Most are sessile and have three embryonic layers: endoderm, ectoderm and mesoderm.

Class: Ascidiacea Ascidiacea is a filter feeder. Its common name is the "sea squirt."

Subphylum: Cephalochordata Have a **notochord**,[279] as a feature throughout life. It is referred to as the "**first backbone.**" A representative animal is the lancelet (amphioxus). Amphioxus has a small soft body about 7 cm long.

Subphylum: Craniata (Vertebrata) Has a **skull**,[280] cartilage *or* bony skeleton. Includes all fish, amphibians, reptiles and mammals – chordates with backbones.

Class: Sarcopterygii (Coelocanths). Lungfish. Lungfish are living fossils.

Class: Chondrichthyes (sharks, rays and skates) have a two chambered heart
     Ectothermic:[281] fish with cartilage skeletons, jawed and paired fins.

Class: Osteichthyes (bony fish) have a two chambered heart
     Ectothermic– Bony skeletons.

Class: Amphibia: toads, frogs and *caecilians* (*legless frogs*). Caecilians look like worms, snakes or salamanders. Amphibia have a three-chambered heart.
     Ectothermic – lives in water and on land.

Class: Reptilia lizards, snakes, turtles, crocodiles, alligators and dinosaurs (extinct). Reptiles have a three-chambered heart.
     Ectothermic – breathe air and lay eggs.

Class: Aves (birds) have a four-chambered heart, vertebrates that have feathers.
     Endothermic,[282] wings, bipedal and lays eggs. Birds evolved from
     dinosaurs and carry on the dinosaur lineage.

Class: Mammalia (mammals) have a four-chambered heart.
     Endothermic, female of the species has two pectoral, functional mammary glands, male not functional, hair, middle ear bones, most have sweat glands and specialized teeth. Size range: bat (3 cm) to blue whale (33 m).

Order: Monotreme (duck-billed platypus): *egg-laying mammal.*

Infraclass: Marsupialia (kangaroos, wallabies, possums, koala, wombats). Marsupials carry young is a pouch.

---

278   Dorsal, from the Latin, *dorsum* meaning "back."

279   All embryos of chordates have a pliable rod-shaped structure called a notochord. It may continue throughout life in some chordates and in some vertebrates becomes a vertebral column.

280   A brain, organs of smell, sight, and hearing enclosed by this structure.

281   Cold blooded, will depend on heat present in the environment.

282   Warm-blooded animal that maintains a constant body temperature.

Infraclass: Eutheria (**placental**[283] **animals**)
Superorder: Euarchontoglires (rodents, lagomorphs, tree shrews, primates, including humans).
Order: Artiodactyla: Are even-toed ungulates (deer, camels, pigs, cows, sheep).

Order: Perissodactyla: Are odd-toed ungulates (horses, rhinoceroses)

Order: Carnivora: carnivores: (cats, bears, weasels)

Order: Cetacea: (whales, dolphins)

Order: Chiroptera: (bats)

Order: Insectivora: insect-eaters (hedgehogs, moles, shrews)

Order: Lagomorpha: (rabbits and hares). Rabbits bear their young in underground burrows, lacking fur and eyes closed, while hares bear their young in shallow nests above ground fully furred with eyes open.

Order: Primates: (apes, monkeys, lemurs, humans)
Eyes face forward, eye socket surrounded with bony ridge, cranium is dome-shaped, most have a large brain, each limb has five digits, fingernails, most have opposable thumbs, the collar bone is part of pectoral girdle.
Family: Hominidae (hominids)
Genus: Pongo (orangutans)
Tribe: Hominini[284]
Subtribe: Homininae[285]
Genus: Gorilla (gorillas)

Genus: Australopithecus
 1. Species *A. africanus* (*A. africanus* was named "Lucy." She is an early hominid that lived 3-2 Million (**Ma**)[286] years ago.

Genus: Homo

 1. Species: *H. habilis*: (extinct) lived 2.3- 1.4 Ma ago.
 2. Species: *H. erectus*: (extinct) lived 2 Ma-70,000 years ago.
 3. Species: *H. neandertalis*:(extinct) lived 130,000 to 30,000 years ago.
   In Europe, *Homo neanderthalensis* encountered *Homo sapiens. May have*
     interbred with H. sapiens.

 4. Species: *H. sapiens*: (humans) The only primate that is considered a **person.**[287] *H. sapiens* appeared 200,000 to 150,000 years ago. Some migrated out of Africa about 70-60,000 years ago.

---

283   An organ that permits nourishment of the embryo and fetus before birth.
284   Includes humans, common chimpanzee and the bonobo.
285   Erect bipedal locomotion. Exclusively seen in *Homo sapiens*.
286   Megaannum equals 1,000,000 years or in scientific notation 10^6 years.
287   Ability to have citizenship, equality, liberty, legal rights and responsibilities.

Chapter 5 The Divisions of Natural Science

Knowledge

The human **mind**[288] likes to group things in order to understand where it is and where it is going. The mind thinks, has emotions, perceives, remembers and imagines. *The mind is not a physical structure* to be confused with the brain. The brain can be directly observed and be dissected, but the mind cannot. The mind can be directly accessed by the individual that possesses it. It can be accessed only indirectly by others.

The mind wants **to know**.[289] It desires to understand things. It classifies physical objects and phenomena in its environment, thus leading to the construction of general principles called theories and laws.

The mind studies itself and how it acquires knowledge as well as studying the natural world around It. *The human mind learns how it learns.* **Epistemology**[290] is the science of what knowledge is and how knowledge is acquired. Humans say, "*I know*," but what is it "to know?" What is known? Humans define things. Definitions set boundaries that can be pushed further into "knowing" a subject. The circle below represents the total of what is "known."

None of the following schemes are all inclusive. These schemes simply give an idea of what humankind's knowledge base includes. In addition, there is an overlapping of knowledge between different "branches" of knowledge.

All human knowledge can be divided into the *arts* and *natural sciences*. See figure 5.1. The arts and sciences are society's collective creative spirit of literature, poetry, fine arts and exploration of causation of natural phenomena.

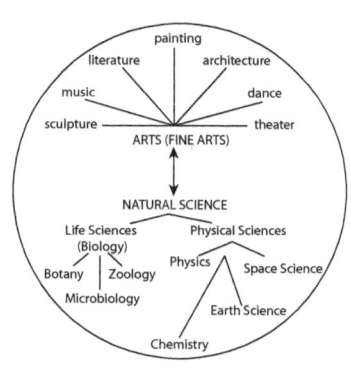

Figure 5.1. The "world" of knowledge.

The arts and sciences influence each other. For example, Italian artists dissected cadavers and added to the knowledge of anatomy as well as representing of the human form during the period we call the Renaissance. Their dissections and art inspired physicians and philosophers of the time to think differently about humankind and the world in general. Their collective cultural efforts resulted in a rebirth of learning.

---

288   Three theories deal with mind/brain: *dualism*, *materialism* and *idealism*.
289   To know is to have factual, descriptive knowledge, skills through experience or education.
290   Epistemology is the study of theories of knowledge.

The Natural Sciences

Scientists uncover *causes of events in nature*. Scientists organize their discoveries to see patterns in nature. Natural science is divided into two general areas of study: *biological science* and *physical science*. See figure 5.2.

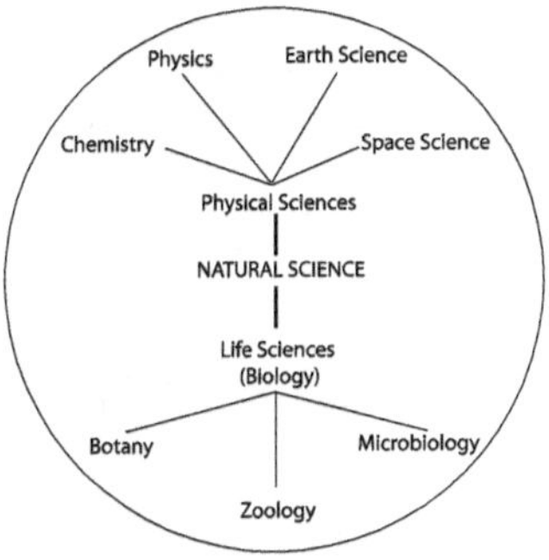

Physics   Earth Science
Chemistry   Space Science
Physical Sciences
NATURAL SCIENCE
Life Sciences (Biology)
Botany   Microbiology
Zoology

Figure 5.2. A representation of branches of natural science.

There are no strict boundaries between individual sciences. For example, The science of *biochemistry* studies chemical reactions that take place in an organism. Biochemistry requires knowledge of biology and chemistry. *Biophysics* requires knowledge of physical laws that aid in the understanding of biological systems.

The Life Sciences

Biology studies living things. Living things possess unique characteristics nonliving things do not exhibit. See chapter 2. Biology is divided into three branches, *zoology*, *botany* and *microbiology*. These sciences consist of smaller specialized areas of study. See figure 5.2.

1. *Zoology* studies animal life. Animals are defined as eukaryotic multicellular organisms lacking chlorophyll. Animal nutrition is *heterotrophic*. Heterotrophs feed on other life forms. Some major branches of zoology are **anatomy**,[291] **physiology**,[292] **histology**,[293] **ornithology**,[294] **ichthyology**,[295] *embryology* and *comparative zoology*.

2. Botany studies plant life. Plants are eukaryotic, multicellular photosynthetic organisms. Their mode of nutrition is *autotrophic*. Plants use sunlight to make their food from carbon dioxide and water. A few divisions of botanical science are **phytotomy**,[296] *plant physiology*, **phytopathology**,[297] **agronomy**,[298] *plant ecology* and *plant genetics*.

3. Microbiology studies life forms that require a microscope to be observed. See figures 2.1a, b and c in chapter 2. Microbiology studies *prokaryotes* and one celled *eukaryotes*. See figure 2.3 and 2.4 in chapter 2. Bacteria are prokaryotes. Algae, fungi, protozoa and parasitic worms are eukaryotes.

291	The study of structure of living things.
292	The science that studies the function of living things.
293	The science of tissues.
294	The study of birds.
295	The study of fish.
296	Plant anatomy.
297	Diseases of plants.
298	The study of crop plants such as corn, wheat and other grains.

Virology,[299] *bacteriology*, **algology**,[300] **mycology**,[301] *protozoology and parasitology* are just a few of the many subdivisions of microbiology. See figure 5.3. *The above six subjects are generally considered to define the science of microbiology,* even though some organisms studied are **macroscopic**, such as mushrooms, some parasitic worms and some algae.

The Physical Sciences

Physical sciences explore the non-living world; rocks, water and the atmosphere of the Earth and **celestial**[302] bodies. Physical science includes *chemistry*, *physics*, *earth science* and *space science*.

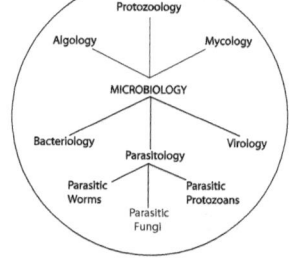

Figure 5.3. Branches of microbiology.

1. Chemistry may be broadly divided into *inorganic chemistry* and *organic chemistry*. Inorganic chemistry studies *simple* substances that *do not contain the element carbon* and are *not associated with living things*. Carbon dioxide (CO_2) is an exception.

Organic chemistry studies molecules that have *complex* structures, *contain carbon* and are *associated with life*.

Biochemistry is a branch of chemistry that overlaps with biology. Biochemistry studies the chemistry of living things. Knowledge of chemistry is essential for understanding biology. Below is an elementary review of some topics in chemistry.

A Brief Overview of the Science of Chemistry

$$2H_2 + O_2 \xrightarrow{\triangle} 2H_2O$$

REACTANTS → PRODUCTS

Figure 5.4. The chemical equation representing the formation of water from hydrogen and oxygen gases.

Chemistry *studies how different substances react with each other.* **Chemical equations**[303] *represent chemical reactions.* The balanced chemical equation in figure 5.4 demonstrates some fundamental concepts of chemistry. The *reactants* side of the equation shows two molecules of hydrogen ($2H_2$) and one molecule of oxygen (O_2). A "+" sign separates the reactants on the left side of the equation. Hydrogen gas burns. Oxygen gas supports burning. When thermal energy (triangle) is added to the mixture of gases, a chemical reaction takes and two water molecules result. Water extinguish fires. Water is shown on the *product* side of the equation. Water

299   Virology is the study of viruses.
300   The study of algae.
301   The study of fungi.
302   Dealing with stars, planets and other bodies that appear in the sky.
303   Using symbols and formulas to represent and describe a chemical reaction.

92

has properties different from hydrogen and oxygen atoms. *The original properties of the reactants are lost* after a chemical reaction takes place.

The same *kind* and *numbers* of atoms are shown on both sides of the chemical equation using symbols, formulas and numbers. According to the *Law of Conservation of Matter*, the same number of and kinds of atoms on the left side of the equation must be present on the right side of the equation. Count the atoms up; four H atoms and two O atoms are on the right side and four H atoms and two O atoms are on the left side.

2. Physics is also divided into many subsciences. Common divisions of physics are **classical mechanics,**[304] gravity, light, lenses, sound, electricity, magnetism and **quantum mechanics.**[305]

A Brief Overview of the Science of Physics

The science of physics studies how *matter* and *energy* are related. *Matter is anything that has mass and takes up space (volume).* **Energy is defined as the capacity or has the ability to do work.**

The most famous example of this relationship of matter and energy is Albert Einstein's famous equation: $E = mc^2$. In this equation **E** represents *energy*, **m** represents *mass* and **c** is equal to the **speed of light.**[306] The validity of this equation was proven when a piece of uranium the size of an apple was converted into energy with the power of 20 thousand tons of TNT. More **benign**[307] examples of the relationship of matter and energy would be the use of **simple machines.**[308] A *lever* is a rigid bar that rotates around an axis on a pivot point called a *fulcrum*. Simple machines, multiply, decreases or reverses the direction of an applied force.

A lever can be a wooden beam or a crowbar (*matter*). A *force* or effort is put in at one end of the lever and multiplied at the other end. The effort force moves a force called the load or *resistance* (R). The movement exhibits the property of *energy*.

Levers can produce a *mechanical advantage* (**MA**). Anyone can attest to this if they have tried to open a container's lid with their fingers. Using a screwdriver offers greater mechanical advantage than fingers. MA of lever is a measure of how much the **effort** (force) is multiplied by a machine. The formula used to calculate MA is $MA = F_e(input)/F_r$ (**output**). The distance from the fulcrum to where the effort is applied is referred to as the *effort arm* (F_e). The distance from the fulcrum to where the resistance is applied is referred to as the *resistance arm* (F_r).

---

304   The study of how large bodies (matter) react when a force is applied to them.
305   Studies the nature of particles, waves and the interaction of matter and energy.
306   In a vacuum it is 186,000 miles per second (300,000,000 meters per second).
307   Not harmful.
308   There are six: lever, screw, inclined plane, wedge, pulley, wheel and axle.

The positions of the **fulcrum**,[309] *load* (**R**) and the applied *force* (**E**) determines the class of a lever. There are three classes of levers: first, second and third.

A *first class lever:* See figure 5.5a. Nodding the head is like a seesaw, a first class lever. It is made up of the *condyles* (flat projections at the base of the occipital bone of the skull) and the first cervical vertebra, the *atlas* (C1). The *atlas* and the *occipital bone* form the *atlanto-occipital* joint.

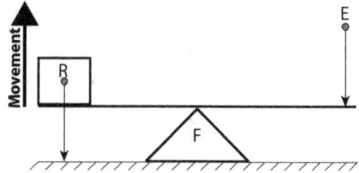

Figure 5.5a. The fulcrum is *between* the applied force and the load (resistance).

A *second class lever:* See figure 5.5b. An example of a second class lever in the human body is raising one's self onto the tip of the toes (*plantar flexion*). Plantar flexion is the increase of the angle of the foot *from* a right angle to more than 90 degrees of the foot with respect to the leg. Lifting a wheelbarrow is another example of a second class lever.

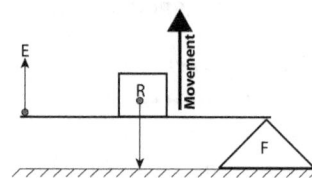

Figure 5.5b. Class 2: The load is *between* the fulcrum and the applied force (effort). A wheelbarrow and plantar flexion of the human foot are examples.

A *third class lever:* See figure 5.5c. *A third class lever is the most common lever in the human body.* **Flexing**[310] *the biceps brachii to raise the forearm is a common example of a third class lever.* Flexion is decreasing the angle of the **forearm**[311] with respect to the **arm**.[312] The fulcrum is the elbow and the resistance is the weight of the fore-

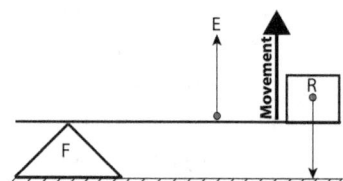

Figure 5.5c. Class 3: The effort force is applied *between* the ful-crum and the load (resistance). In this example, a weight in the hand is being lifted and being brought closer to the arm.

arm. A third class lever is used when a mass is held in the hand and the angle of the forearm is decreased with respect to the arm.

3. Earth science studies nonliving things on Earth: its atmosphere, **hydrosphere**[313] and the **lithosphere**.[314] Geology is the study of the Earth's rocks and the processes that enabled the Earth *to form* and change over time. *Petrology* is the study of rocks.

309   Pivot point around which the lever rotates.
310   To decrease the angle of two articulated bones.
311   Elbow to the wrist.
312   Shoulder to the elbow.
313   The Earth's waters.
314   The Earth's crust.

Humans live on a solid Earth bathed in a gaseous atmosphere that consists of 79% nitrogen, 20% oxygen and less than 1% of other gases. See table 5.1.

Table 5.1. Composition of the Earth's atmosphere.		
Gas or Substance	Symbol or Formula	Percentage
Nitrogen	N_2	78.08
Oxygen	O_2	20.95
Water vapor	H_2O	0 to 4
Carbon dioxide	CO_2	0.038
Argon	Ar	0.93
Neon	Ne	–
Helium	He	–
Krypton	Kr	–
Nitrogen oxides	–	–
Sulfur compounds	–	–
Ozone	–	–

4. Space science studies everything beyond the Earth's atmosphere. Space science examines how the **universe**[315] began, and how the Earth, stars and planets were formed. The universe to the ancients consisted of the visible stars and planets. Today the *universe* is known to be much more. It includes all that exists in space: stars, planets, **galaxies**[316] and solar systems.

The universe is estimated to be about 13.8 billion years old. The universe consists of the planets and all other bodies in space. The Solar System is approximately 4.6 billion years old. The Earth is approximately 4.54 billion years old.

The current theory is that all of space was condensed into a single hot, dense point 13.8 billion years ago. Creation of the universe took place at the moment of expansion termed the **"big bang."** The universe cooled and condensed from the giant cloud of interstellar dust.
All of the galaxies in space are still moving farther away from each other. As galaxies continue to move away from each other, the universe is expanding. The expansion is explained as the "expanding universe theory."

---

315   All of the matter and energy that exists. All of the planets, stars and interstellar dust in space.
316   A huge system made up of stars, the remains of stars, the planets that may rotate around the stars, gases, interstellar dust that is held together by gravitational forces.

Chapter 6 The Scientific Method: The Scientist's Intellectual Tool
Deductive Reasoning – The Beginning of Science
The word *science* comes from the ancient Greek word *episteme* meaning "knowledge" or to "know." *Episteme*, translated into Latin, is s*cientia*, meaning *knowledge* or *wisdom*. The ancient Greeks firmly established the foundation of science. Greek Science was not experimental science, although there is evidence of some experimentation. The Greeks used philosophical inquiry and logic to arrive at truth. Questions about nature were answered by ***argument***. To the Greeks, argument was clarification, exchanging points of view and persuasion through the use of reason and evidence to prove a point. Their arguments employed ***deductive reasoning***. Deductive reasoning is reasoning from a *general premise* that is accepted as true, to a conclusion about a *specific* thing. For example, *"All dogs bark." "Skippy is a dog."* Therefore, *"Skippy barks."* If the premise is accepted as true, then the conclusion that follows will be true. If the premise is false, then the conclusion *may* be false. In this case, the premise is false because not all dogs **bark**.[317]

The Beginning of Modern Experimental Science
The Renaissance began with Italian engineers, architects, anatomists, writers and philosophers of the late 1100s and early 1200s. It was a *rebirth of learning*. The **scientific revolution** was a part of this rebirth. The scientific revolution was a rational way of thinking based on experiment. By around 1500, the scientific revolution was well underway. A German scholar, Albertus Magnus (1193-1280), Pietro Pomponozzi (1462-1525), Galileo (1564-1642), Francesco Redi (1626-1697), Lazzaro Spallanzani (1729-1799) and many others were pioneers of this revolution. This *modern way of thinking* asked *how* things happened, *not why*. Scientific thinking spread to the rest of Europe by the 1700s. People now believe in direct observation of natural phenomena to uncover **causation**.

The Slow and Uneven Evolution of Modern Scientific Thinking
People did not wake up one day and say "I will not be superstitious. I will be a rational, thinking person." Reason *slowly replaced superstition* among individuals. One example of the persistence of superstition is the word *influenza* (It). Influenza means to be under the influence of an unlucky star. The word remains in use today. It was coined during the 1743 influenza outbreak in Europe. The fact that by the mid-1700s superstition is still applied to an event demonstrates the slow movement towards rational scientific explanations. Three hundred years later, almost everyone believes in *cause and effect*, not just the highly educated.

317   A "barkless" dog breed is the African hunting hound called the *Basenji*.

Inductive Reasoning

The English philosopher Francis Bacon (1561-1626) was greatly influenced by Italian Renaissance scientists such as Galileo (1564-1642). Galileo insisted on the use of experimentation and inductive reasoning to explore nature. *Inductive reasoning* is best described as *arriving at a conclusion based on direct observation of specific phenomena*. It is reasoning from a *specific* observation to a *general* conclusion. This is the opposite of the Greek *deductive* method of discovery.

Using the Cell Theory as an Exercise in Inductive Reasoning

Matthias Schleiden (1804-1881), a German botanist, examined numerous samples of plant tissue with a microscope. All samples revealed cell structure. He concluded that all plants are made up of cells. Theodor Schwann (1810-1882), a German physiologist, did the same thing with animal tissue. He concluded that all animals were made up of cells. In 1839, both conclusions were combined into the **cell theory** – *"all living things are composed of cells,"* even though neither scientist examined *all* animals or *all* plants. The cell theory continued to be confirmed by other scientists' observations.

Using the Cell Theory with the Modern Model of Scientific Thinking

The modern model of scientific thinking is not as simple as induction versus deduction. The **hypothetico-deductive** method is the modern model of the scientific method. *Hypothetico* refers to a hypothesis. *Deductive* refers to deductive reasoning. Today, scientists have large electronic databases that contain records of past experiments that are useful to find out if a hypothesis has been proposed previously.

Francesco Redi (1628-1697) hypothesized that **all cells come from pre-existing cells**.[318] From the 1600s to the 1800s, scientists used microscopes to examine many things. Leeuwenhoek published descriptions of what he called **animalcules** [319] in scrapings from his teeth, in pond water and **hay infusions**.[320] Leeuwenhoek, **Hooke**,[321] Malpighi and others reported seeing structure in living things.

Schleiden and Schwann knew of these discoveries and may have led them to *theorize* that *all* plants *will* have cell structure, and all animals will have cell structure. Both are **general statements** that allow for testing new, specific hypotheses. "Leaves have cell structure." "Stems have cell structure." It is reasoning from the general to the specific.

318   A dictum made famous by Rudolf Virchow (1821-1902).
319   Leeuwenhoek termed these microscopic living things "little animals."
320   Hay infusion is dried grass in water.
321   He examined cork. The contents of which were no longer present. He named these dead remnants "cells" because the structure reminded him of monks' cubicles.

Exploring the Experimental Method

The experimental or the scientific method is the scientist's tool for extracting answers from nature about *how* things work. The scientific method may have variations, but it can be distilled down to *five steps*.

1. Question – A statement that reveals a specific cause for an event.

2. Hypothesis – A **tentative** [322] answer to a question.

3. Experiment – A duplication of conditions used to test a hypothesis.

4. Results – Data collection. A written record of observations.

5. Conclusion – The reasoning process of judging if the hypothesis is true or false based on experimental results.

Different Experimental Approaches

A scientist may *ask an original question* of nature, one that has never been asked before, as did Gregor Mendel. Another scientist may *challenge* a commonly held belief because he or she believe it to be incorrect, as Galileo did. Another scientist may wish to **verify**[323] another person's experiment by repeating his or her work. *Whatever the approach, the problem solving process always starts with a question or problem.*

1. Question

The scientific method always begins with a properly formulated question. If the wrong question is asked, the answer will be useless. *The question must be formulated so that one **variable** is isolated. A variable is something that can be changed.* If the amount of fertilizer, water, light or oxygen given to plants is changed, then each item is a variable.

Gregor Mendel wondered how traits are passed from one generation to another in pea plants. See chapter 13. Columbus asked, "Can the east (China) be reached by sailing west?" Emile Roux believed a toxin, not the physical bacterial cell, caused diphtheria. In each case, a single variable was examined.

A *literature search* of previously published work is done before finalizing a question. The search may reveal that a question was answered in the past. The *rediscovery* of Mendel's discovery of the laws of heredity is a classic example. Mendel published his findings in an **obscure**[324] botanical journal in 1866, only to be forgotten. Mendel's laws of heredity were rediscovered by Hugo deVries, Carl Correns and Erich von Tachermak in 1890 during their literature searches. All three gave Mendel credit for his discoveries, even though he died years before. Mendel was the first to publish. He had established **priority**.[325]

322   Tentative means temporary.
323   To confirm. Repeating a scientist's experiment should give the same results.
324   Little known about or unknown.
325   The first published work demonstrating a discovery.

Gathering Information about the Problem or Question

If a researcher has a question to investigate, a literature search has to be performed, and the material must be read thoroughly. Careful reading of past experiments allows one to select the best hypothesis, the *working hypothesis*. The investigator then follows up on the working hypothesis.

Locating records of experiments done in the past used to take a lot of time. Today, electronic *literature searches* find published works quickly. If the literature search reveals that a question has been answered in the past, then the investigator may choose to confirm the work. *Confirmation* is essential to maintain the integrity the body of scientific information. When other scientists repeat another researcher's work and get the same results, then the conclusions of the original scientist are trusted until someone repeats the original experiment and gets different results.

2. Hypothesis

After becoming familiar with the literature that deals with a problem, the scientist formulates a *hypothesis*. An excellent example is Columbus' hypothesis that the Earth is round and, therefore, one can sail west around the world and arrive at his starting point. Some people thought the same in times past, but he was the only one to carry out an experiment based on his *firm* belief in his hypothesis. He sailed west to reach India and China in the east. In doing so discovered the New World. He just did not know about the two continents between Europe and Asia.

Another example of an elegant hypothesis is the French physician Emile Roux's belief that a toxin secreted by *Corynebacterium diphtheriae* causes diphtheria, not the organism itself. Roux's belief was based on **Friedrich Loeffler's**[326] observation that the diphtheria organism stayed at the injection site of experimental animals. Evidence indicated that something is given off by the bacteria that caused diphtheria. He brilliantly hypothesized that the watery mix bacteria grew in might contain this disease-causing substance. He separated the bacteria from the watery mix and injected only some of the watery mix into an experimental animal. The animal died of diphtheria, proving that it was a toxin produced by the bacterium that caused diphtheria.

3. Experiment

An experiment is a *method* of gathering information about an observable event in nature. An experiment is a *test* of a hypothesis. *The experimental design must force a situation to occur that will physically duplicate conditions as they occur in nature*. The design must control for, and *isolate only one variable* and eliminate all others.

---

326   Discovered the causative agent of diphtheria, *Corynebacterium diphtheriae*.

Aristotle believed that a heavy object, when dropped from a height, will hit the ground before a light object. His argument was that an object will move up or down in order to find its "natural place." He also believed an object's speed is directly proportional to its weight. In other words, the heavier the object, the faster it will fall. His ideas became **dogma**[327] for 2000 years. Galileo challenged Aristotle's conclusions. His challenge was with experiment, not philosophical argument.

Galileo asked the question: "Do objects of different masses *accelerate* at the *same* speed or *different* speeds when dropped at the same time from the same **height?**[328] He hypothesized that mass is independent of gravity. If mass is independent of gravity, he reasoned, then different masses will hit the ground at the same time when dropped at the same time from the same height. To slow down the "falling" of masses, he used long inclined plane. Galileo used his pulse as a timing device as perfectly round brass balls of different masses rolled down the inclined plane. He carefully recorded and analyzed the **data**.[329]

4. Results

Galileo's results demonstrated that balls of *different masses* arrived at the bottom of the inclined plane at the same time. Both bodies had to have the same rate of **acceleration**[330] for this to happen. *He had experimentally proved that mass is independent of gravity. Equally important, his work was reproducible.* Thus, a 2000 year old belief was disproved by experiment; not argument.

5. Conclusion(s)

Based on his careful measurements, Galileo concluded that the mass of a body is independent of gravity as it falls to Earth. It made no difference if the mass was big or small. Gravity did not affect the speed of a falling mass. Galileo had to determine if his hypothesis was correct or not after thorough analysis of his the *data*. Reasoning as to the validity of a hypothesis is a *conclusion*. His **criticism**[331] of the experimental procedure supported his conclusion. A thorough investigator tries to destroy a hypothesis to see if it continues to hold true. After his self-criticism, Galileo published his findings to communicate his work to the scientific community and for other scientists to review, repeat and verify his work.

---

327   A belief that is not to be doubted. An established truth not to be challenged.

328   The dropping of two balls of different weights from the Leaning Tower of Pisa, Italy, is probably a myth.

329   Data (*pl.*) is either a quantity or qualities. Typically data are quantities expressed as measurement. Datum is singular or data.

330   Coined by Galileo, "acceleration." From Italian, accelerare *"to go faster."*

331   Criticism is from the Latin, *criticus* a judge. Criticism is a good and leads to improvement in a work or creation.

A Simple Experiment Illustrating the Experimental Method

Scientists deal with *cause* and *effect*. For example, an investigator may wonder about how to make plants grow faster. Having knowledge of nitrogen as an essential element for plant growth, discovered by a German chemist, Justus von Liebig (1803-1873) and the farmers common use of manure to increase crop yield, will help to form a question.

1. Question and Hypothesis Formulation

A *question* must be clear and lend itself to an objective assessment. The question must *only* deal with the effect of *fertilizer* (the *independent variable*[332]) on plant *growth*. Plant *growth* is the *dependent variable*. For example, a literature search may reveal three *possible hypotheses*: **a)** Fertilizer will have a positive effect on plant growth, **b)** Fertilizer will make plants "greener," or **c)** Fertilizer will make plants grow faster. *All are all valid questions, but not all are good questions.*

A "positive effect" is too *general* and color is too **subjective**.[333] "Fertilizer will make grass grow faster" is objective and easily quantified. *Height is easily measured. Growth is an increase in height over time.* Refining the question further for clarity yields: "Fertilizer will increase the *rate* of growth of grass." The species of grass must be specified in order for the experiment to be repeated properly. Kentucky bluegrass, *Poa pratensis,* is a well known species of grass with an average rate of growth. The *working hypothesis* may now be expressed as, "Fertilizer increases the rate of growth of Kentucky bluegrass, *Poa pratensis.*"

2. The Experiment

Two groups of identical *Poa pratensis* are needed. The *experimental group* will get fertilizer, and the *control group* will not. Both groups are observed over **time**[334] under the same conditions. A **table**[335] of three columns and six rows may be constructed to record data. Entering information in a table ensures orderly and accurate recording of **raw data**.[336] Table construction helps reveal experimental design flaws. See table 6.1.

Column I is a record of *time* the experiment took to complete. **Column II** lists the heights of experimental plants (fertilizer). **Column III** lists the heights of control plants (no fertilizer). *Constants* are the same variables common to both groups. Light, oxygen, heat, soil, water and species of grass are constants. The data collected from the *experimental group* is **compared** to the data collected from the *control group*.

---

332   An **in**vestigator changes the **in**dependent variable is an easy way to remember what the independent variable is.

333   An observation that can be influenced by a person's personal opinions.

334   Time is the measurement of the interval between two events.

335   A table is rows and columns of data.

336   Data taken from some source that has not been mathematically processed.

3. Experimental Procedure

Harvest a 60 cm by 30 cm portion of mature Kentucky bluegrass from a plot of soil. Cut the plot in half to make two identical **30 cm by 30 cm**[337] plots of grass. Cut the top of the grass evenly and run several level strings across the top of both plots of grass. The evenly cut tops of grass is *height zero*. Measure the increase in height above the strings. Measure the growth at each corner of both plots and in the middle. The areas to be measured are designated: *a* -upper right, *b* -lower right, *c* -lower left, *d* -upper left and *e* -is the center. Read the ruler at *eye level* to avoid error. Average the five points together. The average of the five areas is height number 1 in the experimental and control groups. The experiment begins at *time zero*, when 10 ml of fertilizer is added to the experimental plot of grass. See table 6.1. The experimental procedure is clearly stated to make the experiment easily repeated for *verification*.

4. Results

Experimental *results* are *observed outcomes*. Raw data are re-corded in a *data collection* **table**.[338] See table 6.1. Daily measurements are taken of the plant's *height,* the *dependent variable* (effect) in areas designated as I *a,b,c,d e* and II *a,b,c,d e*. An observed increase in height may be the result of fertilizer, the proposed cause. Each measurement of grass height is a **datum**. Each datum is entered in the correct day (time) column in table 6.1. Each *height* is paired with a *time* to yield an *ordered pair* of data. The ordered pairs are entered as **data points**[339] on a *graph*. The data points may reveal cause and effect.

Graphs

Graphs are a way to **visualize**[340] data and *make it easier to draw conclusions* from the data. See figure 6.1 on page 104. Table 6.1 is a mass of numbers that makes it difficult to draw conclusions. It is valid to summarize data to make it easier to interpret. See table 6.2. A *summary* of data in table form (see table 6.2) will suffice for publication. **Raw data** is always kept by the investigator as a record. Raw data is not always presented in published experimental reports.

The dependent variable in the above experiment is plotted on the y-axis and time is plotted on the x-axis. Growth is expressed in centimeters per day in the experimental and the control group. Time is expressed in days. A graph may be constructed from the summary of the data in table 6.2.

337   30.45 cm equals 1 ft.
338   A table is rows and columns of data.
339   A data point is a point in a graph. It represents an ordered pair of information.
340   To see or make a mental image.

Table 6.1. Raw data. The Effect of fertilizer on the growth of *Poa pratensis*.

I	II 10 ml of Fertilizer added						III No fertilizer added					
Time (days)	Experimental Group Height in centimeters*						Control Group Height in centimeters*					
	a	b	c	d	e	*Av.*	a	b	c	d	e	*Av.*
0	0.0	0.0	0.0	0.0	0.0	*0.0*	0.0	0.0	0.0	0.0	0.0	*0.0*
1	0.5	0.4	0.5	0.5	0.3	*0.4*	0.2	0.2	0.1	0.2	0.3	*0.2*
2	0.9	0.8	0.8	0.7	0.8	*0.8*	0.2	0.3	0.2	0.3	0.2	*0.2*
3	1.1	1.2	1.2	1.3	1.3	*1.2*	0.5	0.5	0.4	0.6	0.4	*0.5*
4	1.4	1.5	1.6	1.5	1.5	*1.5*	0.7	0.8	0.7	0.6	0.6	*0.7*
5	1.9	1.8	1.7	1.8	1.9	*1.8*	1.0	1.2	1.3	1.0	1.0	*1.1*
6	2.0	2.2	2.2	2.3	2.2	*2.2*	1.2	1.3	1.3	1.2	1.2	*1.2*
Average of daily averages taken at five measurement points over six days.						**1.3**	Average of daily averages of five measurement points over six days.					**0.65**

*a-e refers to the position in the tray the measurement was taken: a=upper right, b=lower right, c= lower left, d= upper left and e=center.

Table 6.2. Summary of Results of the Effect of Fertilizer on the Growth of *Poa pratensis*

I	II	III
Day	Daily Average Height, in cm, of plants (plot A) *with fertilizer.*	Daily Average Height, in cm, of plants (plot B) *without fertilizer.*
0	0.0	0.0
1	0.4	0.2
2	0.8	0.2
3	1.2	0.5
4	1.5	0.7
5	1.8	1.1
6	2.2	1.2
	Average height = 1.3 cm	Average height = 0.65 cm

The Experiment Is Not Over Yet!

Mathematical operations applied to the collection, analysis and interpretation of data is termed *statistical analysis*. Statistical analysis determines if the sample size is large enough to yield valid results or if the data is the result of chance. If the results are not caused by chance, then there is a real relationship between the independent variable and the dependent variable. Statistical analysis answers these questions.

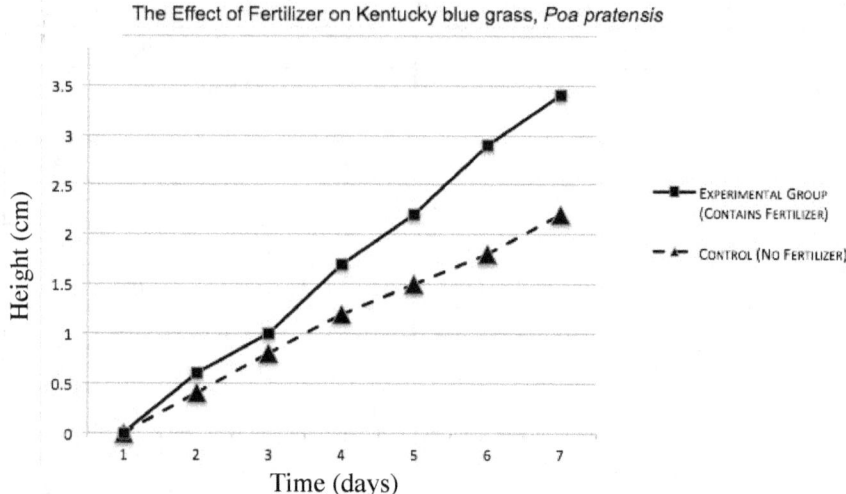

The Effect of Fertilizer on Kentucky blue grass, *Poa pratensis*

Legend:
— ■ — EXPERIMENTAL GROUP (CONTAINS FERTILIZER)
– ▲ – CONTROL (NO FERTILIZER)

Figure 6.1. Fertilized grass compared to unfertilized grass.

5. Conclusion(s)

A conclusion is arrived at by analyzing experimental results to see if there is a relationship between the independent variable (fertilizer) and the dependent variable (height of grass plants). The line representing the growth of the plants in the *experimental group* shows an increase in height compared to the *control group*. See figure 6.1.

A conclusion is arrived at by comparing the working hypothesis to the results – "Fertilizer will increase the rate of growth of Kentucky bluegrass, *Poa pratensis.*" The results strongly indicate *a cause for the effect*. To say that fertilizer is the *cause of the effect* is a reasonable *conclusion* based on critical analysis of the data in tables 6.1, 6.2 and figure 6.1. Fertilizer increases the rate of growth of *Poa pratensis* is now a *fact* and is entered into the body of verified scientific information.

An investigator must *be objective*. At first it may seem simple, but there are many pitfalls. For example, an investigator can favor a hypothesis by measuring plants one way in the experimental group and another way in the control group. The investigator may drop a critical measurement that does not favor the hypothesis. To ignore data is bad science. It is not even science. The experimenter cannot be **biased**.[341] Errors in judgement can be caused by a blind belief in a hypothesis, leading to failure to consider other possibilities.

---

341   A scientist holds a belief based on his or her personal beliefs instead of objective evidence.

Objectivity

Belief in one's own hypothesis *can* lead to erroneous conclusions. A classic illustration of a lack of objectivity is that of the German geologist Dr. Otto Hahn. Hahn microscopically examined meteorites. He found shells, sponges and other life forms imbedded in the meteorite. Hahn published his findings in 1880. Others only saw crystals when Hahn's photomicrographs were examined. Hahn's results clearly were imagined.

An investigator has to be *critical* and *objective*. The investigator should examine his work in the **conclusions** and **discussion** sections of an experimental report. The scientist must honestly ask and answer: "Was the correct question asked of nature?" "Was the hypothesis correctly formulated to answer this question?" "Was the experiment designed properly?" "Were the results collected without bias?" Only after thorough self examination, the report is submitted for publication in a **peer-reviewed journal**.[342] After publication, an experiment is examined by other scientists. If the conclusion of an experiment is accepted as true, it will become a fact in the scientific literature.

Communication With Other Scientists

Scientists must communicate with each other in order for the body of scientific knowledge to grow. Communication is done through peer-reviewed journals, personal communications, letters and lectures at learned societies. Peer review assures the integrity of scientific findings.

The theory of *biogenesis* was an alternative explanation to the theory of *abiogenesis* (**spontaneous generation**).[343] See chapter 2. People believed in spontaneous generation for thousands of years until it was challenged experimentally by *Francesco Redi* (1626-1697), *Lazzaro Spallanzani* (1729-1799) and *Louis Pasteur* (1822-1895).

Francesco Redi offered the first proof for biogenesis in 1668. His belief in "**Omne vivum ex ovo**"[344] led him to hypothesize that maggots *did not* come from rotting meat. Flies, he hypothesized, came from flies. His experiments proved that flies *lay eggs on meat and the eggs develop into* maggots. His findings became a fact of science that fit into the theory of biogenesis. *Spallanzani* challenged the idea that microbes are produced *from* broth. He experimentally proved that microbes enter the broth from the air and grow in the broth. His work was published, entered the literature as a fact that fit into the theory of biogenesis.

---

342   A peer-reviewed journal is published on a regular basis. It is devoted to science in general or to a specific field.
343   Living things could arise from nonliving things by a process termed *generatio spontanea* or spontaneous generation. This theory held sway until the late 1800s.
344   Every living thing comes from a pre-existing living thing.

Louis Pasteur, building on Redi and Spallanzani's work experimentally proved that microbes are in the air. He demonstrated that air held no "vital force." Pasteur's experimental findings entered the literature and fit into the theory of biogenesis. *The three scientists demonstrated that abiogenesis is a theory to be discarded.* Their experimentally proven facts support the general framework called the theory of biogenesis.

Forms of Trustworthy Scientific Information
Theories, Laws and Facts

Theories, laws and facts are subject to peer review and, therefore, are considered to be trustworthy sources of information.

Theories explain phenomena in nature. Theories are broad generalizations that are accepted as true until it is proven false. New discoveries that support the theory of **evolution**[345] are made frequently, but parts of the theory may have to be modified. Each bone is considered a fact.

New information may be strengthened, weakened or modified theories. Examination of a fossilized bone helps explain the theory as a whole. Presently, there is no evidence to indicate that the *theory* of organic evolution is false. Many hypotheses exist within the overall theory of organic evolution. *One hypothesis deals with the disappearance of the dinosaurs.* Walter Alvares (1940-) hypothesized that the impact of an asteroid on the Earth's surface wiped out the dinosaurs. This is the *hypothesis* known as **catastrophism**. Initially rejected, its **adherents**[346] verified his hypothesis with new facts. An alternative view of the appearance and disappearance of life forms on Earth was the *hypothesis* of **uniformitarianism**. This hypothesis stated that changes take place in a population over time and accumulate over generations. **Uniformitarianism** held sway for about 150 years. Evidence did not support this view.

Laws describe natural phenomena. A Law is accepted as true until it is proven false. Laws may have to be modified or discarded as new evidence comes to light. Galileo's law of falling bodies or Newton's laws of motion describes how a masses will act under different conditions. A ball thrown into the air will take a specific path. The ball's position, how long it will take to fall back to Earth and exactly where it will land, can be mathematically calculated.

A fact is the result of an experiment. Hypotheses are proposed and confirmed by experiment. If a hypothesis is proven correct, it becomes a fact. The new fact enters into the body of scientific knowledge.

---

345   Living things present on the Earth today evolved from simpler ancestors. The first to formulate a theory of evolution of living things based on natural selection were Charles Darwin and Alfred Wallace.
346   Individuals that stick or cling to an idea.

Forms of Scientific Information That Are Not Trustworthy
Anecdote

Anecdotal information is a result of casual, nonscientific observations passed from person to person. *It is not peer-reviewed information published by members of the scientific community.* For example, in the 1950s, many people did not send their children to public swimming pools because parents heard children could get polio there. "Johnny had friends that went to public swimming pools and some got polio." This statement comes under the heading of hearsay. It could be true or not true. It is not statistical proof. It happened to be **true**.[347] It was later proved that the transmission of the polio virus was a fecal-oral route.

Science is a search for truth, but some people engage in behaviors that are in opposition to the scientific community's ideals and beliefs. Below are behaviors that put at risk the advancement of science, jeopardize the body of scientific information and destroy scientific **integrity**.[348] If there is a breach of integrity, behavior may fall into one of two general categories: *research misbehavior* and *research misconduct*.

Research Misbehavior

Research *misbehavior* is a serious infraction and is a reflection on a principal investigator (PI). The PI bears responsibility for the conduct of scientists that do research in his laboratory. In 2011, a senior author on a research paper was found to be responsible for scientific research misbehavior. The PI trusted his colleague, but did not properly supervise the research or the writing of a 2005 paper that bore his name. The paper was later proven to be fraudulent. Those responsible for the research were guilty of *misconduct* and dismissed from their positions. The PI was guilty of misbehavior for not properly supervise the work.

Research or Scientific Misconduct

A **standard code**[349] exists to guide scientists in their work. Violation of this code is termed *scientific misconduct*. Misconduct can take three forms: *plagiarism, falsification* and *fabrication*. Plagiarism, falsification and fabrication are a **blatant**[350] disregard for the ethical standards of the scientific community.

---

347   It was later proven that one could contract polio by exposure to water in swimming pools because the polio virus can be shed from the lower gastrointestinal tract of an infected individual into the water and then ingested by others.
348   Keeping moral and ethical principles in any endeavor; soundness, honesty and having moral character.
349   As defined by the National Science Foundation (NSF).
350   Obvious, clear.

Plagiarism

Plagiarism is *one of the most common forms of misconduct.* *Plagiarism is to take credit for writing that belongs to someone else.* An author may "forget" to cite another person's work. This should not happen if the investigator is *honest* and *thorough.* In the great majority of cases it is a purposeful act. An investigator in India was found to have plagiarized at least 70 research articles between 2004 and 2007.

Falsification

Falsification is to hide or modify something to present it as true. For example, a graduate student at a medical college in New York State published a paper in 2004 claiming to have seen expression of certain **immunoglobulins**[351] in *Borrelia burgdorferi*, the bacterial agent of Lyme disease in humans. The investigator used a blank slide as a control and photocropping to show what he wanted to show, not the real results, thus falsifying photomicrographs of the experimental results.

Fabrication

Piltdown Man is one of the most widely known scientific *fabrications* of data in the history of **anthropology**.[352] The **fabricator**,[353] a well-respected scientist, combined a jawbone of an orangutang[354] with the skull of a modern human to approximate the "missing link" between apes and man. He then planted the "find" in a gravel pit in Piltdown, England in 1912. The perpetrator of the hoax then led a group of respected anthropologists to the site and allowed them to "discover" the greatest find of all time. Modern scientific techniques exposed the hoax in 1953.

The *Cardiff Giant*, another famous fabrication occurred in 1869. A George Hull was an atheist that wanted to humiliate a friend who believed the Bible should be taken literally. The Bible said that giants once roamed the Earth. Hull had a mason make a ten foot high statue resembling a human. He aged the stone and buried it in Cardiff, New York. He claimed it was a petrified giant human that once roamed the Earth. Many were fooled and paid to view it. It was exposed as a fraud in 1870.

A serious *fabrication of scientific data* occurred in 1974 at a prestigious research institute in the United States. An investigator claimed to have transplanted tissue that resulted in black patches of fur in the coat of a white mouse. In fact, the investigator used an indelible marker to color patches of fur black on a white mouse. When it was discovered what he had done, he was dismissed and banned from doing scientific research.

351   Immunoglobulins are antibodies.
352   Anthropology is the study of humanity; its past and its present.
353   Make something with the intent to deceive.
354   A found only in Borneo and Sumatra. Classified in the genus Pongo.

Making False Claims by Pretending to be Scientific
Pseudoscience

Pseudoscience or false science is characterized by vague claims and a lack of scientific verification. Sometimes the person making claims for a remedy might try to link the fraudulent product to a famous person.

The medical field has many examples of **pseudoscience**[355] because a lot of money is made by selling "miracle drugs." One of these concoctions which supposedly cured all illness, came from a book of homeopathic medicines from the 1880s and went as follows: mix powdered starfish, secretions from a skunk, powdered coal and human urine. These concoctions were taken in small amounts and were generally harmless. Homeopathic treatments are a form of alternative medicine. Much of it is based on pseudoscience.

The Federal Food and Drugs Act of 1906 and the creation of the Food and Drug Administration (FDA) addressed many of the abuses of pseudoscience and **pseudomedicine**[356] in the U.S. In 1906, President Theodore Roosevelt signed The Federal Food and Drugs Act of 1906 or **"Wiley Act"**[357] into law. The Federal Food and Drugs Act forbade "adulterated" food to be transported across state lines. Food or drugs were considered adulterated if they were reduced in quality or strength by the addition of fillers or had substances added that could cause injury.

In 1990 Federal Food, Drug, and Cosmetic Act was amended to declare Human Growth Hormone (HGH) a controlled substance. In the United states a prescription is required from a physician. HGH is only to be used for the treatment of a disease condition. There are several serious side effect from using HGH. HGH has been used to treat obesity, fibromyalgia, Crohn's disease and ulcerative colitis.

---

355   False science.
356   False medicine.
357   Harvey W. Wiley (1844-1930) was instrumental in passing the establishment of the FDA. He was a crusader for the regulation of food and drugs in the nation.

Chapter 7 Measurement

A Standard

Measurement is the process of *matching* a *standard quantity* to an *unknown quantity* of a *physical something*. A **standard** represents **magnitude**[358] or "how much" that everyone agrees to use. For example, if a 1 kilogram mass is placed on one side of a double pan balance and salt is poured onto the opposite side of the balance, eventually the salt will balance the 1 kg mass. The 1 kg of salt is now said to be **massed**.[359] A *known magnitude* has been matched to the *unknown physical "something"* – salt.

Ancient Standards and Systems of Measurement

Standards of measurement are not new. Peoples of ancient China, Egypt, Mesopotamia, and the **Indus**[360] Valley used precise standards of measurement for commerce. The Egyptian cubit, a unit of length, was the distance from the elbow to the tip of the furthest digit. A standard cubit was made of stone and copies were made of wood.

The Romans traded with many provinces beyond Italy, all countries agreed upon and adopted Roman standards. The standard Roman mile for civilian and military transportation was 1,000 paces of a Roman soldier. The English word "mile" is a contraction of *mille passus* (L.), 1,000 paces of a Roman soldier. The inch is derived from L. *uncia*, or "thumb" represented a twelfth of a Roman foot, the *pes*.[361] The *libra* (lb) [362] about **329 g**,[363] was the Roman unit of mass. The mile, inch, foot and the **pound** are remnants of the Roman system of measures.

Modern Standards and System of Measurement
The Metric System

The French devised a method of weights and measures to simplify commerce and break with the pre-revolutionary order. The French wanted new standards of measurement based on something natural. Named the metric system (L. *metricus,* to measure), the result was an easy to use decimal system. **Base**[364] **units** are multiplied or divided by 10. The meter was defined as the distance of one ten millionth (1/10,000,000) of the Earth's circumference from the North Pole to the equator. The meter was the starting point used to derive base units. The kilogram and the liter were constructed (derived) from the meter.

---

358   Magnitude refers to a quantity or an amount.
359   "Massed" because an unknown mass is compared to a known a kilogram mass.
360   A region bordering the Indus river in India (2500 -1800 B.C.).
361   Almost equal to the U.S. foot.
362   From the Latin, *libra* means "balance."
363   A little over seven-tenths of a pound (lb).
364   Basic units, commonly referred to as standards in the SI system.

The original *metric system* used the centimeter (cm) as the base unit of length, the gram (g) as the base unit for mass. See figure 7.1. The *centimeter* and the *gram* were *derived* from the meter. See figure 7.2. Derived units are combinations of base units.

The International System of Units – The SI System

By 1875, the modern metric system, the **International System of Units** (*Système International d'Unités*), or the **SI system**, was agree upon as an almost universal system of measurement.

The *kilogram (kg) replaced the gram* as the base unit of mass, and the *meter (M) replaced the centimeter* as the base unit for length.

As international trade developed, proliferated in use. The SI system fulfilled the need for *uniformity of measurement* for trade and scientific communication. Scales of all laboratory instruments are in SI units in *all laboratories* throughout the world and most market places.

Measuring devices have a *scale* imprinted on them. *A scale is a line divided into equal parts.* Each division on a scale represents a unit of measurement. A scale can be very fine or it can be crude. If the scale is finely made, then the *precision* of the device will be increased. *Accuracy* is the result of careful measurement. A plastic metric ruler usually has a scale marked off in centimeters with the centimeters divided into millimeters. A *unit* is a defined amount of a physical quantity. The units of 1 millimeter, 1 meter and 1 kilometer are defined quantities.

Units of the SI System - Base Units

The modern metric system (SI) consists of seven base units or standards. Each base unit is equal to the quantity 1.0 and can be subdivided into smaller units or multiplied to make larger units. Every country of the world has copies of each SI standard. The seven base units of the SI system are the agreed upon standards against which everything is measured. Base units in the SI system are the: **meter (m)**[365] (length), **kilogram (kg)** (mass), **second (s)** (time), **kelvin (K)** (temperature), **mole (mol)** (amount in chemistry), **ampere (A)** (electrical current) **and candela (cd)** (light intensity). Multiples and fractions of base units are preceded by a **prefix**.[366] The most commonly used base units in the SI system are the *meter, kilogram* and the *second*. *One liter of pure water at 4 °C has a mass of 1 kilogram*. See figure 7.4.

The liter (volume) and the degree centigrade (temperature) are not base units. Both are derived units commonly used in every marketplace

---

365   Lower case "m" is the abbreviation for meter *and* mass the context will determine the correct abbreviation.

366   A prefix, when placed in front of a word, changes the word's meaning. For example, **kilo**gram is 1000 grams, but **deci**gram is 1/10th of a gram. Bold face is a prefix.

around the world, except in the U.S. The United Kingdom and a few other countries use non-metric measures along with metric measures.

Fractional parts of the meter such as centimeters and millimeters (mm) are used to measure something smaller than a meter. See figure 7.1b and c. Multiples of the meter such as a decameter or kilometer are used to measure things larger than a meter. Multiplication of the centimeter by 100 results in 1 meter. Division of the meter by 1000 results in 1000 smaller units called millimeters. Base units are preceded by prefixes that indicate a fraction or a multiple of one standard unit. Commonly used prefixes are listed in table 7.2.

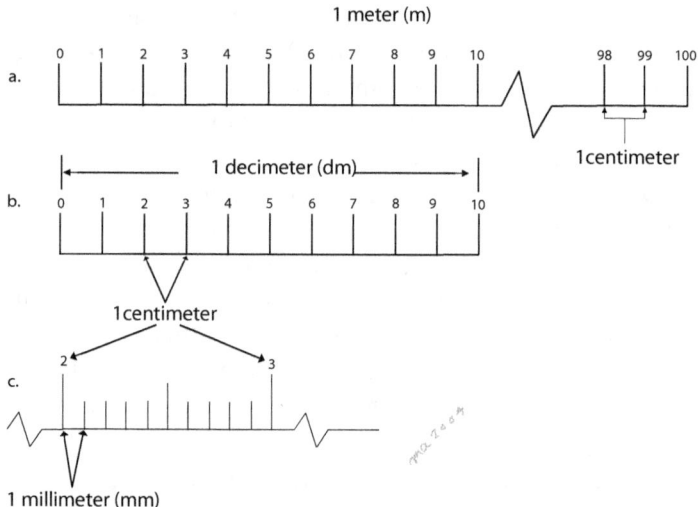

Figure 7.1. The standard meter above (a) consists of 100 centimeters. Ten centimeters (b) is equal to a decimeter or 1/10th of a meter. Each decimeter consists of 10 centimeters and each centimeter (c) is 1/10th of a decimeter.

1. *The Meter - Length*

The standard **meter**[367] is used to measure length. *Length* is defined as the distance from one **location**[368] (A) to another location (B). The meter is divided into 100 cm and each centimeter is divided into 10 mm for everyday purposes. Therefore, 1000 mm are in 1 m. See figure 7.1.

The meter was originally agreed upon as 1/10,000,000 of the distance from the North Pole to the Earth's equator. With advances in the science of measurement (***metrology***), the definition of the meter has become precisely defined as the distance light travels in a vacuum in 1/299,792,458 seconds. This distance is the official meter.

367   The meter is 39.37008 inches
368   From the Latin, *locus,* meaning place.

Estimation, Precision and Accuracy
Estimation

Measurement in the laboratory commonly involves using rulers, graduated cylinders and balances. None of these devices are perfect. Therefore, at times, estimates of some measurements are made. *Estimation* is an approximation of a measurement. Division marks (units) on an instrument are matched to a standard, such as the standard meter. *A measurement that lies between two lines on a scale has to be estimated.* However, the estimated number cannot exceed the limits of *precision* of the instrument being used. See figure 7.2. Accuracy to a centimeter is the only possibility since centimeters are the only divisions on this scale.

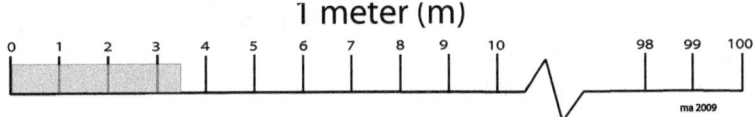

Figure 7.2. If the edge of the object being measured is midway between centimeter 3.0 and centimeter 4.0, then we can validly estimate the number to be 3.5 cm. Since there are no further divisions smaller than a centimeter we can not be sure of a measurement beyond 3.0 cm. The number 3.0 is a significant digit, but 5 is not because it cannot be read directly from the ruler. It is an estimated number. See page 119 for further information.

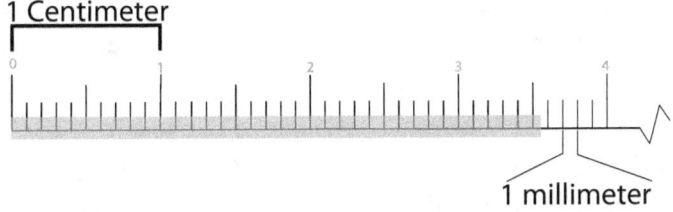

Figure 7.3. A metric ruler with centimeters divided into millimeters.

Precision and Accuracy

Precision and accuracy are very different to the scientist, even though both are used interchangeably in everyday usage. *Precision* is primarily function of the measuring device. *Precision is best defined as the ability to reproduce the same measurement each time it is performed with the same instrument.*

The ruler shown figure 7.3 illustrates how each centimeter is divided into 10 mm. Clearly, the ruler in figure 7.3 is *more precise* than that in figure 7.2. The edge of the object measured with the ruler in figure 7.2 appears to be midway between centimeter 3 and centimeter 4, but there is *no certainty* in between 3 and 4. The ruler in figure 7.3 allows for direct measurement to the nearest millimeter. There is *certainty* that the object is 3.5 cm (35 mm) long. The length of 3.55 cm can only be *estimated* because the object appears to be halfway between millimeter 5 and 6 or 3.55 cm.

Accuracy is how close an individual comes to the accepted or true value of what is being measured. Accuracy is a mostly a function of the person doing the measuring. If the person doing the measuring is careless, then inaccurate measurements will result. If the instrument is crude, then no amount of care will produce accurate measurements. Accuracy depends more on skill and care of the person doing the measuring. Accuracy can be improved with practice. No amount of practice will increase accuracy with imprecise instruments.

2. The Liter - Volume a Derived Unit of Measurement in the SI System

The *liter*, is a derived unit of *volume* in the SI system. *Volume is the amount of space an object occupies.*

The liter is not a base unit because it is mathematically derived.

Calculating the Volume of a Regular Object

If an object is **regular**[369] in shape, like a cube or a rectangular box, then the volume is calculated by using the formula: $V = L \times W \times H$ (V= volume, L= length, W= width and H= height)

If L, W and H are measured in **meters**   V will be expressed in m^3
If L, W and H are **decimeters**   V will be expressed in dm^3
If L, W and H are in **cm**   V will be expressed in cm^3

For example: $10 \text{ cm}^1 \times 10 \text{ cm}^1 = 100 \text{ cm}^2$
$100 \text{ cm}^2 \times 10 \text{ cm}^1 = 1{,}000 \text{ cm}^3$

When multiplying exponents, add them. See example above. Do not write "cm^1" because it is understood that **cm** is equal to one centimeter raised to the power of 1. It is written here as cm^1 for instructional purposes only.

The cubic meter, 1m x 1m x 1m, is not a practical unit to use in the laboratory or the marketplace. It is too big. The smaller cubic decimeter (dm^3) is much more convenient to use. Figure 7.1a.

The volume of a liquid can be conveniently measured with a graduated cylinder. A graduated cylinder is made by pouring 1 liter of pure water at a temperature of 4 °C at 1 atmosphere from a standard liter container into a glass cylinder. Placing a mark on the graduated cylinder at the highest water level equals 1 liter (1 dm^3 or 1,000 cm^3). If the cylinder is divided into 1,000 equal divisions, each division will be 1 millilter (mL) or 1 cubic centimeter (cm^3). See figure 7.1 a-c.

---

369   A regular object has symmetry. A cube has for sides of equal length.

A slight downward bend of the upper surface of a column of water, called the *meniscus*, can be seen in the graduated cylinder. See figure 7.4 c. The number of ml should be read at the *bottom of the meniscus*.

Deriving the ml (cm³) from the meter

If a cube measures 1 dm x 1 dm x 1 dm *then* the cube has a volume of 1 dm³. One dm³ is equal to 1,000 cm³ because a dm contains 10 cm. *If* 1/1000th of 1,000 cm³ equals 1 cm³ *and* it was decided to call 1,000 cm³ 1 liter (L), *then* 1/1000th of a liter equals 1 milliliter (mL).

$$\text{Simply put: } 1\text{ L} = 1\text{ dm}^3 = 1{,}000\text{ cm}^3$$
$$\text{Therefore: } 1\text{ cm}^3 = 1\text{ mL}$$

Calculating the Volume of an Irregular Object

A rock does not have a regular shape. Its volume is measured using the water displacement method. This is easily done by filling a graduated cylinder to a known level, for example 50 mL. After dropping the rock in the water, the water level will rise. If the water level rose to 54 mL, then the volume of the rock is 4 mL. (If 1 cc = 1 mL then 4 cc = 4 mL).

3. The Kilogram - Mass

The kilogram is a base unit of mass in the SI system. *Mass (m)* is the amount of matter contained in an object. If the matter is a piece of iron, then its mass is equal to the amount of iron atoms. If the mass is a liter of water, then its mass is the amount of water molecules present.

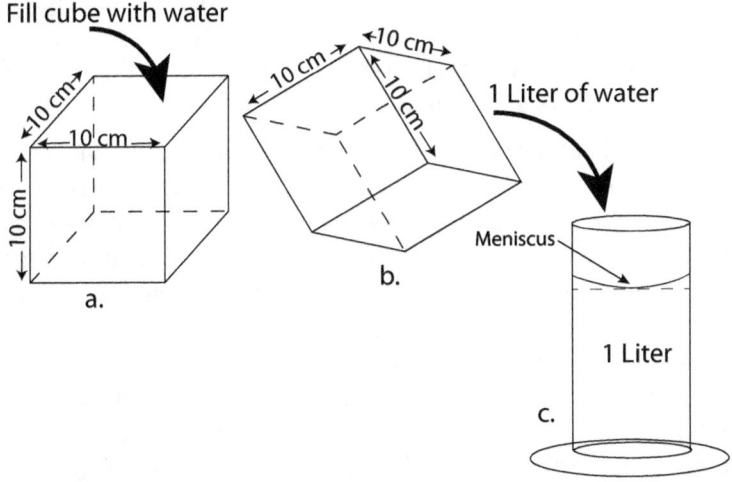

Figure 7.4. Constructing or deriving the liter (L) from the meter.

When one uses a balance, the object is being massed, not weighed. Massing an object is done by putting an object on a balance in order to *compare* the mass of the unknown object to a known mass, the *kilogram*.

A container that measures 1 dm x 1 dm x 1 dm has a volume of 1 dm³, 1000 cm³ or 1 liter. The standard kilogram mass is defined as a liter container filled with pure water at a temperature of **4° C** and air pressure equal to 1 **atmosphere**.[370] See figure 7.4 a.

Mass and Weight

Mass and weight are often confused. *Mass* is the *amount* of matter in an object. *Weight* is a measure of the *pull of gravity* on an object. *Mass* does not change anywhere on the surface of the Earth or in space. The number of atoms or molecules remains constant in an object.

An object will *weigh* less as it is moved further away from the center

100 Pounds ☐ ☐ 1 kilogram

North Pole

☐ 1 kilogram
☐ 97 pounds

Equator

South Pole

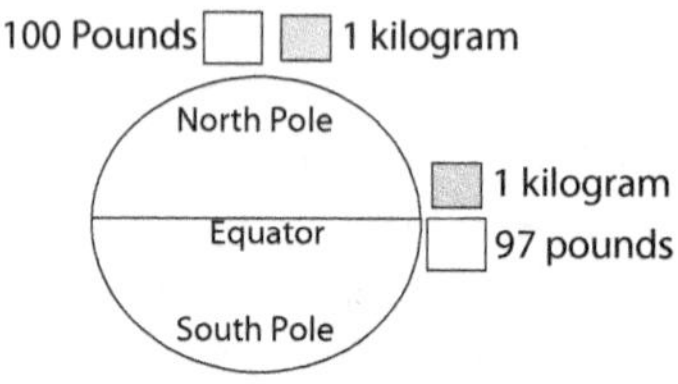

Figure 7.5. Weight will be less at the equator compared to the North Pole.

of the Earth. Gravity changes slightly with position on the **Earth**[371] and therefore, so will weight. If a 100 lb weight is moved from the North Pole to the equator, it will weigh about 3% less at the equator because the pull of gravity is less at the equator. An object is slightly further away from the center of the Earth at the equator. See figure 7.5.

4. The second (s) - Time is a measurement of how long an event lasts or the interval between events such as the interval between waves striking the shore. The *second* (s) is the base unit of time in the SI system.

Multiples and divisions of the second are arrived at by using the prefixes in table 7.1. A kilosecond, for example, would be *1,000 seconds,* and a microsecond would be *1,000,000th of a second.* Minutes, hours, days, weeks and years are commonly used in science.

5. Kelvin (K) - Temperature *is a measurement of the average kinetic energy of the atoms or molecules in a substance. **Kinetic energy** is the random motion of atoms or molecules* It is a physical property of matter commonly interpreted as "hot" or "cold." Hot objects have a greater **thermal energy** content than objects that feel cold because the total number of its molecules in a hot object move faster than the total number of molecules in an object with less thermal energy.

All atoms and molecules of **solids, liquids, gases** or **plasma** are in constant motion due to their kinetic energy. The faster atoms or molecules move, the further apart will be spaced. Plasma is a fourth state of matter, but is not an ordinary state of matter.

---

370   1 atmosphere is the pressure exerted by the atmosphere on objects at sea level.
371   Since the Earth is flattened slightly at the poles and bulging slightly at the equator, a person at the equator will be further from the center of the Earth and will weigh slightly less than at the poles.

Solid, *liquid* and *gas* are three *ordinary* states of matter. Water molecules in the *solid state* (ice) *move slower than molecules of liquids*. Water molecules (liquid) *move faster* than molecules of water in the ice state. Water molecules in the *gaseous* (steam) *state move faster than*

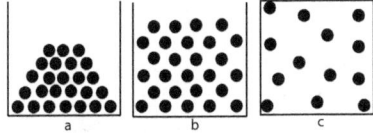

a b c

liquids and solids. As the temperature in a substance rises, its atoms or molecules increase in kinetic energy and move further apart. As the temperature goes lower, the atoms or molecules move closer together. See figure 7.6. When a nurse takes a patient's temperature, the thermometer is measuring the average kinetic energy of molecules in the liquids it comes in contact.

Figure 7.6. Representation of particles in: (a) solids, (b) liquids and (c) gases.

Comparison of Kelvin, Centigrade and Fahrenheit Temperature Scales

The **kelvin (K)** is the basic unit of temperature in the SI system. The term "degree" is omitted. The kelvin scale is based on absolute zero or the cessation of all molecular motion. Absolute zero is 0 K. 0 **K** equal to -273.15 °C or -459.67 °F. In theory, all molecular motion ceases at absolute zero where atoms have no kinetic energy and therefore no thermal energy. Absolute zero has never been achieved.

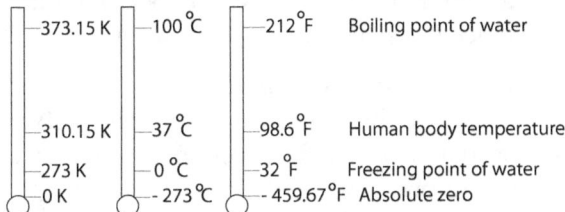

Figure 7.7. A comparison of three temperature scales: the Kelvin, Celcius and Fahrenheit. Notice boiling, body temperature, freezing and absolute zero on all three scales.

The kelvin scale (K) is used in physics laboratories. The **Celsius**[372] (°C) scale is in everyday use in *most* countries of the world to report daily temperatures and universally in all laboratories. See figure 7.7

Thermal Energy and Heat

Thermal energy is the total energy possessed by a body. Thermal energy is the sum of the kinetic energy possessed by the atoms or molecules in an object. The thermal energy in a kilogram of iron is the total kinetic energy of all the iron atoms. The thermal energy in a kilogram of ice is the total of kinetic energy of all water molecules in the block of ice.

372   Anders Celsius (1701-1744) was a Swedish astronomer. The Celsius temperature scale was known as the centigrade scale prior to 1948.

Heat is the transfer of thermal energy from one system to another. Heat is sensed if there is a transfer of thermal energy. If a 1 kg mass of iron at a temperature of 50 °C is placed on a 1 kg block of ice (0 °C), then *heat* will be transferred from the 1 kg mass of iron to the 1kg block of ice. Melting of the ice will take place.

5. The mole (mol) - is an amount of something. It is also called **Avogadro's number (N$_A$)**. The mole is a chemical unit of mass. It expresses amount. A mole is the number of atoms in 12 grams of carbon 12. Using scientific notation, Avogadro's number is 6.02 x10^{23} or 602,000,000,000,000,000,000,000 in ordinary numbers. One mole of any substance has a mass in grams. It is equal to its molecular (atomic mass). The mole is reproducible and can be converted to grams, molecules or atoms. For example, 1 mole of H$_2$O has 6.020 x 10^{23} atoms. On the periodic table hydrogen is noted as *atomic mass unit (amu)* 1 and oxygen has an amu of 16. Water has are two atoms of H and one atom of O. Add the amu of two H atoms and the amu of one O atom. Together the amu of two H atoms and one O atom total 18 amu per mole or 18 grams.

6. The ampere (A) - a unit of electric current. It is the amount of electric charge passing a point in an electric circuit per unit time. The unit is named after the French physicist André-Marie Ampère (1775 -1836).

7 The candela (cd) - the standard unit of luminosity or brightness. Luminosity is a measure of the amount of power that is given off by a particular source of light in a specific direction. A standard candle gives off the light intensity of one candela.

Scientific Notation

Scientific notation is a shorthand for writing numbers that are too large or small to be easily written. Numbers are written in the format: *a* x **10^b** using scientific notation. The lowercase *"a"* represents a number that is **greater than or equal to 1, but less than 10**. Stated symbolically: $1 \leq a < 10$. For example, 53,000 is 5.3 x 10^4 using scientific notation. To represent *a*, move the decimal point between the 5 and 3 to get the number 5.3. The number 5.3 is greater than 1, but less than 10. To get b, count the places from the last "0"on the original number to the decimal point between 5 and 3. It is four places. Therefore, 53,000 is 5.3 x 10^4. See example in table 7.1. The exponent is positive when the standard number being put in scientific is greater than zero. The exponent is negative when the standard number being put in scientific notation is less than zero.

Table 7.1. Examples of scientific notation			
Very Large Numbers		Very Small Numbers	
Standard Form	Scientific Notation	Standard Form	Scientific Notation
30,000,000	3×10^7	0.003	3×10^{-3}
1,000,000	1×10^6	0.000001	1×10^{-6}
13,060,000,000	1.306×10^{10}	0.00037	3.7×10^{-4}

Significant Digits

Precision of a measurement depends upon the number of *significant digits* that can be achieved. Significant digits are the numbers that contribute to a number's precision. Zeros are exclude if used as placeholders, or if a mathematical calculation is far beyond the precision of the instrument. For example, 005 has two zeros called leading zeros that can be dropped and the number is still five. Leading zeros on a check are used to prevent anyone from altering the number. The number 5.00 is still five if the last two numbers, referred to as trailing zeros, are dropped.

Precision is only to the level of the nearest *centimeter* in figure 7.8. There is only one significant number in the measurement. The ruler clearly shows the object to be more than 5 cm. How much more can only be *estimated*. The object can be estimated to be 5.5 cm because the object is close to the midpoint of 5 and 6. See figure 7.4 on page 115 for further information.

12 centimeter ruler

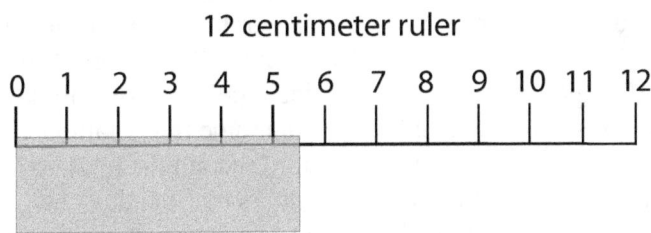

Figure 7.8. A metric ruler divided into centimeters.

In figure 7.9, the level of precision is to the nearest *millimeter*. The numbers that result from the measurement of the same object will have more numbers that have true meaning. These numbers are termed "significant" figures. The object measures 5.5 cm. The object is estimate to be about one half of the next millimeter. Therefore, the measurement is 5.55 cm, but only the first two fives are significant. The last five is an estimate and not read directly on the ruler.

119

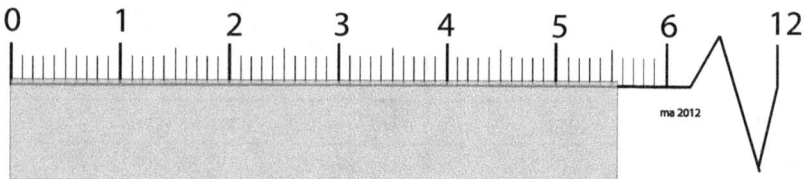

Figure 7.9. A twelve centimeter metric ruler. Each centimeter is divided into 10 millimeters. Only the first six centimeters is shown in detail.

Additional Derived Units of Measurement of the SI System

The agreed upon seven standards of the SI system of measurement are listed on pages 111 through 118, but there are over 20 approved *derived units* of measurements besides the *liter*. Derived units are combinations of base units. Unit names are always written as lower case. Abbreviations are usually written in lower case except if it is a person's name. The newton (N), volume (V) and density (d) *for elementary presentations)* are just three of the many derived units.

The *newton* (N) is a derived unit of *force*. It is made up of kilograms (kg), meters (m) and seconds (s). The newton is defined as the force needed to accelerate a mass of one kilogram at the rate of one meter per second squared (s^2). Symbolically written as $\mathbf{1N = 1kg \cdot m/s^2}$. *Force* is a push or a pull that accelerates a mass. A force can be the pull of gravity (g) or an externally applied force such as a push or pull.

Forces are measured in newtons (N). The abbreviation of a newton is capitalized and written as N, while the word "newton" is not, unless it is at the beginning of a sentence. For example, if g represents the force of gravity (9.8 m/s), then after one second a falling body will have a velocity of 9.8 m/s. After the 2nd second the same freely falling body will have a velocity of 19.6 m/s. After the third second the body will have a velocity of 29.4 m/s until terminal velocity is reached due to air friction. The body is uniformly falling faster per unit of time. This is *uniformly accelerated* motion.

Density (d) is a derived unit that expresses the ratio of mass (m) of an object to its volume (V). Density of an object is calculated using the formula $\mathbf{d = m/V}$.[373] If cubic centimeter of water is massed, it will be found to have a mass of 1g. Therefore:

$$d = m/V$$
$$d = 1g/1cm^3$$
$$d = 1g/cm^3$$

---

373   The Greek letter P (rho) is often used as an abbreviation for density. $P = m/V$.

Table 7.2. Most frequently used prefixes in the SI system of measurement.

Prefix	Number	Fraction	Decimal Equivalent	Scientific Notation
fempto-	100,000,000,000,000	1/100,000,000,000,000	0.000000000000001	$1x10^{-15}$
pico-	100,000,000,000	1/100,000,000,000	0.00000000001	$1x10^{-12}$
nano-	1,000,000,000	1/1,000,000,000	0.000000001	$1x10^{-9}$
micro-	1,000,000	1/1,000,000	0.000001	$1x10^{-6}$
milli-	1000	1/1000	0.001	$1x10^{-3}$
centi-	100	1/100	0.01	$1x10^{-2}$
deci-	10	1/10	0.1	$1x10^{-1}$
Standard Unit 1 Meter, 1 Liter and 1 Kilogram ($1 = 1x10^{0}$)				
deka-	10	na	na	$1x10^{1}$
hecto-	100	na	na	$1x10^{2}$
kilo-	1,000	na	na	$1x10^{3}$
mega-	1,000,000	na	na	$1x10^{6}$
giga-	1,000,000,000	na	na	$1x10^{9}$
tera-	1,000,000,000,000	na	na	$1x10^{12}$

Chapter 8 Physical Tools of the Scientist

The scientific method is the intellectual tool that is crucial to science, but physical tools are needed too. Most students are familiar with common scientific equipment and instruments. See figure 8.1. Scientific equipment is used and stored in a ***laboratory***. Early scientists used a kitchen as a laboratory. The "kitchen laboratory" eventually evolved into a dedicated laboratory for scientific inquiry. Robert Hook (1635-1703) carried out many of his investigations in his kitchen laboratory. Hooke made and used compound microscopes in his laboratory and is remembered for coining the word "***cell***." He also made great contributions to the science of physics.

Equipment may be as simple as an inclined plane that Galileo used to answer his questions about acceleration. Some equipment may be as complex as an electron microscope or a **myriad**[374] of other specialized kinds of instruments. Equipment may be categorized as *glassware*, *hardware* and *instruments*. Below are some common examples of laboratory glassware and hardware students should be familiar.

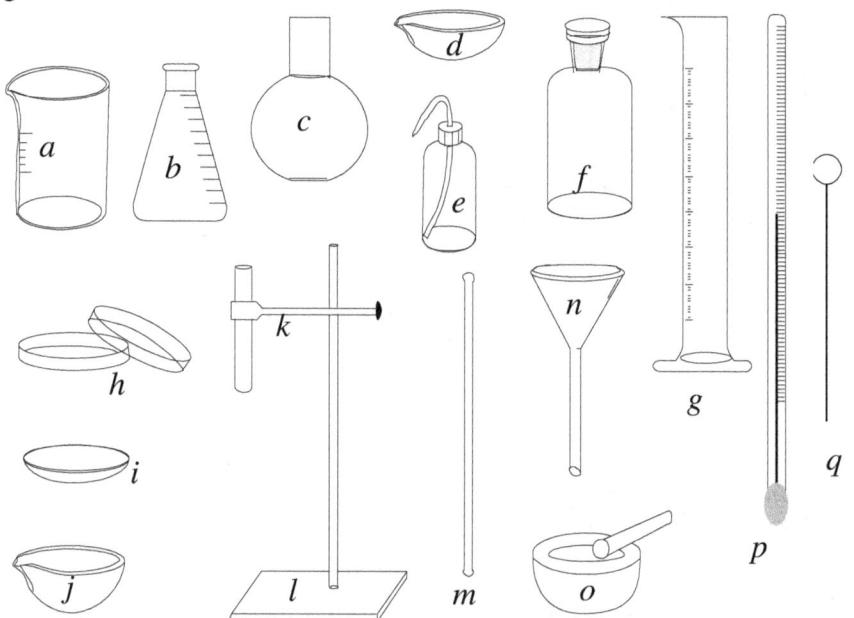

Figure 8.1. Basic glassware used in most laboratories.
(*a*) beaker, (*b*) Erlenmeyer flask, (*c*) Florence flask (also known as a boiling flask), (*d*) crucible, (*e*) plastic drop bottle, (*f*) glass stoppered stock bottle, (*g*) graduated cylinder, (*h*) Petri dish, (*i*) watch glass, (*j*) evaporating dish, (*k*) clamp holding a test tube, (*l*) ring stand, (*m*) glass stirring rod, (*n*) funnel, (*o*) mortar and pestle, (*p*) thermometer and (*q*) bacterial transfer loop.

374   A great number.

The Microscope Makes the Invisible World Visible

The Italian physician, Girolamo Fracastoro, described a compound microscope and its ability to enlarge very small things in 1538 many years earlier than others. Galileo learned of the compound microscope and built an improved version. He called his microscope an *ochiolino,* Italian for "little eye." His friend Giovanni Faber (1574–1629, a German physician, suggested the name "microscope" from the ancient Greek "*micro*" small and "*scopos*" to see.

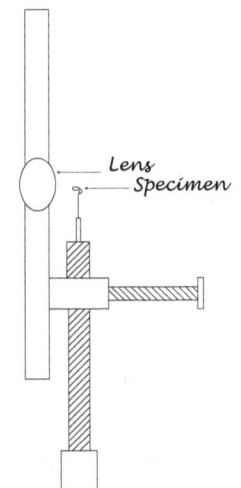

Figure 8.2. Diagram of Antonie van Leeuwenhoek's simple microscope.

Hans Lippershey (1570-1619), a German-Dutch eyeglass maker claimed to have invented the microscope. He was denied a patent on the grounds that it was too easy to construct, and so many people knew about it.

The real inventor of the microscope may never be known. However, there is no doubt that the Dutch tradesman, **Antonie van Leeuwenhoek** (1632-1723), with his *simple microscope*, founded the science of microbiology, earning him the title of "Father of Microbiology." Leeuwenhoek's *simple microscope* could achieve a magnification of about 300 diameters. A *simple microscope* consists of *one lens*. See figure 8.2.

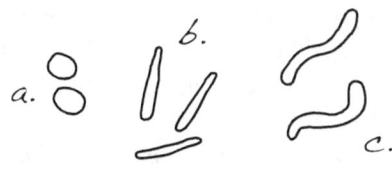

Figure 8.3. Representation of van Leeuwenhoek's drawings of bacteria: (a) cocci, (b) bacilli, (c) spirals.

Leeuwenhoek was able to see bacteria and protists. He called these microscopic living things his "little animals." He kept accurate records and made clear drawings of what he saw. He was the first to record and publish his microscopic observations. See figure 8.3.

Robert Hooke built *compound microscopes*. Compound microscopes have two or more lenses. Hooke published his observations in a book entitled *Micrographia* in 1635. In it he described a fly's foot, bird feathers, a bee's stinger and "cork" cells. The name "cell" remained as a description of living cells even though Hooke's observations and drawings were of the walls of once living cork cells.

The Compound Light Microscope

The compound light microscope is an **icon**[375] of science. The micro-scope makes it possible to see things that are too small to be seen with the unaided eye. It is also known as a *compound*, *optical*, *bright field* or *light micro-scope*. Presently, there is a vast array of different kinds of microscopes, each with a specialized purpose. See table 8.2 on page 132 for the different kinds and specialized uses of these microscopes.

The compound light microscope con-sists of two or more lenses that magnify a specimen. A specimen must be thin in order for light to be **transmitted**[376] through it. Magnification is achieved by the bend-ing or **refraction**[377] of light rays by lenses. See figure 8.4.

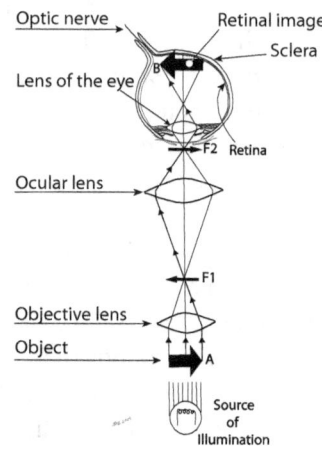

Figure 8.4. Tracing the path of light waves from a source of illumination, through a speci-men to the retina.

Parts of the Compound Light Micro-scope

A microscope is made of many parts. Each part has a specific function. See figure 8.10 and table 8.1. The *ocular lens* is located in the uppermost part of a hollow *body tube*, and the *objective lens* is in the lowermost part. The arrange-ment allows both lenses to be raised or lowered at the same time. The *coarse and fine adjustment knobs* are used to raise or lower the body tube. The coarse adjustment is used for large movements of the body tube. The fine adjustment is used for very small movements. *The coarse adjustment should be used only with the lowest power objective.*

The *field of view* appears as a bright circle of light in the ocular lens. As the mag-nification of the objective lenses is increased, the diameter of the field of view *decreases* (see figure 8.5) and light intensity is lessened.

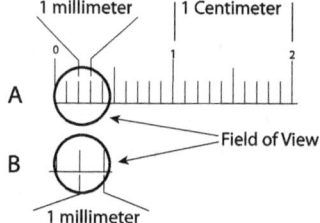

Figure 8.5. Illustration of field of view. (**A**) is a plastic ruler under 40x. About 4.5 mm of the ruler is visible. (**B**) The field of view is decreased under 100x. Only 2 mm are now visible of the same ruler.

---

375   A symbol.
376   Referring to light that passes through a substance like glass without distortion.
377   Refraction is the bending of light waves as they pass from one medium to another.

The *stage* is a platform below the objective lens upon which the specimen is placed. An opening in the center of the stage admits light from a *source of illumination*. A dimmer switch on the base of the microscope slides back and forth to increase or decrease light intensity in the field of view.

The *iris diaphragm* regulates the amount light passing through the specimen. Opening the iris diaphragm admits more light. Closing it restricts light just like the iris of the human eye. The iris diaphragm can also increase contrast between the structures in a specimen if properly adjusted. Increasing contrast with the iris diaphragm takes practice. The diaphragm is connected to the bottom of a *substage condenser (Abbe condenser)*. As the name implies, the condenser concentrates light from a source before it enters the specimen. A control on the condenser body moves it up and down, increasing or decreasing the diameter of the beam of light entering the specimen. *Each time an objective lens is changed from one lens to another, the condenser should be adjusted* by raising or lowering it, thereby adjusting the diameter of the beam of light entering the objective lens.

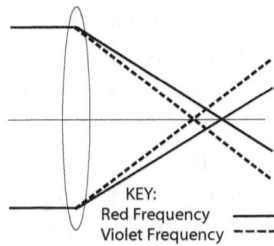

KEY:
Red Frequency ——
Violet Frequency ----

Figure 8.6. Diagram illustrating the principle of chromatic aberration. If the red and violet frequencies are not focused at the same point, chromatic distortion occurs.

An *achromatic condenser* corrects for *chromatic aberration* (chromatic distortion). Microscope lenses do not focus all frequencies of light at the same point (focal point). Higher light frequencies refract more than lower frequencies. Violet, a higher frequency than red, bends more than red. As a result, violet and red wavelengths focus do not focus at the same focal point. This causes color distortion. See figure 8.6. If chromatic aberration is not corrected, spikes of color will appear at the edges of the field of view. *Spherical aberration* is another distortion that can occur with lenses. Spherical aberration occurs when light rays are bent more at the edges of a lens than at the center of a lens. As a result, the image will be distorted or blurred at the edges. See figure 8.7.

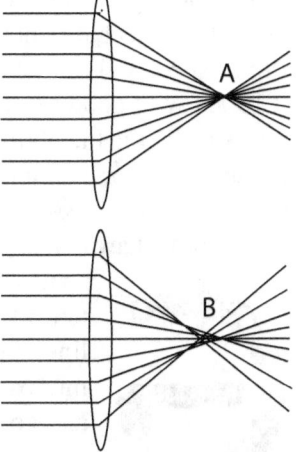

Figure 8.7. Diagram showing the proper focusing of light waves by a lens (A) and distortion of light waves (B).

125

Calculating Magnification of a Combination of Lenses

The ocular lens, usually marked 10x, magnifies the image of the objective lens 10 times or 10 diameters. *The total magnification of a combination of lenses is found by* **multiplying the magnification power of the ocular lens** *by the* **magnification power of the objective lens**.

Objective lenses are at the lower end of the body tube in a *revolving turret*. The revolving turret simplifies changing magnification by rotating different power lenses into position directly under and in line with the ocular lens. The magnifying power of an objective lens is marked on its side. Magnifying powers of objective lenses are commonly 4x, 10x, 40x and 100x. In addition to a printed number, objective lenses are color coded; red is 4x, yellow is 10x, blue is 40x and white is 100x. The length of an objective lens housing becomes longer as the power of the objective lens is increased. The lowest power is the shortest lens, and the highest power is the longest lens. The lowest power lens is the *scanning lens*. The scanning lens is usually 4x. This lens is used to *locate and center* the specimen in the field of view.

If a **4x objective lens** is in place with a **10x ocular lens**, the total magnification is forty diameters or 40x. If the 40x objective is in place; then the total magnification is 400x. The lens marked 100x is an immersion oil lens. Immersion oil prevents the loss of light intensity. A drop of immersion oil has the same refractive index as glass placed between this lens and the object. The light intensity from the source of illumination, through the object and then to the objective lens remains virtually undiminished. Oil immersion lenses will magnify an object 1000x with good resolution.

Resolution is the ability to distinguish one object from another. For example, if two lines are drawn next to each other (figure 8.8A) on a piece of paper and the paper is moved away from an observer, there would be a point where the observer could not tell if there were two lines or one line. See figure 8.8B. A good microscope has high magnification and good resolution.

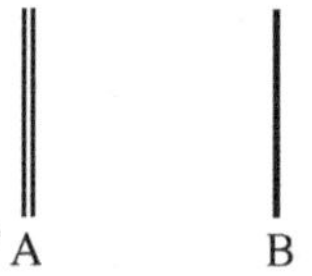

A   B

Figure 8.8. Resolution. The human eye is unable to distinguish or resolve the two lines at a distance.

Apparent Movement of an Image

Figure 8.9. The letter "e" under low power (40x).

If the letter "e" cut from a newspaper, is placed right side up on the microscope stage it will appear *upside down* and *inverted right to left* in the ocular lens. See figure 8.9. When the letter "e" is moved upwards on the stage, the image in the ocular lens appears to move

126

downwards. When the letter "e" is moved downwards, the image appears to move upwards. When the object is moved to the right, the image will appear to move to the left. If the object on the stage is moved to the right, the image appears to move to the left in the ocular lens.

Table 8.1. Parts of the compound light microscope and their functions.	
Part	Function
Ocular lens or eyepiece	Usually magnifies an image 10x
Revolving turret or nosepiece	A rotating holder for the objective lenses
Objective lens	Magnify object 4x, 10x, 40x, or 100x
Stage	A platform under the objective lens with an opening through which light passes
Sub-stage condenser	A lens beneath the stage that can *help* increase or decrease the intensity of light and prevent refraction of light
Iris diaphragm	Regulates the amount of light passing through the stage opening.
Iris diaphragm lever	Regulates the opening and closing of the iris diaphragm
Base	Supports the working parts of the microscope
Arm	Supports the body tube and objective lenses above the stage
Mechanical stage	Moves the slide vertically and horizontally
Mechanical stage controls	Moves the mechanical stage.
Coarse adjustment	Used to make fast, large movements of the objective lenses
Fine adjustment	Used to make slow, very small, almost imperceptible movements of the objective lenses
Light intensity control	Regulates the voltage to the source of illumination
On/off switch	Turns on source of illumination

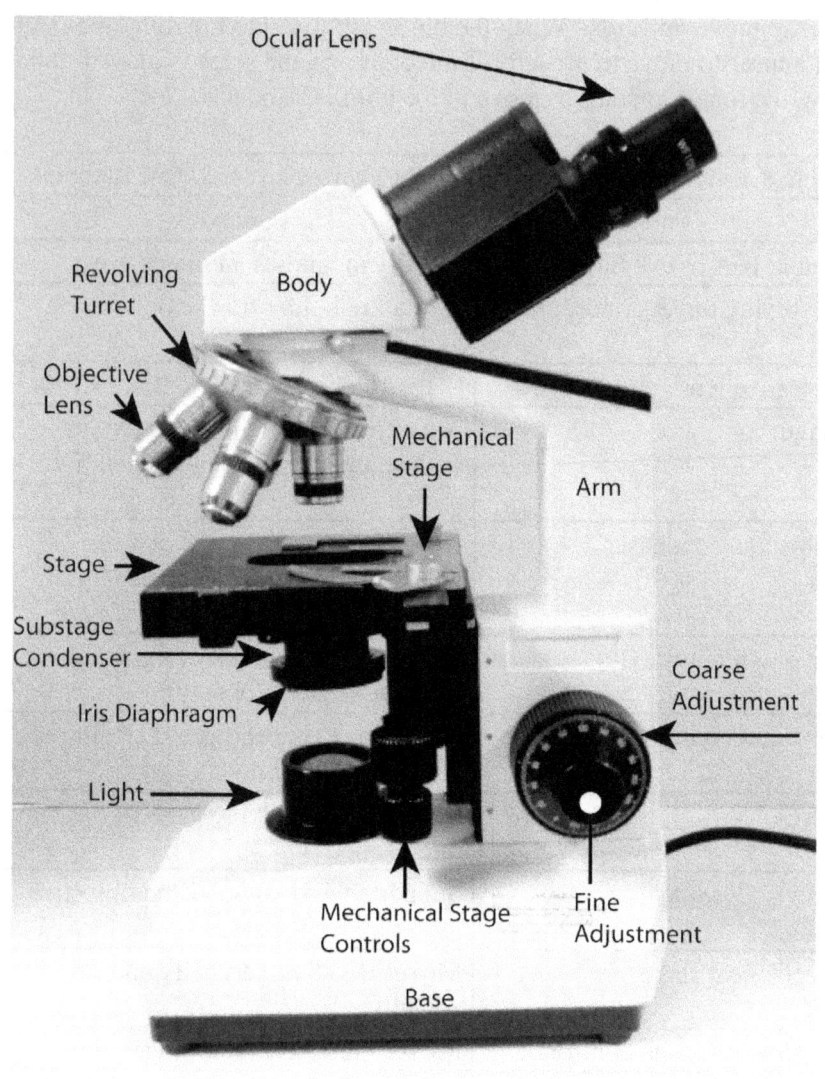

Ocular Lens

Revolving Turret

Body

Objective Lens

Mechanical Stage

Arm

Stage

Substage Condenser

Iris Diaphragm

Coarse Adjustment

Light

Mechanical Stage Controls

Fine Adjustment

Base

Figure 8.10. A typical compound light microscope.

The Nature of Light

The science that studies *light* is a branch of physics called *optics*. *Light waves* travel in straight lines in all directions from its source. Light waves can be *transmitted*, *refracted* or *reflected*.

All electromagnetic radiation, including visible light, is composed

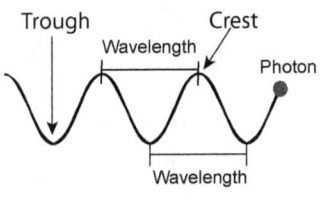

of particles called *photons*. A photon is a *small* packet or *quantum* of light energy. Photons are particles that move as *transverse waves*. These are waves where the photon moves at right angles to the direction of the wave that carries it. A photon moves up and down as the wave that carries it moves horizontally. See figure 8.11.

Figure 8.11. Representation of a photon.

Visible light is composed of transverse waves of different frequencies. Transverse waves are *electromagnetic waves* or electromagnetic radiation. Frequency of transverse a wave refers to the number of time a crest, or trough passes a given point in one second.

There are many kinds of transverse waves. The *electromagnetic spectrum* below is a way of categorizing them. See figure 8.12. As the frequency increases, the wavelength gets shorter.

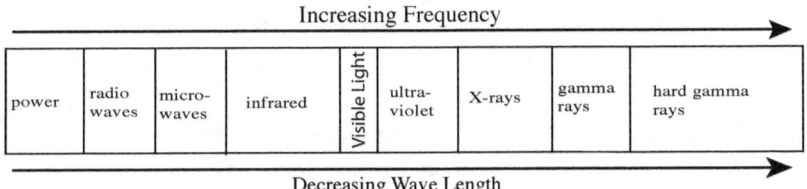

Figure 8.12. The electromagnetic spectrum.

Electromagnetic radiation can travel through a medium like glass or air, but it can also travel through a vacuum, just as light from the Sun travels through space to reach the Earth. Living things are surrounded by electromagnetic radiation. Electrical wiring in a house, power lines and motors give off small amounts of electromagnetic radiation. All three are long wavelengths and range from 1 to 300 feet.

Radio waves are long waves. Radio waves carry energy for cell phones, radios and television. Stars emit radio waves and are among the longest radio waves, about a mile long. *Microwaves* are about a foot to fractions of an inch in length. Microwaves carry enough energy to boil water and cook food. Radar devices use shorter microwaves. Radar images shown in weather reports are created by microwaves bouncing off clouds, rain or snow.

Infrared waves generate thermal energy. Infrared lamps are often seen keeping food warm in restaurants. Standing in front of a fire feels warm because of invisible infrared waves the fire gives off. Infrared radiation from the Sun generates the warmth we feel on Earth.

Visible light (white light) is the only part of the electromagnetic spectrum humans can see. Visible light is one small part of the electromagnetic spectrum. In a vacuum, where there are no particles of matter present to slow it, light travels at 186,000 miles per second. Light will move at different speeds in different media. In air, light slows down a tiny bit and in glass it slows down more. The more particles present in a medium, the slower light waves move.

The eye detects the visible spectrum of light, but not above the violet frequency, nor below the red frequency. Visible light is made up of many different frequencies. Structures in the retina of the eye called cones, detect each frequency of light as a different color.

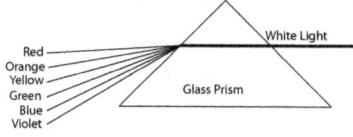

A glass **prism**[378] separates the electromagnetic frequencies called color. Individual wavelengths of light slow down as they pass from air to glass. Red wavelengths bend the least, and violet wavelengths bend the most. The change in speed causes light waves to bend, thus separating light into the colors of the spectrum. See figure 8.13. A rainbow

Figure 8.13. A glass prism bends violet wavelengths more than the red. Violet is a higher frequency than red.

results from water droplets in the air that act as prisms.

Each observed color of an object is a reflected wavelength. All other colors are absorbed by an object except the one that is the reflected wavelength. A red flag reflects the red frequency of visible light, and it absorbs all the other frequencies.

Ultraviolet light is not visible and has shorter wavelengths compared to visible light, but it carries more energy than frequencies in the visible spectrum. Ultraviolet light is not sensed and can cause severe sunburns.

X-rays are the next shortest wavelength and are used extensively in diagnostic medicine. X-rays penetrate soft tissue but not bone. The radiation that passes through soft tissue exposes photographic film, but dense objects absorb the X radiation preventing the exposure of the film. This is why soft tissue appears black and hard objects appear white.

Gamma rays are among the shortest wavelengths and possess the greatest amount of energy. Gamma rays are given off by the nuclear decay of atoms in stars.

---

378   A optical material like glass consisting of three sides that refracts light waves.

The Behavior of Light

Transmission of light refers to light passing *through* an object. The light waves are not absorbed or reflected. See figure 8.14.

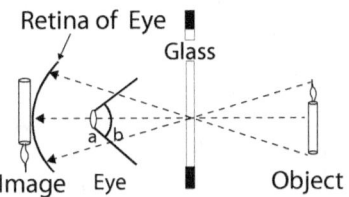

Figure 8.14. Light waves pass through the glass. A clear image can be seen by the viewer. The lens is a. The cornea is b.

Light waves striking clear glass are transmitted through the glass. An image can be seen through the glass because most of the light passes through something that is *transparent*. Transparent objects have little ability to reflect or absorb light waves. Light waves pass directly through the object. If light waves are scattered, light can be seen, but a clear image is not visible through the glass; the glass is said to be *translucent*. If an object is *opaque*, all of the light is absorbed. Therefore, no light will pass through the object. Solid walls are opaque.

Refraction is the bending of light rays. Refraction is caused by a change in the direction of light rays. The change in direction is due to a change in the speed of a light wave. In figure 8.15, there are three types of media: air, water and glass. Each medium has a different index of refraction. Air has a refractive index of 1.0003, water is about 1.33 and glass is about 1.5. As the index of refraction increases, light bends more.

Figure 8.15. Refraction of light waves reflected by a pencil in a glass of water.

Reflection occurs when light rays strike a surface and bounce off. A mirror can be defined as a metallic-coated surface that changes the direc-

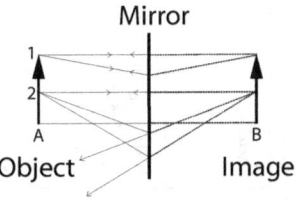

Figure 8.16. Reflection of light from a plane (flat) mirror. Only two points on object A are considered, numbered 1 and 2.

tion of a light wave. A *plane or flat mirror* is shown in figure 8.16. Light waves that strike the mirror at (A) are called *incident rays*. Light waves that are reflected back to the observer are called *reflected waves*.

Incident rays are shown bouncing back to form a *virtual* image (B). If the light ray strikes the mirror at 90 degrees, it goes straight back. If the light wave hits the mirror at an angle, it is reflected back at the same angle. Reflected images appear to be behind a mirror and erect.

131

Table 8.2. Selected types of microscopes.

Type of Microscope	Description	Staining	Source of Electromagnetic Spectrum	Maximum Magnification	Resolution	Mounting Specimen	Changing Magnification
Simple	Single lens	Staining of specimen is not required.	Natural or visible light	about 300x	about 1 micrometer (1 millionth of a meter)	in front of lens	none
Compound Light or Bright Field	Two or more lenses	Specimen requires staining	Natural or visible light reflected from a mirror or an incandescent electric bulb	1500-2000x	about 0.2 micrometers	On a platform (stage) that can be raised or lowered	Revolving turret with objective lenses of different magnification
Dark Field	Two or more lenses	Staining of specimen is not required.	Reflected natural or visible light. Background is dark and object is illuminated from above	Same as compound above	Same as compound above	Same as compound above	Same as compound above
Phase Contrast	Two or more lenses	Staining of specimen is not required.	Indirect visible light, manipulated wavelengths	Same as compound above	Same as compound above	Same as compound above	Same as compound above
Electron Microscope	Optico-magnetic coils	Special staining	Beam of electrons	10 million x	50 trillionths of a meter	Ultra-thin specimens	Voltage regulation

Common Laboratory Hardware
Balances

An equal arm balance consists of a beam with a pan suspended at each end. This system has been in use for thousands of years to compare unknown masses with agreed upon standard masses. See figure 8.17.

One pan holds a *known mass* and the other pan holds the *unknown mass*. When the balance comes to rest in a horizontal position, the unknown mass is equal to the known mass. The correct term is *massing* a substance, not weighing it. Instead of putting fractions or multiples of a

Figure 8.17. An equal arm balance depicted on an Egyptian papyrus c. 2000 BC.

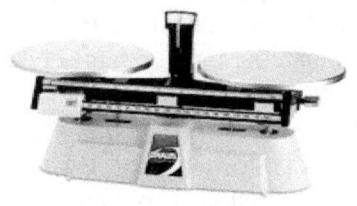

Figure 8.18. A single pan balance.

kilogram on a pan, modern balances have sliding masses that can be moved incrementally until the unknown mass is balanced against the sliding mass. See figure 8.18. Electronic massing devices are used more often in modern laboratories.

The Bunsen Burner

Most laboratories need flame for heating and combustion of materials. Bunsen burners are used for these purposes. Bacterial transfer loops must be sterilized by passing them through a flame prior to and after use to avoid cross contamination between cultures. Flaming will only be effective if the transfer loop is placed in the hottest part of the flame. See figure 8.19. This is a routine procedure in any microbiology laboratory.

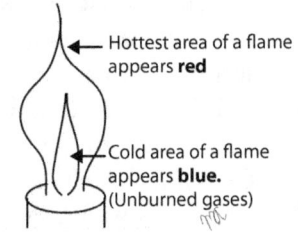

Hottest area of a flame appears **red**

Cold area of a flame appears **blue.** (Unburned gases)

Figure 8.19. Illustration of the parts of a flame.

Centrifuges

Centrifuges are machines that will spin extremely fast to multiply the force of gravity. A mixture is put in a cup in a centrifuge and "spun down." The faster the cup spins, the faster particles suspended in a mixture will settle out. Spinning a mixture containing suspended particles will cause the particles to settle out faster than if the mixture is left standing. Centrifuges can be simple tabletop units or large units in which temperature can be controlled.

Chapter 9 Techniques[379] Used to Study Living Things

Microbiology

Microbiology is the study of *viruses, archaea, bacteria, algae, fungi, protists* and parasitic *helminths (worms)*. Most are microscopic and have to be *cultured* in the laboratory in order to be studied.

Studying Viruses

Virology is the science of viruses. *Viruses are not alive, but they are known to infect all living things.* Viruses are commonly grown in *tissue culture*. Tissue culture is growing viruses in single cells (chicken eggs), tissues, organs or in an organism such as a non-human primate.

Edward Jenner (1749-1823), an English physician, knew that women who milked cows did not get smallpox after they had a case of cowpox. Jenner hypothesized that a cowpox infection might prevent smallpox. He tested his hypothesis by **vaccinating**[380] a child with material from a cowpox pustule. He later injected smallpox material into the same child. The child did not get smallpox. *Immunity* was conferred on the child. Jenner's discovery earned him the title of the "Father of Immunology."

Jonas Salk (1914-1995) developed the first polio vaccine in 1955. His used inactivated polio virus to vaccinate large numbers of children. Salk's vaccine confirmed immunity against the polio virus.

Studying Bacteria

Early bacteriologists grew bacteria on the material bacteria grew on naturally. If bacteria were found growing on potatoes, then slices of potato were used to culture them. In the 1880s, broths containing beef extracts began to be used. Later, gelatin was added to a broth to make it semisolid like **nutrient agar**.[381] Modern microbiologists use hundreds of different kinds of liquid and semisolid media. See table 9.1.

Undefined media are like broths. The exact composition is *unknown*. It has water, salts, glucose and beef extract. *Nutrient* agar or *nutrient media* are examples. *Defined media* have a *known* composition. The pH of media has to be adjusted for proper growth of bacteria. A *Selective* growth medium allows the growth specific types of bacteria or fungi. *MacConkey agar* and *Eosin methylene blue (EMB)* selects for growth of *Gram-negative bacteria*. In general, bacteria like a slightly acid pH. *Sabouraud's medium* is selects for fungi and inhibits bacterial growth. *Sabouraud's medium* has a pH of 5.6 and contains dextrose, peptones and agar. Fungi grow best at a neutral to a slightly alkaline pH.

379   A way of doing something, doing a scientific procedure.
380   From the Latin, *vacca* meaning cow.
381   A gelatin-like substance obtained from red algae. It is used as thickeners in foods and microbiological media.

Culturing Bacteria

Less than one percent of all known bacteria have been cultured in the laboratory. Nutrient agar in a Petri dish is commonly used for culturing bacteria. A *mixed culture* is produced by exposing the sterile agar in a Petri dish to room air. See figure 9.1. The culture contains many different kinds of microbial colonies. Each bacterial cell and fungal spore that falls on the surface of the agar from the air will reproduce asexually by binary fission. Bacteria will continue to reproduce and pile up on each other until a visible *colony* is produced. Colonies are composed of descendants of one cell. All members of the colony are genetically identical. All are *clones* of one another.

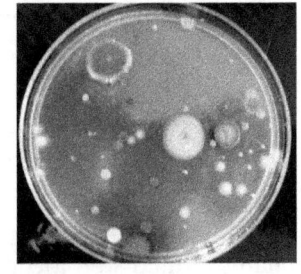

Figure 9.1. A mixed culture of many different kinds of bacteria on a nutrient agar plate.

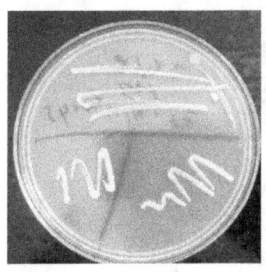

Figure 9.2. A streak plate showing a pure culture.

A *pure culture* is obtained by using a sterile transfer loop to removing a small sample of one bacterial colony growing in a mixed culture and gently moving the loop across the surface of a sterile agar plate. This process produces a *streak plate* that contains only one kind of bacterium. See figure 9.2.

Temperature Requirements of Bacteria

Bacteria, like all living things, have an *optimal growth temperature*. An optimal growth temperature is the temperature at which enzymes function best. *Psychrophiles* are bacteria that grow best at low temperatures, 32 °F to 68 °F (0 to 20 °C). Most bacteria are *mesophiles*. Mesophiles grow in a more temperate range of 68 °F to 104 °F (20-40 °C). Some bacteria are *thermophiles*. Thermophiles like it hot, 104 °F or higher (above 40 °C). An enzyme possessed by the thermophile *Thermus aquaticus* is used to amplify small amount of DNA into large amounts of DNA in the laboratory. Some bacteria are *hyperthermophiles*. Hyperthermophiles like it very hot, up to 252 °F (122 °C). Hyperthermophiles can live and reproduce in boiling water. *Extremophiles* are those bacteria that live in extreme conditions of high temperature, pressure, salt, ionizing radiation or toxic elements.

Making Bacteria Visible – Simple Staining

Simple staining is the use of one color to stain bacteria. It is a quick and easy way to make bacteria visible. Simple staining reveals shape, arrangement and helps in identification. If a thin, dried film of **heat-fixed**[382]bacteria is covered with drop of purple stain, bacteria will appear dark blue. There are a few biological stains that allow living cells to be observed; they are called *vital stains*. Most stains kill bacteria.

Shape and Arrangement of Bacteria is Revealed by Simple Staining

There are three basic forms of bacteria: *cocci* (round), *bacilli* (rod) and *spiral* (curved). Although, there are several variations of each form, round, rod and curved forms are easily identified.

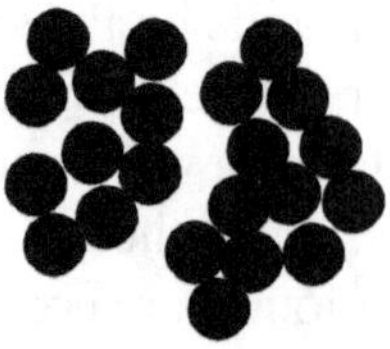

Cocci are round bacteria that can occur as single cells, pairs of cells or grape-like clusters called staphylococci. See figure 9.3. Cocci also are found as chains termed streptococci. See figure 9.4. *Staphylococcus aureus and Streptococcus pneumonia* are two important examples of these two arrangements. Some cocci will demonstrate formations called *tetrads* (figure 9.5a) or *sarcinae* (figure 9.5b). Tetrad formations occur when cocci divide on two planes and fail to separate, forming groups of four cocci. Sarcinae results when cocci divide on three planes and fail to separate, forming groups of eight cocci.

Figure 9.3. Staphylococci.

Figure 9.4. Streptococci and pairs (diplococci).

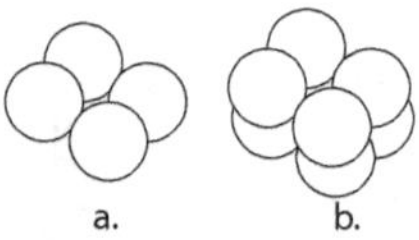

Figure 9.5a tetrad and 9.5b sarcinae.

Bacilli are rod-shaped bacteria. When bacilli occur as long chains, they are termed *streptobacilli*. See figure 9.6. Notice *diplobacilli* (two bacilli).

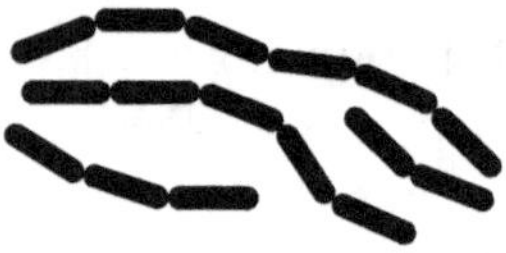

Spirals occur in three forms: *curved (vibrio), spirillum (spiral)* and *spirochaetes (tightly coiled)*. *Vibrios* are comma shaped spirals, as in *Vibrio cholerae*. Some spirals may be thick walled and rigid such as *Spirillum minus,* the causative agent of rat bite fever.

Figure 9.6. Typical streptococci formation.

Spirochaetes are flexible and more tightly coiled. Spirochaetes may be corkscrew shaped as in *Campylobacter jejuni*, or helix shaped like *Helicobacter pylori* or highly coiled like *Treponema palidum palidum*. See figure 9.7.

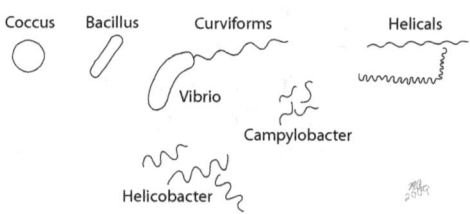

Figure 9.7. Common shapes of bacteria.

―――――――――――――
382   Bacteria washed off a glass slide if heat is not applied to the bottom of the slide.

Differential Staining
The Gram Stain
Gram staining uses differences in chemical composition of bacterial cell walls to place bacteria into one of two groups; Gram-positive or Gram-negative. The Danish bacteriologist, Hans Gram developed this staining method to distinguish the agent of streptococcal (lobar) pneumonia, *Streptococcus pneumonia*, from *Klebsiella pneumoniae* pneumonia.

A purple *primary stain* is placed over a heat-fixed bacterial smear on a slide and rinsed off with water. Iodine is placed on the smear and washed off with alcohol. A pink *counterstain* is applied. *Gram-positive bacteria* will appear *purple*. *Gram-negative bacteria* will appear *pink*.

Gram staining helps in the selection of a proper antibiotic. For example, pneumonia has two major bacterial causes; Gram-positive *Streptococcus pneumoniae* or Gram-negative *Klebsiella pneumoniae*. If the causative agent of the pneumonia is *S. pneumoniae,* one of several *penicillins* may be prescribed. If the agent is *K. pneumoniae, then ampicillin/sulbactam* may be prescribed. Each year it is estimated that about 175,000 people are hospitalized for bacterial or lobar pneumonia. Some vaccines are effective against 23 strains of pneumococcus.

Acid-Fast Staining
Acid-fast staining, also known as the *Ziehl-Neelsen stain* is a test for members of the genus *mycobacteria*. Mycobacteria have a high *mycolic acid* content in their cell walls that holds onto a red dye. *Mycobacterium tuberculosis* is *acid-fast* because it appears *pink* after an acid/alcohol wash. After the wash, purple counterstain is placed over the smear. If the bacterial smear appears pink, the bacteria are acid-fast. If the bacteria appear purple, the bacteria are not acid-fast. For treatment see page 138.

Serious Disease Caused by Gram-Positve Bacteria
Staphylococcus aureus is a normal inhabitant of the skin that can cause serious skin infections. See pages 60 and 185. *Streptococcus pyogenes* is found in the throat and skin and can cause *Streptococcal pharyngitis* or "strep throat." If untreated, this condition can become *scarlet fever*, a serious systemic infection. A **sequela**[383] of scarlet fever is *rheumatic fever*, an inflammatory disease that causes damage to heart valves, joints, skin and brain.

Serious Disease Caused by Gram-Negative Bacteria
Neisseria and *Klebsiella* are two genera of Gram-negative cocci. *Neisseria gonorrhoeae* causes **gonorrhea**. *Neisseria menigititis* causes **cerebrospinal meningitis**.

---

383   From the Latin, meaning that which follows. Pleural is *sequelae*.

Serious Disease Caused by Acid-Fast Bacteria

*Mycobacteria cause two serious human diseases; **tuberculosis** (TB) **and leprosy.** Mycobacterium tuberculosis* causes tuberculosis, and *mycobacterium leprae* causes leprosy. Both diseases are curable with a series of antibiotics taken for a prolonged period of time. Isoniazid (INH), rifampin (RIF) and rifapentine (RPT) are used to treat TB. Leprosy is treated with dapsone and rifampicin for six months.

Other Differential Staining Methods

There are many differential chemical stains that make **subcellular structures** stand out from the rest of the cell. Just a few examples are: methylene blue makes nuclear material (DNA) stand out, iodine stains starch dark purple and malachite green stains fungal spores green.

Koch's Postulates: A Specific Organism Causes a Specific Disease

Robert Koch (1843-1910), a German physician, was *the first to devise a method to associate a specific bacterial disease with a specific bacterial cause.* He isolated and discovered the bacterial agents of anthrax, tuberculosis and cholera. His method came to be known as Koch's postulates. An outline of his method is described below:

1. **Isolate** the organism from the diseased animal (primary isolate).

2. **Culture** the suspected organism in pure culture.

3. **Infect** a healthy test organism with a sample from the pure culture. The test animal **should**[384] develop the disease.

4. **Re-isolate** the organism from the test animal and compare it with the primary isolate.

Exceptions to Koch's Postulates.

Postulate 1. Postulate 1 originally required the organism be found in great quantity, but **asymptomatic**[385] carriers of bacterial diseases such as typhoid fever and cholera will not harbor great quantities of the organism. **Typhoid Mary**[386] is a famous example of a carrier of typhoid fever who never got the disease, but started several outbreaks.

Postulate 2. Some infective agents such as prions cannot be grown in culture.

Koch's Postulates and Molecular Biology

1. *A nucleic acid sequence* or pathogen should be found in abundance in the infected individual.

2. Copies of the *nucleic acid sequence* should not be found in people without the disease. If a disease is resolved, no sequences will be found.

---

384   Most test animals will get sick, but not all.
385   Does not show any symptoms or subclinical infections.
386   Mary Mallon (1869 -1938). First identified asymptomatic carrier of a pathogen.

General Types of Nutrition

It is important to know the nutritional requirements of bacteria in order to grow and identify them from isolates taken from patients. Some bacteria are *not fastidious* and will grow on a variety of nutritional substrates. Bacteria that are not fastidious will grow on a *general microbiological nutrient growth media*. Media are used in liquid or semi-solid form. The semi-solid from is usually referred to as *nutrient agar* (see page 134).

Bacteria that are *fastidious* require specific nutrients, without which the bacteria will not grow and reproduce. Examples of fastidious bacterial genera are *legionella*, *brucella*, *bordetella* and *leptospira*. Examples of fastidious species are *Francisella tularensis* and *Borrelia burgdorferi.* These bacteria require selective media. See table 9.1.

Table 9.1 Selected microbial growth media.

Name	Ingredients	Use
Nutrient Agar (general microbiological medium)	Undefined	Non-fastidious bacteria.
MacConkey agar (MAC) selective	Bile salts, crystal violet dye, neutral red dye, lactose and peptone.	Allows Gram-negative bacterial growth and inhibits the growth of Gram-positive bacteria.
Blood agar	Sheep blood, meat extract, added tryptones and sodium chloride, agar.	Shows: 1. Beta (ß) hemolysis as a complete digestion of red blood cells leaving a clear ring around the bacterial colony and 2. Alpha (∂) hemolysis is a partial hemolysis of red blood cells that produces a greenish color in the agar.
Thayer-Martin agar (TM)	Chocolate, agar.	Isolation of *Neisseria gonorrhoeae*.
Chocolate agar (CHOC)	Heated blood in agar.	Used to grow fastidious respiratory bacteria, (*Haemophilus influenzae*).
Sabouraud Dextrose agar	Dextrose, L peptone, agar, pH 5.6.	Fungi
Mannitol Salt Agar (MSA)	NaCl, mannitol, the indicator phenol red.	Differential medium to identify B-hemolytic streptococci.

Studying Protozoa

Hay, wheat or rice in spring water ("hay infusion") is commonly used to culture protozoa. Spring water avoids chlorine and trace metals in tap water that may be toxic to protozoans.

Nutrition of paramecia and amoeba is *holozoic*, a method that involves the ingestion of whole particles of food. Paramecia and amoeba feed on bacteria, algae, smaller protists and yeasts.

Studying Fungi

Fungi grow on almost everything. *Sabouraud dextrose agar* (SDA) at a pH of 5.6 is the medium of choice to grow fungi in the laboratory. The pH is low enough most bacterial growth is inhibited. In general, bacteria like a pH of 6 to 6.8, although some thrive at pH ranges of 1-3.5 SDA contains *gentamicin*, an antibiotic that prevents the growth of Gram-negative bacteria. Many fungi form symbiotic relationships with plants, animals and algae. *Lichens* are fungi and algae that have a symbiotic relationship. The fungus provides support and protection for the algae and the algae provide food for the fungus.

Cornmeal Agar (CMA) is used to grow many kinds of fungi also, especially the **fungi imperfecti**.[387] *Malt Extract Agar* (MEA) is a good growth medium for fungi that are normally found in soil or fungi isolated from wood. *Potato Dextrose Agar* (PDA) is suitable for growing a wide range of fungi.

Studying Plants

Key:
N - Nucleus
CW - Cell Wall
Chl - Chloroplasts
V - Vacuole
CM - Cell Membrane

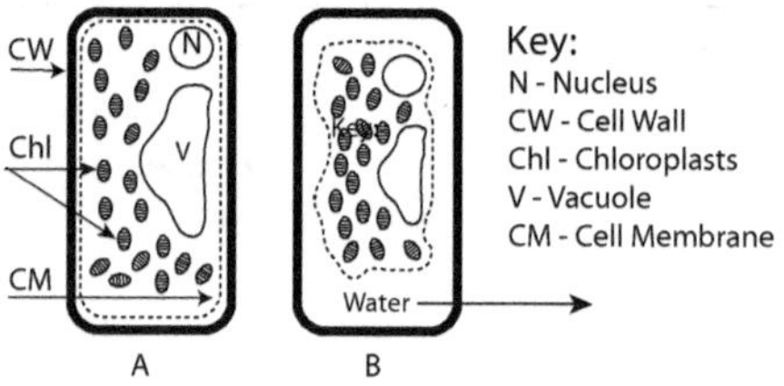

Figure 9.8. Illustration of plasmolysis in a plant cell.

Green plants use sunlight, water, nutrient rich soil, oxygen and carbon dioxide to live and grow. Plants convert carbon dioxide from the air and water from the soil into simple sugars. Chlorophyll is the organic catalyst that transfers energy of sunlight to chemical bonds of sugar molecules. Plants carry on photosynthesis during the *day* to make food. Cell respiration is carried on *day and night* to supply energy needs.

Water is necessary to generate *turgor pressure*. Turgor pressure keeps the plant erect. Turgidity is lost when plant cells undergo *plasmolysis*. Plasmolysis is the shrinking of a cell's cytoplasm caused by the loss of water by *osmosis* out of a cell, causing a visible gap between the cytoplasm and the cell wall, causing the plant to wilt. See figure 9.8.

---

387   Do not fit neatly into common classification schemes. Classification based on morphology of reproductive structures. Only asexual, no sexual reproduction present.

Studying Animals

Vertebrates and invertebrates are two large groups of animals. The two groups are not officially recognized, but it is a simple way to divide the animal kingdom. Vertebrates have a *dorsal nerve cord* protected by bony vertebrae of a spinal column. Invertebrates do not have either. Lower vertebrates such as fish, amphibians, reptiles and birds are discussed to some extent here. Humans are considered in more detail.

1. Anatomy

Anatomy includes observation of an organism's surface or internal structures. In general, anatomy refers to surface and internal structure of plants, animals and humans. *Macroscopic* anatomy, sometimes referred to as *topographical* or *gross* anatomy, is viewing structures that are observable with the unaided eye. *Microscopic anatomy* studies cells or structures requiring magnifying devices. Microscopic anatomy involves knowledge of *histology and cytology*. Histology the study of tissues and cytology is the study of cell structure.

2. Dissection

Dissection is *not* cutting and removing structures from an organism. *Dissection is observing the relationship of one structure to another in cadavers.*[388] It is a method of exploring structures beneath the surface of an organism.

Ancient Egyptians had a great working knowledge of anatomy, mostly related to their practice of embalming their dead. Galen (A.D. 131-200), a Roman physician, was the authoritative and most trusted source of anatomical knowledge for over 1,000 years. He was a surgeon in the Roman army, a physician to gladiators and emperors. Galen used nonhuman primates and pigs for his studies assuming both to be the same as humans. This assumption introduced errors in his writings.

During the late Middle Ages, Mondino de' Luzzi (1275-1326), an Italian physician and anatomist, rather than defer to Galen's authority, relied on his own observations. He did his own dissections, instead of having an assistant do them, as other anatomists did at the time. He was the first to perform a public dissection to instruct medical students. His work paved the way for Andreas Vesalius.

Andreas Vesalius (1514 -1564), a Belgian physician and anatomist, wrote the classic text entitled, *De humani corporis fabrica libri septem* (On the Fabric of the Human Body in Seven Books) in 1543. This book became a standard up to the modern era. Vesalius reorganized the study of anatomy. He found many errors in Galen's work and effectively ended Galen's authority and is the "Father of Modern Anatomy."

---

388   A dead body.

3. Comparative Anatomy

Comparative anatomy is the examination of structures between species of animals to discover similarities and differences between them. Similarities of structure between species may indicate a common evolutionary ancestor. Edward Tyson (1650-1708), an English scientist and physician, is regarded as the founder of modern comparative anatomy. Tyson examined the anatomy of pygmies, monkeys, apes and humans and concluded that chimpanzees are anatomically closer to humans than monkeys. He also discovered that porpoises are mammals. In 1555, Pierre Belon (1517-1564), a French naturalist, showed that the skeletons of humans and birds have the same bones, and the bones are connected in much the same way.

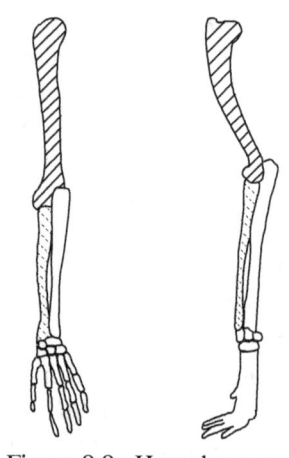

Figure 9.9. Homologous structures. The humerus a human compared to a humerus in a dog's limb.

A **homologous**[389] *structure is a structure that was present in an ancestor. The gene for the structure was inherited, and it descended to related species.* The forelimbs of humans, dogs and the bones in the flippers of whales are examples of homologous structures. See figure 9.9. They have body structures similar to each other, even though they do not have the same function.

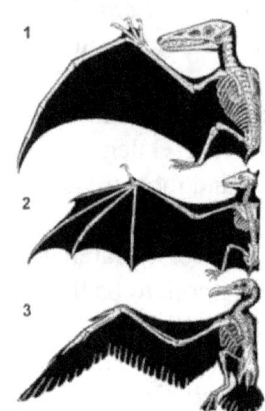

Figure 9.10. Analogous structures. Comparison of bones of: 1. a pterosaur, 2. bat and 3. bird.

Analogous[390] **structures** look similar in unrelated species. *These structures were not inherited from a common ancestor.* Analogous structures may have a similar function, but each evolved separately. Such structures evolved in similar environments and may serve the same or a similar purpose. The wing of a bat and that of a bird are used for flying. See figure 9.10. The bat's wing and the bird's wing have the same function. The structures do not have a common ancestry, therefore, not passed down to offspring. The eye of an insect and the eye of a human function as organs of sight, but each evolved separately. They too, do not have a common ancestor.

---

389   Refers to having similarities in structures that have common origins.
390   Analogous refers to a resemblance, but no common ancestry or relationship.

4. Embryology

Embryology is the study of an organism's development from a fertilized egg (zygote) to a fetus. An animal zygote undergoes a series of cell divisions called *cleavage*. After a number of cell divisions, a *solid* ball of cells termed a *morula* forms. The morula develops into a *hollow* ball of cells called a *blastula*. Soon, the blastula develops into a three layered *gastrula*. A gastrula has three *primary germ layers*: *ectoderm*, *mesoderm* and *endoderm*. Each primary germ layers gives rise to different tissues and organs. See page 195, figure 12.7.

5. Histology

Histology studies microscopic structure of tissues. Animals have *four basic types of tissues*: (1) *epithelial*, (2) *connective*, (3) *muscle* and (4) *nervous* tissue. *Tissues are a group of similar cells that perform the same function and are derived from the same embryonic origin.*

Tissues have to be chemically **fixed**[391] to preserve structure for microscopic study. Tissues must be imbedded in wax and sliced into thin sections with a **microtome**.[392] The exceedingly thin section is stained to achieve contrast of tissue structure. If the tissue looks diseased, it is examined by a *histopathologist*. *Histopathology* is the study of microscopic specimens of tissue that show signs of disease.

a. *Epithelial tissue* consists of closely epithelial cells packed cells arranged in one or more layers. Epithelial tissue protects structures, secretes and absorbs substances. Epithelial tissue lines body cavities and, like the skin, covers the surface of structures. Skin is composed of two major layers: epidermis and dermis. The epidermis is **avascular**[393] and consists mainly of **keratinocytes**.[394] Below the dermis lies the *hypodermis*. See page 247.

b. *Connective tissue is the most abundant tissue in the body.* Connective tissue: (1) *connects* body structures with tendons and ligaments, (2) *protects* vital organs with bone and (3) *stores energy* in fat cells. Composition of connective tissue: *cells*, *fibers* and *ground substance* (extrafibrillar matrix). Ground substance is colorless, has fibers and cells imbedded in it to support the cells and fibers in connective tissue. **Fibroblasts** make extra cellular matrix and collagen; they and they play a major role in wound healing.

---

391   Preserved in such a way as to maintain structural integrity.
392   A razor-sharp device, hand held or a machine, that is used to make extremely thin sections of tissue.
393   Does not have a blood supply.
394   The epidermis is 95% keratinocytes. Keratinocytes make keratin, a waterproofing material that is a barrier against microorganisms, ultraviolet rays, heat and water loss.

c. *Muscle tissue* enables movement. *A muscle cell (muscle fiber)* is the structural and functional unit of muscle tissue. A muscle cell contains many *myofibrils*. Myofibrils are composed of *thick and thin filaments*. *Thick* filaments are made of *myosin* molecules. *Thin* filaments are made of *actin* molecules. *Thick myosin filaments bind to thin actin molecules.* Thin actin molecules are pulled over the stationary thick myosin molecules. *The molecules slide over each other to cause contraction of the sarcomere.* The action is expressed as the *sliding filament mechanism.* A sarcomere extends from Z-line to Z-line. Z-lines are visible when viewed through an electron microscope. See figure 9.11. Sarcomeres are found in cardiac and striated muscle cells, but *not* in smooth muscle cells.

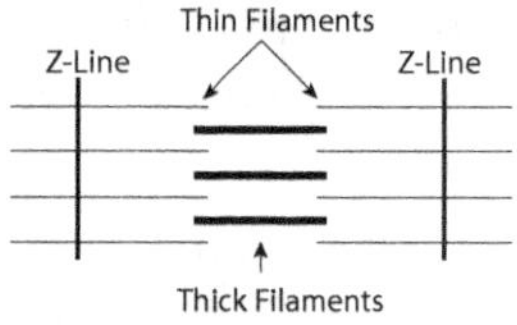

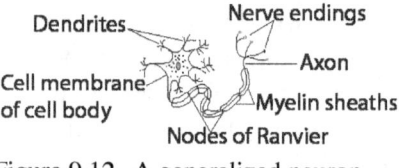

Figure 9.11. A sarcomere.

d. *Nervous tissue* is capable of receiving and transmitting electrical impulses rapidly. The structural and functional unit of nervous tissue is *neuron* (nerve cell). Neurons consist of a *cell body, dendrites, axons* and *axon terminals* (synaptic bulbs). The long axons are covered with a fatty, insulating layer called the *myelin sheath*. Portions of the sheath are marked by indentations called *nodes of Ranvier*. See figure 9.12.

Figure 9.12. A generalized neuron.

6. Cytology

Cytology is the study of cells. Cells are the structural and function units of living things. A single cell is the smallest unit of life. It can be an entire living thing such as a bacterium, amoeba, paramecium, or a multicellular organism that is made up of trillions of cells. The *cytologist* studies the structure, function and chemistry of cells.

7. Molecular Biology

a. Polymerase Chain Reaction (PCR).

The *Polymerase chain reaction* makes millions of copies of DNA molecules from a sample as small as one DNA molecule! DNA contains the code of life expressed as a sequence of four *nitrogenous bases*: *adenine (A), thymine (T), guanine (G) and cytosine (C)*. The four bases make up genes. The *genome* is the total number of genes an organism has. The human genome is estimated to be about 35,000 to 50,000 genes. The genome of every organism is different from each other, except for clones. Archaea, bacteria, protists and human eggs that have divided into identical halves are clones of each other. PCR and gel electrophoresis reveals the numbers and types of genes of an organism.

144

The polymerase chain reaction is an **exponential**[395] doubling of a sample of DNA. The sample is heated and cooled in a thermal cycler machine. See figure 9.13. Heating DNA to about 95 °C unwinds (denatures) the double stranded DNA molecule until the sample contains only single-strands of DNA. The enzyme ***DNA polymerase*** and millions of **A**s, **T**s, **G**s and **C**s are added to the mixture of single stranded DNA. The mixture is cooled to about 55 °C. The **A**s, **T**s, **G**s and **C**s ***anneal*** (stick) to the single stranded DNA to make copies of double stranded DNA molecules. Within hours, millions of copies of the original DNA sample are produced. PCR makes a small sample of DNA large enough to run on a "gel" where DNA fragments are sorted out, compared and identified.

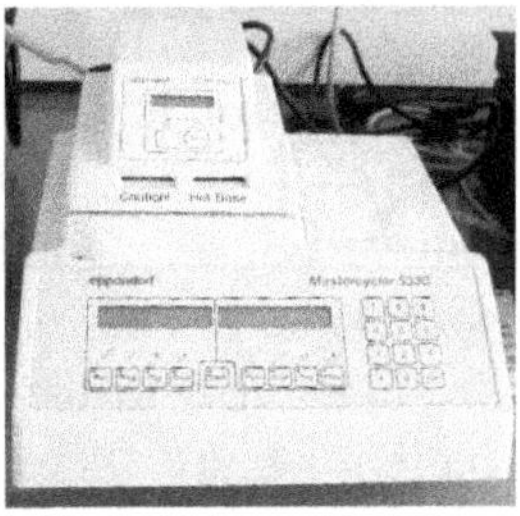

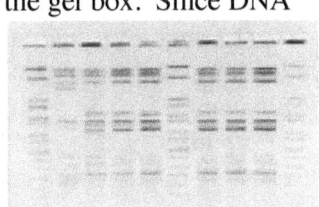

Figure 9.13. A PCR or thermal cycler machine (DNA amplifier).

b. Gel Electrophoresis

A gel is a semisolid material that acts as a sieve. A small sample of DNA is not visible on a gel. PCR provides a large enough sample of DNA, typically $1\mu l$ (one millionth of a liter). One microliter is visible and can be "run" on a "gel." Electricity is passed through the gel in the gel box. Since DNA molecules carry a negative charge, the molecules migrate towards the positively charged electrode in the gel box. Lighter fragments travel further than heavier fragments.

If the bands of DNA on the gel are compared side by side, a visual inspection can determine if two samples of DNA come from the same person. All samples migrate down the gel at the same rate. If the DNA bands match up, the DNA sample is from the same person. If the bands do not match up, the DNA is not from the same person. See figure 9.14.

Figure 9.14 Gel electrophoresis produces a DNA fingerprint.

Noninvasive Diagnostic Methods of Diagnosing Human Disease

Diagnosis is a cognitive process of attempting to identify the nature and cause of a disease. ***A disease is any departure from the normal condition***. Sir William Osler (1849-1919), a Canadian physician, often referred to as the "Father of Modern Medicine," insisted that medical students learn from seeing and talking to patients. Osler noted the importance of taking a good medical history. One of his best known sayings is: "Listen to your patients, they are telling you the diagnosis."

395   Extremely rapid growth.

Physicians use many methods to obtain information about the body. A diagnostic methods is *noninvasive if the method does not produce breaks in the skin or have contact with mucosal linings*. The noninvasive methods of *interrogation*, *inspection*, *auscultation*, *palpation* and *percussion* have been used to help diagnose disease since ancient times.

Interrogation is asking the patient questions. The doctor often gets an idea of what is wrong from the patient's description. *A travel history is crucial*, especially for patients that have come from countries where diseases such as malaria and Ebola hemorrhagic fever are endemic.

Inspection is to make visual descriptions of a patient with the unaided eye. The descriptions include *nutritional state, frequency and volume of breaths* during, **gait**,[396] manner of *speaking, discoloration, swelling, odor, tremors* or *visible masses*. The ancient Greeks described the crab-like appearance of blood vessels surrounding a tumor near the surface of a patient's body. They called the tumor *karkínos,* meaning crab in Greek. The word *cancer* is the Latin translation of the Greek *karkínos*.

Auscultation is listening to sounds made by the heart and lungs. A stethoscope is used to diagnose the health of the circulatory and the respiratory system. In a normal heart, "lub" is the S_1, the first heart sound. S_2 is the "dub," the second heart sound. "Lub-dub" is the sound of valves closing. Auscultation is important to discover heart murmurs or a gallop. The gallop contains *additional sounds of the* S_3 and S_4. The S_4 occurs *before* S_1, and S_4 right after S_3.

Palpation assesses tenderness, swelling, abnormal lumps. The use of pressure or stretching may be used to provoke pain that should not normally occur.

Percussion is tapping on the surface of the body. Tapping assesses the condition of the respiratory system and the abdomen. Reflected sound could be categorized as normal or abnormal.

Noninvasive Diagnostic Methods That Use *Ionizing Radiation*
Radiography

a. *X-rays* are high speed photons. Wilhelm Röntgen (1845-1923), a German physicist is credited with their discovery. He used X-rays for the first time to visualize bones of the hand. X-rays use differences in density to view structures in the body. The object to be X-rayed is placed over a detector that produces a digital image in modern X-ray machines. Bone appears white, and hollow organs like the lungs appear black. See figure 9.15.

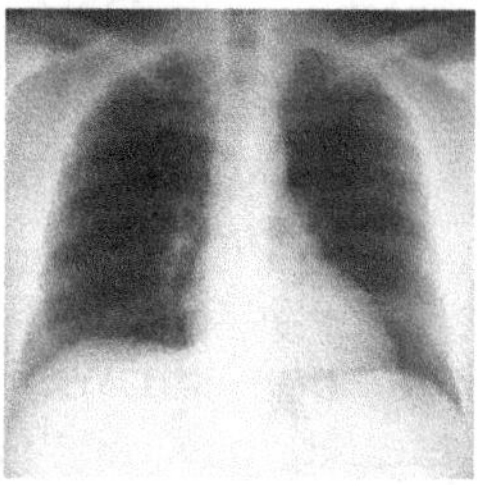

Figure 9.15. Normal x-ray of the human chest.

396   Walking pattern: fast, slow, steady or unsteady.

Hollow organs and soft tissue appear black because the X-rays easily penetrate soft tissue and expose the detector to radiation. Bone or tumors appear white because X-rays cannot penetrate them, thus preventing the detector from exposure to radiation. Low dose X-rays are used for mammography and bone density visualization.

b. *Angiograms* are useful to image the flow of blood in vessels and heart. An opaque material called a *contrast medium*, a special dye, is injected into the circulatory system. The X-rays used in angiograms cannot penetrate the opaque material, and an image of blood flow is formed.

c. *Computerized axial tomography* (**CAT**), or **computed tomography (CT)**.[397] CT scans utilizes X-rays to produce high resolution, three-dimensional images of specific *planes* of soft tissues. See figure 9.16.

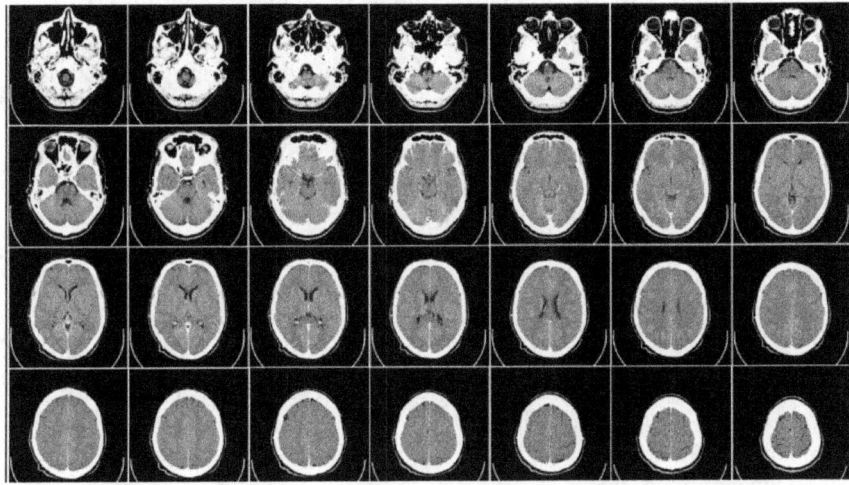

Figure 9.16. Computed tomography (CAT) scan of the human brain.

Sections are superimposed one upon another by manipulating the computerized data. This process is useful if the structure of interest is hidden by other structures that conventional X-rays cannot clearly visualize, such as lung cancers, coronary artery disease and kidney cancers.

d. *Photon emission tomography* (**PET**) utilizes gamma radiation and radioactive molecules commonly called "tracers" that are injected into the bloodstream of a patient. The tracer is short lived and becomes part of a naturally occurring molecule in the body that functions in tissues under investigation. The radioactive tracer gives off **positrons**.[398] The positrons interact with electrons to give off a burst of light that is registered by a detector.

---

397   An X-ray based technique that gives high contrast resolution of soft tissue.
398   A positive electron.

Noninvasive Diagnostic Methods That Do Not Use Ionizing Radiation

a. Sonography

Sound waves are *compressional waves*, also called *longitudinal waves*. Sound waves are produced when particles are caused to move close together (compression) and then move further apart (rarefaction). The gas molecules in a medium such as the air, for example, are compressed and then move apart. The compressions and rarefactions move in the same direction as the wave. See figure 9.17.

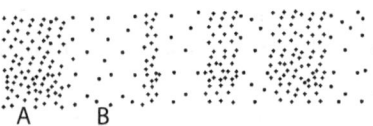

Figure 9.17. Representation of a sound wave. Compressions of air molecules (A) and rarefactions (B) strike the eardrum to produce sound.

Diagnostic *sonography* (ultrasonography) uses sound frequencies beyond the range of normal hearing, between 2 and 18 **megahertz**[399] (MHz). Diagnostic sonography involves the generation of sound waves that travel through the body and are then reflected back to a detector. The reflected image is usually good for viewing *soft tissues* such as muscle, tendons and bladders. See figure 9.18.

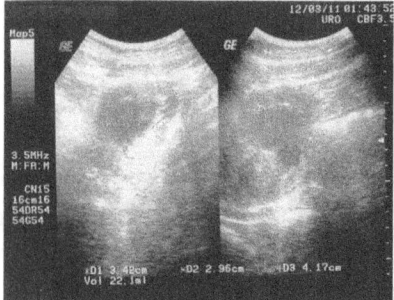

Figure 9.18. A sonogram of a human bladder. Anterior view at left, lateral view at right.

Diagnostic sonography is commonly used for prenatal checkups: monthly observations for the first 28 weeks, every two weeks from 28 to 36 weeks and weekly until delivery.

b. Infrared imaging or thermal imaging utilizes methods that detect infrared radiation. Images that result from thermal imaging are called *thermograms*. Any object that has a temperature above absolute zero (K or -273 °C) gives off detectable thermal energy. As the temperature increases, the intensity of the individual variation of thermal energy increases. Cooler temperatures will appear black, violet and purple. Hot areas will appear red and yellow. The hottest areas will appear white. Breast thermography utilizes this technique.

Magnetic resonance imaging (**MRI**) uses a very powerful magnetic field to make the nuclei of atoms visible on a scanner. This technique makes it possible to see fine differences in soft tissue. It is frequently used to detect artery clogging plaque and abnormalities in blood flow to the brain.

---

399   A unit in the SI system of measurement used to describe frequency.

c. Reflected Light

Dermoscopy or *epiluminescence* microscopy is a dermatological technique. Dermatologists project polarized light onto a patient's skin to better distinguish cancerous skin lesions such as melanomas from benign lesions.

d. *Electrocardiogram (ECG or EKG)*

Willem Einthoven (1860-1927), a Dutch physician, used a type of galvanometer to record *electrical activity of the heart* over time.

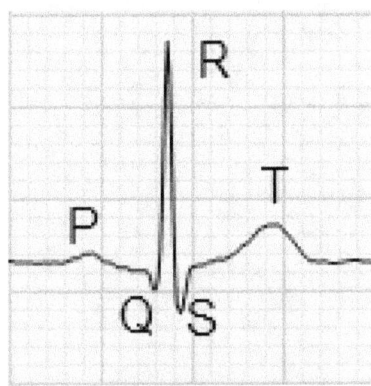

Einthoven labeled the peaks and valleys of the tracing P, Q, R, S and T. Departures from the normal tracings of these points were associated with abnormal heart activity. **Electrodes**[400] attached to the surface skin of the chest detect electrical impulses generated by the

Figure 9.19. PQRST tracing.

heart. Every time the heart beats electrical impulses travel from the upper chambers (atria) to the lower chambers (ventricles) and determine the rhythmic beating of the heart.

A properly functioning heart delivers the right amount of blood in a regulated manner to the cells of the body. Gener-

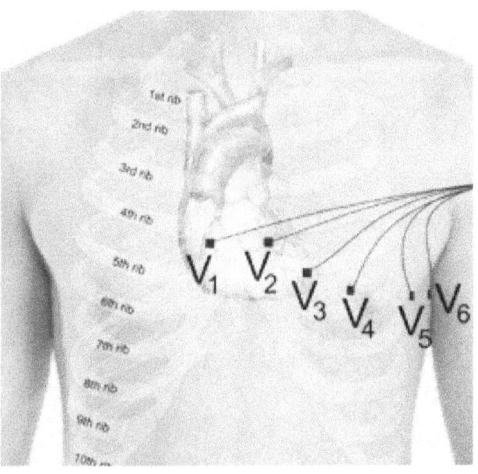

Figure 9.20. Placement of electrodes on a patients chest.

ally, four electrodes are placed on the limbs and six electrodes are placed on the chest (see figure 9.20). The right leg (RL) electrode serves as the electrical ground for the ECG recording machine. The remaining limb electrodes pairs are displayed as leads I, II, and III of the ECG. They are electrically combined to serve as the reference electrode (the modified Wilson's central terminal) for the remaining nine leads of the standard twelve lead ECG.

---

400   An electrode is a substance that can detect electrical energy on a nonmetallic surface and conduct it to a device that will magnify it.

In a very broad sense, an ECG tracing can be looked at as three events:

1. P-wave reflects the movement of an electrical impulse from the SA node to the AV node and from the right atrium to the left atrium.

2. Q, R and S waves are three waves but considered one event. Right and left ventricles depolarize.

3. The T-wave represents the repolarization of the ventricles. The ventricles recover.

Chapter 10 Introduction to Chemistry

Physical and Chemical Properties of Matter

Chemistry studies physical and chemical properties of matter and physical and chemical changes matter undergoes. A *physical property* is a measurable, observable characteristic. **Density**[401] *boiling point, melting point, solubility, color, odor* and *hardness* are physical properties. A *physical change* is a change in a substance's *appearance*. For example, water exists in three states, solid, liquid and gas. A change from one state to another is a temperature dependent physical change, but all are water.

$$2H_2 + O_2 \xrightarrow{\Delta} 2H_2O$$

Figure 10.1. The formation of water from hydrogen and oxygen gases.

A *chemical property* allows a *chemical change* to take place. When substances *react chemically a new substance* is produced. It has different physical and chemical properties compared to the original substance. The balanced chemical equation in figure 10.1 shows the chemical reaction of oxygen gas (O_2) with hydrogen gas (H_2) to form liquid water (H_2O). Oxygen supports burning and hydrogen burns. Water does not burn or support burning. Water puts fires out. *The properties of hydrogen and oxygen are lost.*

Hydrogen and oxygen combine because of their unique chemical properties. The outer energy level of an oxygen atom *needs two electrons*. Two hydrogen atoms can *each* share *one electron* with an oxygen atom. This fills oxy-

$$\xrightarrow[\text{of certain enzymes}]{\text{Enzymatic breakdown}} H_2O_2$$

Figure 10.2. Formation of hydrogen peroxide in the human body.

gen's outer energy level. See figure 10.2. For this reason, two atoms of hydrogen will *always* combine with one atom of oxygen in a *ratio of 2:1*. If the atoms do not combine in a 2:1 ratio, the substance is *not* water. If the ratio of H to O is 1:1, the substance is hydrogen peroxide (H_2O_2). See figure 10.2. Water is an oxide of **hydrogen.**[402] Oxygen combines with metals to form familiar oxides. The orange red rust on iron (Fe) is iron oxide (Fe_2O_3). The dull coating on zinc (Zn) is zinc oxide (ZnO). See figure 10.3.

Divisions of General Chemistry

$$Fe + O_2 \longrightarrow FeO$$
$$Ag + O_2 \longrightarrow Ag_2O$$
$$C + O_2 \longrightarrow CO_2$$

Figure 10.3. Oxidation metals and nonmetals.

Chemistry studies *inorganic* and *organic substances*. *Inorganic chemistry* studies molecules that are simple, not associated with life and have little carbon and hydrogen. *Organic chemistry* studies carbon and organic compounds. Organic compounds are large, complex, associated with living things and has carbon and hydrogen as main components.

---

401   Density is a comparison or ratio of mass to volume of a substance. d=m/V
402   Hydrogen behaves like a metal chemically.

Elements, Compounds and Chemical Reactions

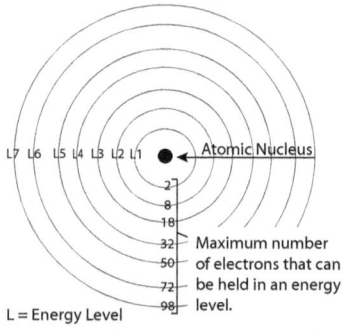

L = Energy Level

Figure 10.4. Maximum electron number in an atom's energy levels. "L" stands for level.

An *element* is *a substance that* cannot be broken down into simpler substances by ordinary chemical means and still retain the properties of the element. *Elements consist of only one kind of atom.* Elements have the same atomic number.

Atoms are made up of smaller par-ticles, *protons*, *neutrons* and *electrons*. Protons have a positive electrical charge, electrons have a negative electrical charge and neutrons do not have an electrical charge. Neutrons are neutral. Protons and neutrons are located in an atom's nucleus. Electrons circle the nucleus in defined *energy levels*. Each *energy level can hold a maximum number of electrons, but energy levels may hold fewer electrons*. See figure 10.4.

Elements

A *symbol* represents an *element*. Some common symbols are C (carbon), H (hydrogen), O (oxygen), N (nitrogen) and Ca (calcium). If a symbol consists of one letter, the letter is capitalized, as in C for carbon. If a symbol consists of two letters, as in **Ca** for calcium, the first letter is capitalized, and the second letter is *always lowercase*. Some names of elements are taken from another language. *Ferrum*-**Fe** (iron), *plum-bum*-**Pb** (lead) and are taken from Latin. Hydrogen is derived from the ancient Greek language, **H**-*hydro* + *gen* (water former). **Bk**-Berkelium is named after a place, Berkley, California, and **Es**-Einsteinium-after a person, Albert Einstein. *Each box in the periodic table contains the ele-ment's symbol, atomic number (proton number) and atomic mass.* Each atom has a unique number of protons. *If the atom contains one proton, the atom must be hydrogen and no other element.* If an atom has two protons, the element is helium and no other atom. The *unnamed ele-ments* have a number only, written in Latin. See table 10.4.

Compounds

Formulas represent compounds. *Compounds* are composed of two or more elements *chemically combined*. *Compounds cannot be sepa-rated by ordinary means physical separation* such as boiling, heating, magnetism, gravity or filtering. NaCl is the formula for sodium chloride (table salt); H_2O is the formula for water and H_2O_2 is the formula for hydrogen peroxide. *Compounds can be separated by chemical means.*

Writing Formulas

Formulas consist of symbols and numbers. Chemists agree to write symbols for *metallic* atoms *first* and symbol for the *nonmetallic atoms second* when writing chemical *formulas*. For example, the formula **NaCl**, shows the metallic element Na written first. The nonmetal Cl appears second. The formula H_2O shows hydrogen written first because it is a gas that behaves like a metal chemically. The nonmetal oxygen appears second in the formula.

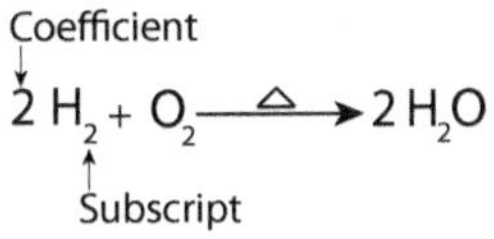

Coefficient

$$\overset{\downarrow}{2}\,H_2 + O_2 \xrightarrow{\triangle} 2\,H_2O$$

$\uparrow$
Subscript

Figure 10.5. Coefficients and subscripts

Chemical Reactions

Chemical equations consist of symbols, formulas and numbers that describe what happens in a *chemical reaction*. In a chemical equation, *reactants* appear on the *left side*, and *products* appear on the *right side*. *Reactants* are separated by a "+" sign. An *arrow* reads as "*yields*," separates reactants from *products* in an equation. See figure 10.5. The balanced chemical equation in figure 10.5 shows the formation of two water molecules from four hydrogen molecules and two oxygen molecules. After energy is added (triangle) to the mixture gases, a rapid reaction takes place yielding water. A *coefficient* indicates *how many atoms or molecules participate* in a reaction. The *subscript* below and to the right of an element tells how many atoms are *present in a molecule*. The *coefficient* "2" before H_2 in figure 10.5 shows the reader that *2 hydrogen molecules* combine with *1 oxygen molecule*.

Atomic Theory

Democritus (460-370 B.C.), an ancient Greek philosopher, proposed the first theory explaining the nature of matter. He reasoned that if a sample of matter is repeatedly divided, a point must be reached where the substance cannot be divided again and still be the same substance. He called it the "indivisible," *átomos* in Greek, "*atom*" in English. Democritus hypothesized that matter is composed of atoms of different sizes, cannot be destroyed and are mostly empty space. The Greeks got it right using deductive reasoning 2500 years ago!

John Dalton (1766-1844), an English chemist and teacher, proposed an atomic theory that explained the composition of matter and how different kinds of matter react with each another. Dalton's theory has many similarities to Democritus' atomic theory. *He proposed that atoms*:

1) Make up all matter.
2) Cannot be divided into smaller particles.
3) Cannot be **destroyed**.[403]
4) Make up elements having identical masses and properties.
5) Different kinds of atoms unite in definite ratios to form compounds.

---

403   Today it is known that atoms can be split to yield different atoms.

Joseph Thomson (1856-1940), a British physicist, discovered the electron. His atomic model is a positive sphere surrounded by electrons. See figure 10.6.

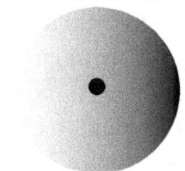

Niels Bohr (1885-1962), a Danish physicist, proposed the *planetary model* of the atom with *protons and neutrons* located in a central *nucleus* and *electrons* orbiting the nucleus in specific paths called energy levels. See figure 10.7.

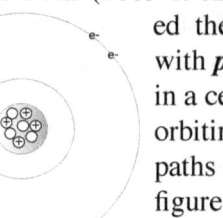

Figure 10.6. Thomson model of the atom.

Figure 10.7. The Bohr of the atom.

The current model is the *electron cloud model*. See figure 10.8. The electron cloud model places electrons "everywhere" around an atom's nucleus, giving the appearance of a "cloud." Only indirect proof for atoms existed until recently, but in 2013, a hydrogen atom was *directly observed* for the first time with the *quantum electron microscope*. *An atom can now be seen*. The first direct observation was of a hydrogen atom. It looks like the electron cloud model.

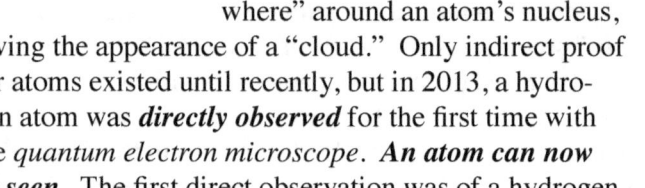

Figure 10.8. The electron cloud model.

The Periodic Table of the Elements

Dmitry I. Mendeleev (1894-1907), a Russian chemist, organized the known elements into the *periodic table* in 1869. **Mendeleev's table**, unlike the modern version of his table, arranged the known elements in order of *increasing atomic **mass number**.*[404] The mass number is the combined number of *protons* and *neutrons* in an atom's nucleus. *Mendeleev noticed that properties of the elements in a row, called a **period**, were **repeated** in the row immediately below.* **Periodicity** is the repetition of properties of elements in the periodic table.

The **modern** periodic table lists about *100 naturally occurring elements* and 18 synthetic elements. The elements are arranged by *increasing atomic number.* The table of elements demonstrates properties and relationships among elements. See table 10.4. The table is arranged in *18 columns (**group** or **family**) and 7 rows (**periods**). Each group contains elements with similar physical and chemical properties. Each **row** is a called a **period**. **Nonmetals** are on the **right side** of the table and* **metals** are on the **left side**. Metals tend to be shiny, hard, have high melting points and *donate 1, 2 or 3 electrons* to nonmetals. Nonmetals tend to be soft substances or gases that *accept 1, 2 or 3 electrons* from metals.

---

404   The atomic mass number, relative atomic mass, average atomic mass, atomic weight are often used interchangeably, but each refers to slightly different quantities.

Oxidation state is the *potential* an atom has to *give*, *take* or *share electrons* in order to achieve a *stable electron arrangement*. If a *neutral* metal atom, gives away 1 electron, the atom becomes positive by 1. Its oxidation state is **1⁺**. If a *neutral* nonmetal atom receives 1 electron it becomes negative by 1. Its oxidation state becomes **1⁻**. Noble elements have an oxidation state of 0. Oxidation states are *positive*, *negative* or *zero*.

Overview of Groups (Families) of Elements

Column numbers indicate the number of *electrons present* in an atom's outer energy level. *Common oxidation states are listed after the column number in* **boldface**. *Row number* equals an atom's number of energy levels. Atoms in row 1 have 1 energy level. Atoms in row 2 have 2 energy levels, and so on down to row 7. See figure 10.4 on page 166.

Column 1: (**1⁺**) *Alkali metals* **give away** their outermost electron in order to be stable. Example: Lithium is Li $^{1+}$ is very reactive. Hydrogen is in column 1 because it behaves like a metal chemically.

Column 2: (**2⁺**) *Alkaline earth metals* **give away** 2 electrons to achieve a stable electron configuration. All have 2 electrons in their outermost energy level. Example: Calcium (Ca) is Ca^{2+} .

Columns 3 through 12: (**2⁺ and 3⁺**) *Transition elements* **give away** 2 or 3 electrons to nonmetals. Transition elements have multiple oxidation states and are located in the middle of the periodic table. Examples: iron (Fe), copper (Cu), lead (Pb), cobalt (Co), zinc (Zn) and mercury (Hg).

Column 13: (**3⁺**) *The boron group* **donates** electrons. They have 3 valence electrons, but can have oxidation states of **1⁺, 2⁺** and **3.⁺** Examples: boron B^{3+} and aluminium Al^{3+}.

Column 14: (**4⁺ or 4⁻**) *Elements in the carbon group* has four valence electrons. They can have positive or negative oxidation states. Carbon commonly *shares* electrons with other carbons and with other atoms.

Column 15: (**3⁻**) *The nitrogen group* elements want to *accept* 3 electrons from other elements to become stable. They have 5 electrons in their outer energy levels and become 3⁻ (N^{3-}) after accepting 3 electrons.

Column 16: (**2⁻**) *The oxygen group* wants to *accept* 2 electrons to become stable (O^{2-}). Oxygen is the first element in column 16. Those below oxygen are soft nonmetals. One representative is sulfur (S^{2-}).

Group 17: (**1⁻**) *Halogen group* wants to *accept* 1 electron to become stable. Chlorine has 7 electrons in its outer energy level. Chlorine wants 1 electron to becomes 1⁻ (Cl^{1-}). Halogens react with metals to form salts. Examples: Fluorine, Chlorine and Bromine

Group 18: (**0**) *Noble elements* do not react with other elements. Examples: Helium (He), neon (Ne), argon (Ar), krypton (Kr) and radon (Rn). Radon is produced by radioactive decay of uranium present in soil.

The *staircase* feature on the right side of the periodic table separates *metals* from *nonmetals*. See page 166. Elements such as silicon, aluminum (Al) and antimony (Sb) are in contact with the staircase and are called *metalloids*. Metalloids have properties of metals and nonmetals.

Electron Number Equals Proton Number in a Neutral Atom. Electrons in the outermost *energy level* govern how atoms reacts with each other. Each energy level has the potential to hold a maximum number of electrons, but can hold less. See figure 10.4.

All elements in column 1 of the periodic table have one electron in its outer energy level. The loosely held electron will combine rapidly with any element that will accept it. One loosely held electron causes column 1 elements very *reactive*. When column 1 elements get rid of 1 electron in its outer energy level, 8 electrons will be found in the energy level just below. This action satisfies the *octet rule*. *The octet rule states that when atoms undergo a chemical reaction they do so to have eight electrons in their outermost levels, except for hydrogen and helium.* Hydrogen and helium like to have two electrons in their outer shells.

Sodium (Na), column 1, will give away 1 electron to chlorine (Cl), column 17, to form sodium chloride (NaCl). See figure 10.13. Column 2 elements give away two electrons from their outer energy level, to expose 8 electrons immediately below. Elements in column 16 accept two electrons in order to have 8 electrons. Magnesium (Mg) gives 2 electrons to oxygen (O). Mg will have 8, and O will have 8, to form MgO.

Most elements have different forms called **isotopes**.[405] Isotopes of an element are found in the *same place* on the periodic table. Isotopes have the *same proton number*, but *different atomic masses*.[406] Only the most common form is shown. The proton number determines what the element is. The neutron number determines the mass of an element.

The nucleus of an atom can be symbolically described by noting the number of protons and neutrons that appear with the symbol for the element. For example, C has a mass of **12** and a proton number of **6**. Carbon's nucleus is described as $^{12}H_6$. The *superscript* number is the *mass number*, and the *subscript* is the

proton number. *The most common isotope of hydrogen* is ordinary hydrogen. Ordinary hydrogen has **one proton** and **no neutrons** in its nucleus. Ordinary hydrogen's mass number is **1** ($^1H_1$). A hydrogen atom with **1** proton and **1** neutron has a mass number

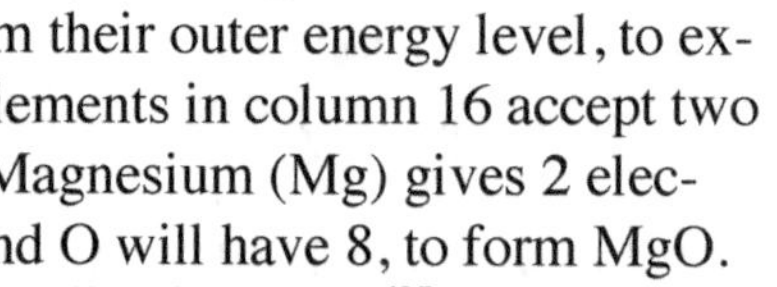

Ordinary Hydrogen   Deuterium   Tritium

^1_1H ^1_2H ^1_3H

Figure 10.9. Isotopes of hydrogen.

---

405   *Isos* (Gk) means the same or equal and *topos* (Gk) means place.
406   Atomic mass is the sum of the protons and neutrons in the nucleus of an atom.

of **2** ($^2H_1$). If a hydrogen atom has a mass of **3**, it is because the atom has **1** proton and **2** neutrons ($^3H_1$). ***Each is an isotope of the other***. See figure 10.9.

Average Atomic Mass

Ordinary hydrogen has 1 proton, therefore, its mass should be 1 on the periodic table, but it is not. It is 1.008. The extra mass is a result of the averaging of the three isotopic masses of hydrogen. Since ordinary hydrogen is the most common isotope, it is weighted more in the average.

Catalysts

Biochemical reactions need to be "helped along" by *catalysts* in order to work. Catalysts speed up or maintain the rate of reactions *without being used up* in the reaction. There are two classes of catalysts: inorganic and organic. *Inorganic catalysts* like nickel (Ni) are used to make margarine from vegetable oils. Hydrogen gas (H_2) is bubbled through the oil with nickel in a process called *hydrogenation* resulting in solid margarine. *Saturated, and trans fats* are produced by this process. Platinum is another catalyst. It is used in "catalytic converters" of automobiles to make the exhaust pollutants harmless.

Enzymes are *organic catalysts that* mediate[407] biochemical reactions. Enzymes lower the amount of energy needed for metabolic reactions to occur. Enzymes work on *substrate* molecules to convert them into *products*. A substrate is a molecule upon which an enzyme acts. The substrate is changed, but the *enzymes do not get used up* in reactions.

Biochemical pathways are composed of specific sets of enzymes. If an enzyme is not present, not folded properly or missing *cofactors*, the pathway is blocked, and a disease condition may result. There are two groups organic enzymes: (1) *simple enzymes* and (2) *holoenzymes* (conjugated) enzymes. Simple enzymes consist of only protein molecules.

The *protein* part of a holoenzyme is called an *apoenzyme*. Apoenzymes are bonded to non-protein cofactors. A cofactor may be an *organic* molecule such as a *vitamin*, called a *coenzyme* or an *inorganic metallic ion* called a *cofactor*. Well known examples of vitamins are vitamins: A, B_1(thiamin), B_3 (niacin), C, D, niacin, and vitamin B_{12}. and biotin (B_7). Examples of metallic cofactors are Zn^{2+}, Fe^{2+} and K^{1-}.

Vitamins and minerals are part of many enzymes. A lack of vitamins in the diet cause enzymes *not to work in metabolic pathways, causing vitamin deficiency diseases*. Well known examples of vitamin deficiency diseases are night blindness, beriberi, pellagra, scurvy, rickets and B-12 deficiency anemia (pernicious anemia), B7 (biotin) deficiency.

---

407   To bring about or to bring forth some effect or result.

Radioactive Elements

Henri Becquerel discovered *radioactivity* in 1896. The French scientist wrapped a uranium compound in photographic film. The result was a blackening of the film due to the radioactive particles given off by the uranium. Soon after Becquerel's discovery, other scientists discovered additional *radioactive elements*. *Pierre* and *Marie Curie* discovered the radioactive elements polonium and radium.

Some elements are *radioactive* because they undergo *spontaneous nuclear decay*. Radioactive decay takes place because of unstable nuclei. These elements give off *ionizing radiation* in the form of particles and rays. Spontaneous atomic decay results in giving off protons, neutrons, beta (ß⁻) particles (electrons), beta (ß⁺) particles (also termed positive electrons or positrons, **alpha (∂) particles** and **gamma (γ) rays**.[408] See Figure 10.10 and table 10.1.

There are three kinds of nuclear decay: *alpha, beta* and *gamma*. *Alpha decay* is the ejection of helium nuclei (two protons and two neutrons). If electrons (ß⁻) or positrons (ß⁺) are ejected from a nucleus, *beta decay* results. *Gamma rays* are a form of energy **emitted**[409] as invisible rays.

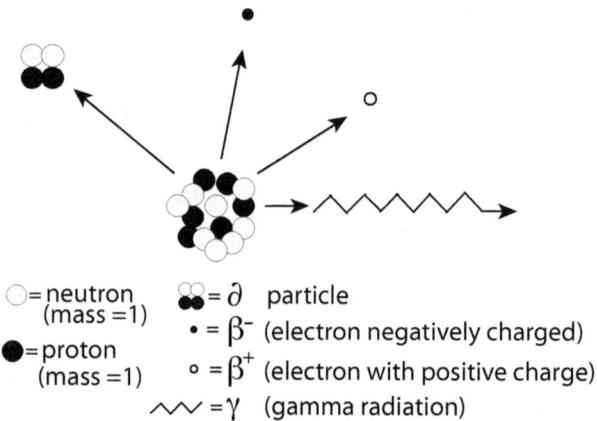

Figure 10.10 Generalized diagram of radioactive decay.

Unknown at the time of their discovery, radioactive elements present serious health risks. People drank "health tonics" containing radium and, as a result, many died. Soon these "tonics" were banned.

Uranium in soil and rocks decay and produce radon. It is estimated that radon causes 20,000 lung cancer deaths a year. All homes should be tested for radon.

---

408   High frequency, short wavelengths of the electromagnetic spectrum.
409   To throw off, to discharge or to send out.

Table 10.1. A partial list of nuclides that undergo natural radioactive decay.				
Element	Nuclide	Atomic Mass Number	Proton Number	Neutrons
Hydrogen	$^3H_1$	3	1	2
Helium	$^5He_2$	5	2	3
Lithium	$^8Li_3$	8	3	5
Carbon	$^{14}C_6$	14	6	8
Nitrogen	$^{16}N_7$	16	7	9
Potassium	$^{40}K_{19}$	40	19	21

A *Nuclide* (see table 10.1) refers to the *nucleus* of an atom. A nuclide refers to the proton number (Z) and the neutron number (N).

Many elements are essential for life. Carbon (C), oxygen (O), hydrogen (H), nitrogen (N), sulfur (S), phosphorus (P) and sodium (Na) are the major elements necessary for living things. There are about 18 more. These elements are supplied by a proper diet. See table 10.2.

Table 10.2. Major elements necessary for life.		
Name	Symbol	Function in the Body
Oxygen	O	Part of carbohydrates, fats, proteins, nucleic acids and water; a final electron acceptor in aerobic respiration.
Carbon	C	Part of structure of carbohydrates, fats, proteins, nucleic acids.
Hydrogen	H	Part of structure of carbohydrates, fats, proteins, nucleic acids; hydrogen bonding
Nitrogen	N	Part of structure of proteins, nucleic acids.
Calcium	Ca	Bone formation, muscle contraction, a cofactor in many enzymes, blood-clotting.
Phosphorus	P	Bone and tooth structure, nucleic acids and ATP.
Potassium	K	Most abundant cation in a living cell. Has to stay in balance with Na. Needed for photosynthesis.
Sulfur	S	Vital to Vitamin B1, insulin and collagen structure and function.
Sodium	Na	Most abundant cation in extracellular fluid. Has to stay in balance with K.
Chlorine	Cl	Regulates water and acid-base balance, transmission of nervous impulses.
Magnesium	Mg	Bone structure, enzyme structure.
Iron	Fe	Hemoglobin in red blood cells.

Chemical Bonds - Forces of Attraction Between Atoms
Covalent Bonding

There are three kinds of bonds of biological importance: 1) *covalent*, 2) *ionic* and 3) *hydrogen bonds*. Covalent bonds are strong bonds that are found in solids, liquids and gases. *Hydrogen and oxygen atoms in a water molecule are held together by covalent bonds*. The formula for water, H_2O, shows that a water molecule contains hydrogen and oxygen atoms. It also shows the ratio of H to O is 2:1, but it does not tell us about electron arrangement. ***Knowing an atom's proton number, and its number of energy levels allows electron distribution to be determined***.

Proton number and the number of energy levels are found in the periodic table. The ***proton number or atomic number*** is the smaller number listed with the element in the periodic table. The atomic number is also equal to the ***number of electrons*** in a neutral atom. The ***row number*** the atom is located on the left hand of the periodic table and equals the ***number of energy levels*** an atom has. Place electrons, beginning with the first energy level, in each succeeding energy level without exceeding that level's maximum number of allowable electrons until no more electrons are available for placement. See figure 10.4 on page 152.

Covalent bonds form when atoms *share electrons*. Hydrogen, for example, has 1 electron orbiting its nucleus. It would like to have 2. Oxygen has 6 electrons in its outer energy level. It would like to have 8. If an oxygen atom comes in contact with two hydrogen atoms, each hydrogen atom can *share* its 1 electron with oxygen. Sharing satisfies oxygen's need for 8 electrons and hydrogen's need for 2 electrons. The outer energy levels of each atom overlap. Overlapping causes oxygen's valence electrons to spend part of their time in the outer energy level of hydrogen, and each hydrogen atom's single electron spends part of its time in the outer energy level of oxygen. Thus, oxygen can "claim" 8 electrons in its outer shell, and hydrogen can "claim" 2. See figure 10.11.

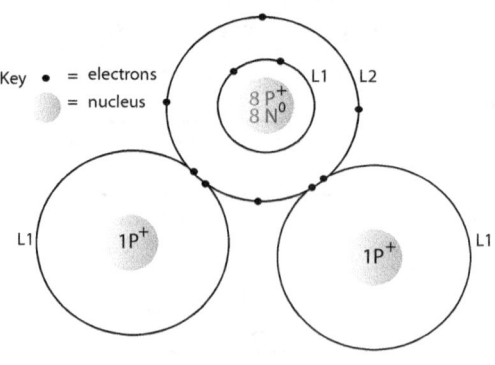

Figure 10.11. Covalent bonding. Oxygen has two energy levels, L1 and L2. L1 has a maximum capacity of 2 electrons. It is full. L2 has a maximum capacity of 8 electrons. It is full because O is sharing an electron from each H atom. Hydrogen's L1 is now full because each H shares an electron from level two (L2) of oxygen.

160

Ionic Bonding

Neutral atoms have equal numbers of protons and electrons. The periodic table shows atoms in their neutral state. *If an atom donates one, two or three electrons* to an atom that *accepts one, two or three electrons,* **an imbalance is created between electrons and protons**. As a result, ions are formed. Ions are atoms or groups of atoms in which the total number of electrons and the total number of protons are *not equal*. The result is a **net positive charge** on an atom forming a **cation** or a **net negative charge** on an atom forming an **anion**. For example, if a neutral Na atom donates 1 available electron to Cl, the previously neutral Na atom becomes positively charged by 1. It is now an ion, written as Na^{1+}. The previously neutral Cl atom becomes negatively charged by 1 and becomes an ion, written as Cl^{1-}. The ions are attracted to each other until repelled by their positively charged nuclei. **Ionic compounds** are formed by the force of attraction between ions. See figure 10.12

Ionic compounds form a lattice-like crystalline structure. Ionic compounds are hard and brittle and have high melting and boiling points. Sodium chloride, calcium carbonate and calcium chloride are examples of ionic compounds. All dissolve in water, a polar solvent.

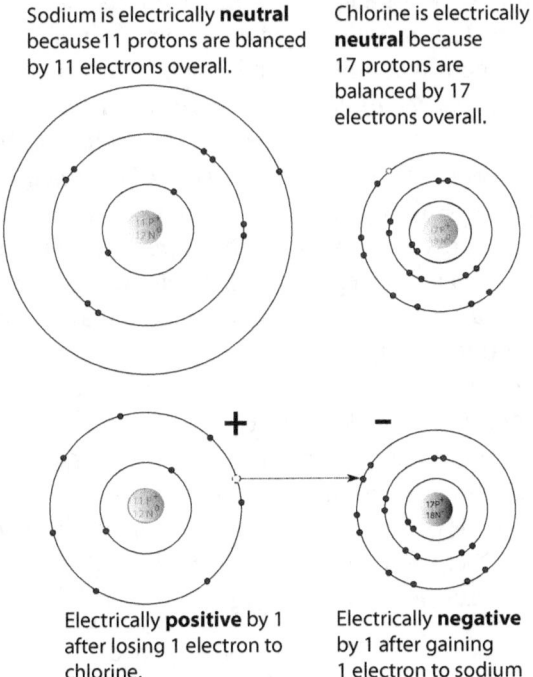

Sodium is electrically **neutral** because 11 protons are blanced by 11 electrons overall.

Chlorine is electrically **neutral** because 17 protons are balanced by 17 electrons overall.

Electrically **positive** by 1 after losing 1 electron to chlorine.

Electrically **negative** by 1 after gaining 1 electron to sodium

Figure 10.12. Ionic bonding of sodium (Na) to chlorine (Cl), arrow, to make the ionic compound sodium chloride (NaCl).

Valence Electrons and Oxidation Numbers

The electronic configuration of a neutral atom determines if there will be a donation, acceptance or sharing of electrons during a chemical reaction. Figure 10.13 shows how a sodium atom and a chlorine atom form an ionic bond. When a neutral sodium atom comes close to a neutral chlorine atom, they react rapidly with each other because sodium has one loosely held electron in its outer energy level and 8 electrons in the energy level below. Chlorine has 7 electrons in its outer energy level and one empty place it wants to fill in its outer energy level.

Sodium has the potential to give away its *single outermost electron* in its **valence**[410] orbital to Cl. When a Na atom gives away its one outer electron, it becomes positively charged by 1 unit to become 1$^+$. Sodium now has 1 less electron compared to its neutral state. For this reason, sodium's oxidation number is 1$^+$. After the Cl atom accepts 1 electron from the Na atom, the Cl atom becomes negatively charged by 1 unit to become 1$^-$. The oxidation number of Cl is 1$^-$. *Na$^+$ attracts Cl$^-$ to form a stable ionic compound.*

Metallic elements give away 1, 2 or 3 electrons from their outer energy levels to achieve a stable 8 electron configuration. After donating electrons, an atom exposes a lower energy level that has 8 electrons. *Nonmetals accept 1, 2 or 3 electrons* into their outer energy levels to attain a stable 8 electron configuration in their outer energy levels. *Donation and acceptance of electrons is typical of ionic compounds.* See figure 10.12. Atoms can share electrons to achieve a stable configuration. *Sharing electrons is typical of covalently bonded molecules.* See figure 10.15. The ability of atoms to react in order to have 8 electrons in its outer energy level is called the *octet rule*. The small atoms of hydrogen and helium are exceptions to the octet rule. Hydrogen and helium follow the *duet rule*, each wants to hold a maximum of two electrons in their outer energy levels.

Hydrogen Bonding

Hydrogen bonds are weak bonds of great biological importance. *Hydrogen bonds are not chemical bonds* as covalent or ionic bonds are. Hydrogen bonds are a *force of attraction* between a *hydrogen atom and a nitrogen or oxygen atom.* See figure 10.13.

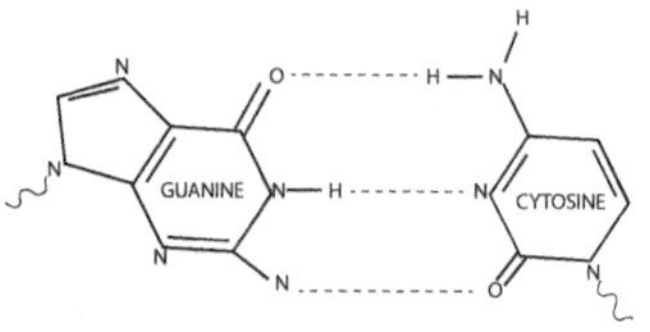

Figure 10.13. Hydrogen bonding between two bases found in DNA. The dashed lines represent hydrogen bonds.

---

410   Combining power. Valence, often confused with oxidation, describes valence orbitals and the number of bonds that can form. Oxidation number is the potential number of electrons that can be given or received.

Hydrogen Bonding and DNA Replication

Replication takes place during interphase of the cell cycle. Hydrogen bonds hold both strands of the DNA double helix together. The double helix has to *"unwind"* for replication to take place. After the double helix is unwound, it must *open* or *"unzip."* Weak hydrogen bonds that hold the strands of the double helix together break. Once separated, each strand can be copied. New bases, the As, Ts, Gs and Cs, attach to each **complimentary**[411] base of an "old" strand. This process is *replication* or making of an *exact* copy of the original DNA double helix. One strand is copied in pieces called *Okazaki fragments*. The fragments are later connected to each other to form a single strand. The other strand is copied in one piece. See figure 10.14.

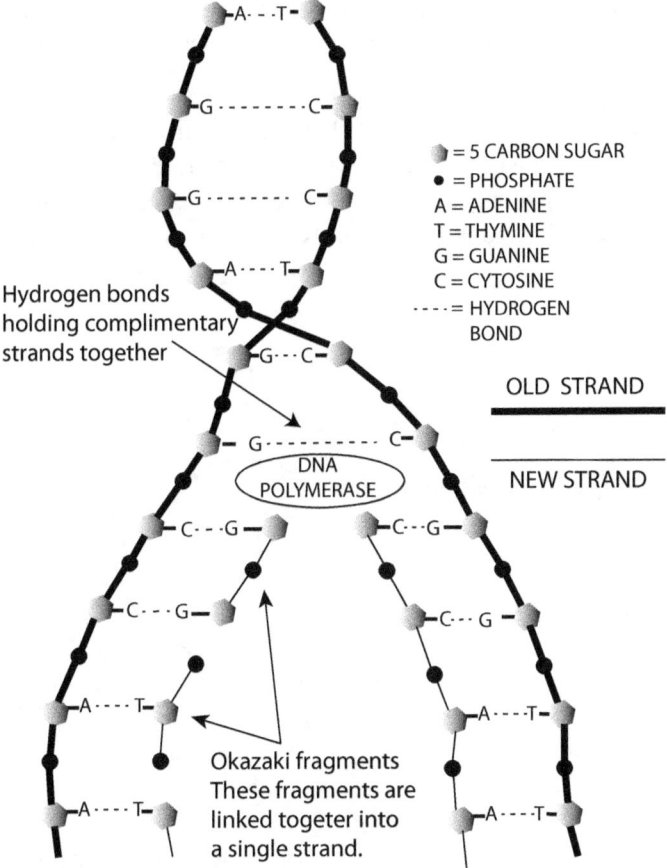

Figure 10.14. Replication of the DNA molecule.

411   Complementarity refers to an arrangement of bases in a DNA double helix so that an **a**denine will pair with a **t**hymine and **g**uanine will pair with a **c**ytosine.

163

Hydrogen Bonding and Protein Synthesis

The *code* or "recipe" for proteins are kept in the nucleus of cells. When a protein such as an enzyme or hormone has to be synthesized, a section of the *DNA double helix* opens to expose the *DNA code* for the gene that makes the protein. The double helix opens easily because weak hydrogen bonds are not strong. A *messenger RNA* molecule (mRNA) molecule is a copy of the DNA code. The copy is called a *codon*. A codon is a sequence of As, Us,[412] Gs and Cs on the mRNA molecule. The short mRNA molecule leaves the nucleus and attaches to a ribosome. *Transfer RNAs* (tRNA) bring *amino acids* to the ribosome. The tRNA molecule has an *anticodon* that insures attachment in the right place on the codon of the mRNA molecule. *Amino acids* are brought to the ribosome one after another like a conveyor belt. Amino acids attach like beads on a string to make a primary protein structure. See figure 13.12.

Polar Molecules

Water is the most abundant substance on Earth. Water occurs as a liquid, solid and a gas. The polarity of water molecules allows it to dissolve almost all substances. For this reason, water is referred to as a *universal* solvent[413] and why it is rarely found in pure form.

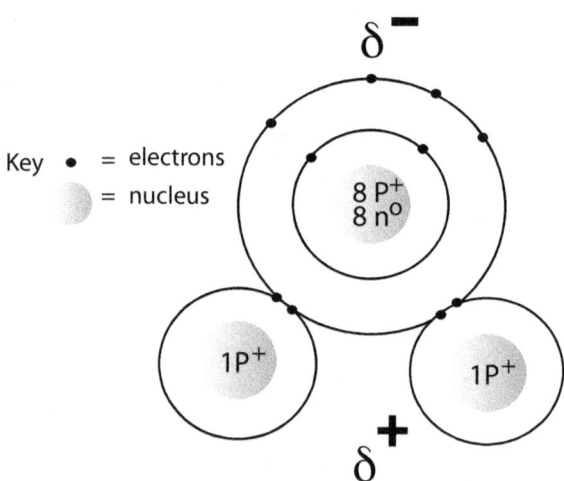

Figure 10.15. Diagram showing electron distribution that causes polarity of a water molecule. Most of the electrons are associated at the oxygen end of the molecule. That side becomes negative and the hydrogen side becomes positive.

---

412   RNA does not contain thymine (T). Thymine is replaced by uracil (U) in RNA.
413   A solvent is the substance that the solute is being dissolved or dispersed.

Water molecules are *polar* molecules because of an *uneven distribution of electrons* around the molecule. See figure 10.15. The oxygen end of the water molecule has a negative charge because there are more electrons at that end of the molecule. The hydrogen side is positively charged because there are fewer electrons at that end of the molecule. Polarity[414] results, and the molecule acts like a tiny magnet. Polarity of the water molecule allows for many of the special properties of water, one being the ability to dissolve almost all substances.

Nonpolar Compounds

Nonpolar compounds have an *even distribution of electrons* over the surface of their electron clouds. Most diatonic molecules such as O_2, N_2 and carbon compounds, such as methane have an even distribution of electrons, and therefore, are nonpolar molecules. See table 10.3.

Table 10.3. A partial list of polar and nonpolar molecules.			
Examples of Polar Molecules		Examples of Nonpolar Molecules	
Name	Formula	Name	Formula
Water	H_2O	Gaseous Oxygen	O_2
Sucrose	$C_6H_{12}O_6$	Carbon Dioxide	CO_2
Ammonia	NH_3	Gaseous Nitrogen	N_2
Sulfur dioxide	SO_2	Methane	CH_4

Diffusion, Osmosis and Active Transport

Diffusion is the movement of atoms, molecules or ions *down a concentration gradient*, referring to the concentration of *solutes* in *solution*. Moving down a concentration gradient occurs when atoms, molecules and ions move from an area of *high concentration to an area of low concentration*. To demonstrate this, place a drop of black ink at the bottom of a glass of still water with a medicine dropper. After some hours, the blob of black ink is gone. The ink particles (solute) spread out among the water molecules (solvent) until each ink molecule is evenly spaced between the water molecules. The kinetic energy of water and ink molecules cause them to bounce off each other until an equilibrium is reached. Molecules can move against a concentration gradient by *active transport*. Active transport requires energy in the form of ATP.

Osmosis is the diffusion of *water molecules* into and out of a cell. See figure 9.8 in the previous chapter. Water molecules will move toward an area of high concentration of water to an area of low concentration.

414   Having the properties of opposing physical characteristics. They may be a North Magnetic Pole and a South Magnetic Pole or a positively charged electrode and a negatively charged electrode.

Periodic Table of the Elements

Families or Groups
1 - 18
(Vertical Columns)

Key

1	→ Atomic or Proton Number
H	→ Symbol
1.008	→ Atomic Mass Number

Periods 1 - 7 (Horizontal Rows)

1	2	3	4	5	6	7	8	9	10	11	12	13	14	15	16	17	18
1 H 1.008																	2 He 4
3 Li 6.94	4 Be 9.01											5 B 10.81	6 C 12.01	7 N 14.01	8 O 16.00	9 F 19.00	10 Ne 20.18
11 Na 22.99	12 Mg 24.31											13 Al 26.98	14 Si 28.09	15 P 30.97	16 S 32.06	17 Cl 35.45	18 Ar 39.95
19 K 39.10	20 Ca 40.08	21 Sc 44.96	22 Ti 47.88	23 V 50.94	24 Cr 52.00	25 Mn 54.94	26 Fe 55.85	27 Co 58.93	28 1 Ni 58.69	29 Cu 63.55	30 Zn 65.39	31 Ga 69.72	32 Ge 72.59	33 As 74.92	34 Se 78.96	35 Br 79.90	36 Kr 83.80
37 Rb 85.47	38 Sr 87.62	39 Y 88.91	40 Zr 91.22	41 Nb 92.91	42 Mo	43 Tc (98)	44 Ru 101.07	45 Rh 102.91	46 Pd 106.42	47 Ag 107.87	48 Cd 112.41	49 In 114.82	50 Sn 118.71	51 Sb 121.75	52 Te 127.60	53 I 126.91	54 Xe 131.29
55 Cs 132.91	56 Ba 137.33	* 57 La 138.91	72 Hf 178.49	73 Ta 180.95	74 W 183.85	75 Re 186.21	76 Os 190.2	77 Ir 192.22	78 Pt 195.08	79 Au 196.97	80 Hg 200.59	81 Tl 204.38	82 Pb 207.2	83 Bi 208.98	84 Po (209)	85 At (210)	86 Rn (222)
87 Fr (223)	88 Ra 226.03	** 89 Ac 227.03	104 Rf (261)	105 Ha (262)	106 Sg (271)	107 Bh (272)	108 Hs (270)	109 Mt (276)	110 Ds (281)	111 Rg (280)	112 Cn (285)	113 Uut (286)	114 Uuq (289)	115 Uup (289)	116 Uuh (291)	117 Uus (294)	118 Uuo (294)

*	58 Ce 140.12	59 Pr 140.908	60 Nd 144.24	61 Pm 144.913*	62 Sm 150.36	63 Eu 151.96	64 Gd 157.25	65 Tb 158.9	66 Dy 162.50	67 Ho 164.93	68 Er 167.26	69 Tm 168.93	70 Yb 173.04	71 Lu 174.9
**	90 Th 232.04	91 Pa 231.04	92 U 238.03	93 Np 237.05	94 Pu (244)	95 Am (243)	96 Cm (247)	97 Bk (247)	98 Cf (251)	99 Es (252)	100 Fm (257)	101 Md (258)	102 No (259)	103 Lr (260)

MA 2012

Table 10.4.

166

Brownian Motion

Brownian motion (*particle theory*) is named after Robert Brown (1773-1858), a Scottish botanist, noticed the random motion of pollen grains in a drop of water under the microscope. Brown could not figure out what caused the pollen grains to bounce around. A pollen grain appears large under the microscope. The motion of pollen grains reflected the motion of the invisible water molecules colliding with them. See figure 10.16.

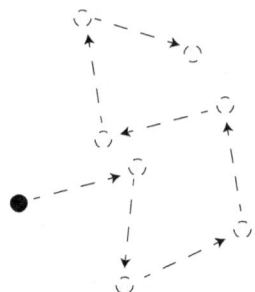

Figure 10.16. Brownian motion of particles.

Mixtures

Mixtures are two or more different substances that are *physically placed together, but they are **not** chemically combined*. Each substance retains its properties. Mixtures can be separated by ordinary means of **physical separation**,[415] boiling heating magnetism, gravity, filtering, magnetism, dissolving or filtering.

Types of Mixtures

Mixtures may be either **homogeneous**[416] or **heterogeneous**.[417] A homogeneous mixture is two or more substances that *appear to be one substance. Mixtures are two or more substances mixed together. **Dissolved particles in a homogeneous mixture cannot be seen**.* Sugar or salt dissolved in water are homogeneous mixtures called *solutions*. The sugar or salt particles cannot be seen, giving the mixture the appearance of being one substance.

Heterogeneous mixtures appear to be different substances. A salad is a heterogeneous mixture. It is easy to identify the different foods and pick each out. Iron filings mixed with sulfur can be separated with a magnet. Iron is magnetic, and sulfur is non-magnetic. The iron filings are pulled out of the sulfur. Each substance keeps its own properties.

Solutions, suspensions, colloids and *alloys are mixtures*. The major differences between each is particle *size*. *Solutions* are mixtures of two or more substances that will not settle out over time. The particle size of the substances in solution, the *solute*, is so small that the kinetic energy of the molecules or atoms of the *solvent* will keep the solute in suspended in the solvent. This is why a solution of NaCl will not settle out and does not require shaking prior to use, as do suspensions and colloids.

---

415   A physical separation can be a simple as using a magnet iron to remove iron filings from a pile of sugar or an iron filing and sulfur mixture.
416   Homogeneous: the same or similar throughout, uniform in nature.
417   Heterogeneous: a varied composition of matter, not uniform.

Sodium chloride (NaCl) is a **solute**[418] that is often *dissolved* in the *solvent* water. NaCl is an ionic compound that **dissociates** [419] into Na⁺ and Cl⁻ ions in water. *Ions are particles that carry an electrical charge.* Since water is a polar molecule, the *positive end* of the water molecule is attracted to the negative chlorine ions (Cl⁻). The positive ends of several water molecules surround the negative chlorine ion (Cl⁻). The *negative side* of the water molecule is attracted to the positively charged Na⁺ ion so that the Na⁺ ion becomes surrounded by the negative sides of several water molecules. See figure 10.17. The vibration of the surrounding water keeps the Cl⁻ and Na⁺ in solution.

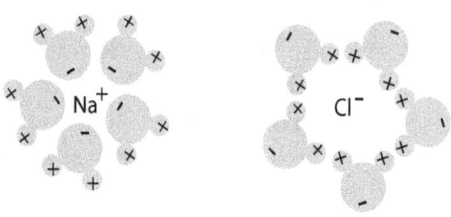

Figure 10.17. Solution theory.

Rate of Solution

Different methods are used to make solutes dissolve faster in a solvent. **Grinding** a solid solute into fine particles will allow it to dissolve faster because more surface area of the solute is exposed to the solvent. For example, powdered sugar dissolves faster than large crystals of sugar. **Stirring** a solution brings the solute in contact with fresh solvent, hastening the dissolving process. **Solutes dissolve faster if the temperature** of the solvent is raised. Increasing the temperature of a solution increases the kinetic energy of the particles causing them to move faster. Solvent molecules that move faster come in contact more rapidly with the solute. *This is true for solids dissolved in liquids, but **not true for gases dissolved in liquids.***

As the temperature of a solution of a gas dissolved in water is raised, gas molecules move further apart decreasing its density and increasing its buoyancy. Raising the temperature forces the gas out of solution. Carbon dioxide (CO_2) dissolved in water, commonly called a "carbonated" drink or soda, is kept cold to keep the dissolved gas in solution.

Saturated Solutions

There are many kinds of solutions. A few are illustrated below. See table 10.5. Solutions can be *dilute* (weak) or *concentrated* (strong). A dilute solution has very little solute compared to the solvent. **Concentrated solution** contains a lot of a solute compared to the solvent. **Unsaturated solutions** are those that have not reached saturation.

---

418   A solute is the substance that is being dissolved in a solvent.
419   Separates into its component parts, sodium and chlorine.

Solutions can be *saturated* or *supersaturated*. A saturated solution *holds as much of a solute as it can at a given temperature*. At this temperature, the solvent cannot dissolve any more solute. If more solute is added to a saturated solution, then a *precipitate* is seen at the bottom of the beaker.

Table 10.5. Types of solutions.		
Solvent	Solute	Example
Liquid	Liquid Solid Gas	Alcohol in water NaCl in water CO_2 in water
Gas	Liquid Solid Gas	Water vapor in air Mothballs (naphthalene) in air Oxygen in Nitrogen (air)
Solid	Liquid Solid	Silver in mercury (dental amalgam) Carbon in iron (steel)

A *supersaturated* solution is one that is forced to hold *more solute than it normally holds* at a given temperature. Carbonated drinks are supersaturated with the solute CO_2 by adding the CO_2 to the solvent water while increasing the pressure and lowering the temperature. *Suspensions* are mixtures of large particles in a fluid such as air or water. The large size of the particles causes them to settle out unless the mixture is stirred. Sand mixed in water is a suspension. If the mixture is not agitated, then the sand particles will quickly settle to the bottom of the beaker of water. This process is termed *sedimentation*. Sedimentation is one of the methods used for the purification of drinking water.

Many medications are made into suspensions if they cannot dissolve in a solvent. Suspensions have to be shaken to keep the particles suspended. The thicker the suspending material, the more it has to be shaken, but it will take longer to settle out. Glycerine is often used as a suspending material. **Amoxicillin**[420] may be in suspension form.

A *colloid* is a type of mixture whereby the substances in the mixture are distributed throughout the mixture on the microscopic level. Fog, shaving cream, **agar**[421] and mayonnaise are all examples of colloids.

Emulsions are usually liquids dispersed in liquids, both of which are **immiscible**.[422] Emulsions are a subcategory of colloids. Milk is an example of an emulsion.

---

420   Taken as tablets, capsules and oral suspension.
421   A gelatin-like substance derived from algae. Used in deserts and microbiological growth media. It is a mixture of agarose and agaropectin.
422   Not able to mix with each other, like olive oil in vinegar.

An *alloy* is a mixture of metals in metals. *Bronze* is a mixture of copper and tin. *Brass* is a mixture of copper and zinc. Mercury will dissolve and mix with almost all other metals. These mixtures are called *amalgams*. One well known amalgam is silver fillings in teeth. It is a mixture of silver and mercury. Most mercury fillings are considered safe, but for those concerned many non-mercury containing substitutes are available. These are a few of the many thousands of alloys.

Acids and Bases

A *neutral* solution has a **pH**[423] of 7.0. Triple distilled water is neutral. It consists of only water molecules. Litmus paper will be unaffected by a neutral substance. Sugar solutions and sodium chloride solutions are neutral when dissolved in pure **water**.[424] *Acids are substances that* have (1) a pH of 0 to 6.9, (2) *turn blue litmus paper red*, (3) have a sour taste, (4) give off H⁺ ions (5) neutralize bases and (6) react with metals to form salts and hydrogen gas.

Bases (1) have a pH of 7.1 to 14, (2) *turn red litmus paper blue*, (3) have a bitter taste, (4) give off -OH⁻ ions, and (5) neutralize acids.

Acids may be *inorganic* or *organic*. Inorganic acids tend to be very strong and dangerous to handle, while organic acids tend to be weak and in most cases common organic acids are consumed as foods. See table 10.6 and table 10.7.

Table 10.6. A partial list of some inorganic and organic acids.			
Inorganic Acids	Name	Organic Acids	Name
HCl	Hydrochloric	CH_3COOH	Acetic
HNO_3	Nitric	$C_6H_8O_7$	Citric
H_3PO_4	Phosphoric	$CH_3- CHOH-COOH$	Lactic
H_2SO_4	Sulfuric	$C_4H_6O_5$	Malic
H_3BO_3	Boric	CH_2O_2	Formic

Table 10.7. A partial list of some inorganic and organic bases.			
Inorganic Bases	Name	Organic Bases	Name
Calcium Carbonate	$CaCO_3$	Atropine	NA
Calcium Hydroxide	$Ca(OH)_2$	Nicotine	NA
Sodium Bicarbonate	$NaHCO_3$	Morphine	NA
Sodium Hydroxide	$NaOH$	Histadine	NA

---

423   pH is an abbreviation for the power of hydrogen. It is a measure of acidity or alkalinity of a substance.
424   Pure lacks minerals and consists of only water molecules.

Neutralization

Acids react with bases to form salts and water. The reaction of hydrochloric acid (HCl) and sodium hydroxide (NaOH) shown below (figure 10.18a) is a **neutralization reaction.** The metallic ion Na^+ of NaOH pairs with the nonmetallic ion Cl^- of HCl and the metallic ion H^+ of HCl pairs with the non-metallic ion OH^- of NaOH. See Figure 10.18b.

a. $$NaOH + HCl \longrightarrow Na\,Cl + HOH$$

b. 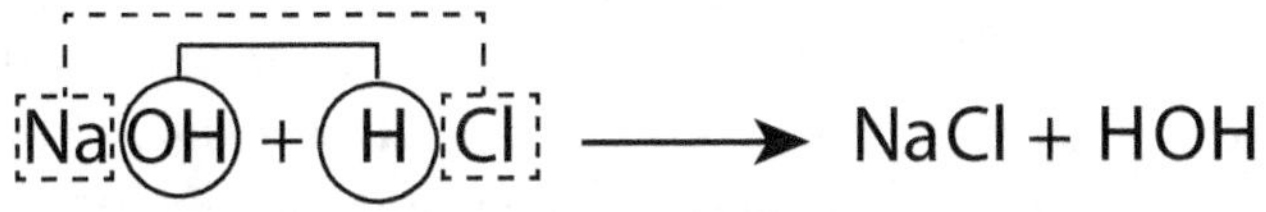 $NaOH + HCl \longrightarrow NaCl + HOH$

Figure 10.18a. The chemical equation for neutralization of an acid with a base. 10.19b. Shows how the metallic ions pair with the nonmetallic ions.

Reactions of Acids With Metals

Acids react with metals to produce ionic compounds and hydrogen gas. This kind of chemical reaction is a **single displacement** reaction. The metal atom displaces the hydrogen of the acid. Free hydrogen atoms unite to form molecular hydrogen gas (H_2). The first reaction in figure 10.19 shows the replacement of hydrogen of HCl on the left side with Fe to produce $FeCl_2$ and H_2 on the right side.

$$Fe + HCl \longrightarrow FeCl_2 + H_2$$
$$Fe + H_2SO_4 \longrightarrow FeSO4 + H_2$$
$$Ag + H_2SO_4 \longrightarrow AgSO_4 + H_2$$
$$Cu + H_2SO_4 \longrightarrow CuSO_4 + H_2$$
$$Cu + HNO_3 \longrightarrow CuNO_3 + H_2$$

Figure 10.19. Examples of reactions of metals with acids.

The reaction of acids with metals are also **oxidation-reduction (redox)** reaction because the oxidation state is changed after an exchange of electrons. Ionic compounds such as $FeCl_2$, $FeSO_4$, and $AgSO_4$ form resulting in the production of hydrogen gas.

Detection of Acids and Bases

Litmus paper is commonly used to test for acidity or basicity (alkalinity) of a substance. It is a quick and reliable test to find out if something is an acid or a base, but it does not tell how acid or how basic something is. Blue litmus paper contains a chemical substance that will turn red in an acid. Red litmus paper contains a substance that turns blue in a base. Some pH papers can give a pH reading from 0 to 14 to give reasonably accurate measurement of pH. Another common way of detecting and measuring the pH of a substance is by using solutions that change color at a specific pH. These are called **indicator solutions.** For very rapid and accurate measurements of acidity or alkalinity (basicity), **pH meters** are used.

Organic Chemistry: The Chemistry of Carbon

The nonmetal element ***carbon*** is represented by the symbol C. Its atomic number is 6, and its average atomic mass is 12. After hydrogen, helium and oxygen, carbon is the next most abundant element in the universe. Carbon has *three naturally occurring isotopes*: C_6^{12}, C_6^{13} and C_6^{14}. Carbon 14 (C_6^{14}) is radioactive.

In pure form, *Carbon exists in three forms*: diamond, graphite and

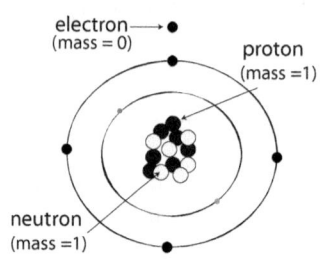

coal. Different forms of the same element are called ***allotropes***. Carbon is at the top of column 14 in the periodic table of the elements. Carbon is ***tetravalent***. Tetravalent means carbon has four electrons in its outer energy level available for covalent bonding, as do all column 14 elements. See figure 10.20. Carbon is the second most abundant element in the human body. Since a carbon atom has 4 electrons in its outer energy level available for covalent bonding, it can combine with other carbon

Figure 10.20. Electron distribution of a carbon atom showing 4 electrons in its outer energy level.

atoms along with other substances to form a great variety of ***organic molecules***. Carbohydrates, fats, proteins and nucleic acids are four organic molecules associated with living systems. ***Hydrocarbons*** are formed when long chains of carbon atoms combine only with hydrogen atoms.

Empirical, Molecular and Structural Formulas

Glucose[425] can be described by the ***empirical formula*** CH_2O. The empirical formula tells the reader are two hydrogen atoms for every carbon atom and one oxygen atom for every carbon atom. The ***molecular formula*** for glucose, $C_6H_{12}O_6$ reveals that there are 6 carbon atoms, 12 hydrogens and 6 oxygen atoms. It is a hexose, a 6 carbon sugar, however it could be any one of 20 other hexoses.

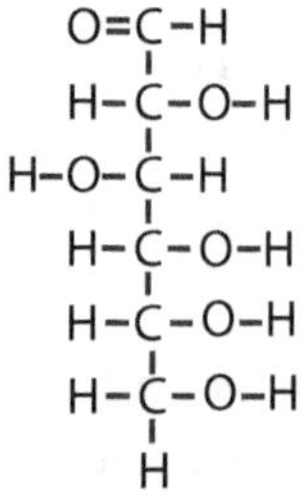

```
O=C-H
H-C-O-H
H-O-C-H
H-C-O-H
H-C-O-H
H-C-O-H
    H
```

Figure 10.21. A chain structural formula for D-glucose.

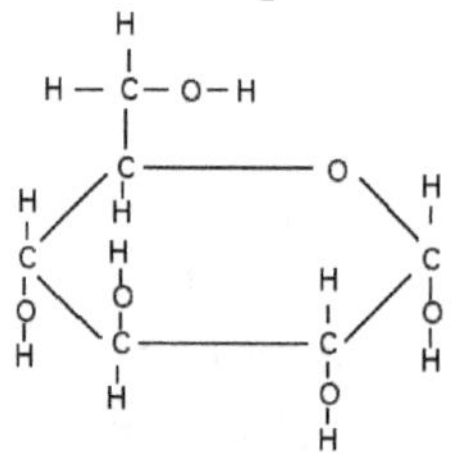

Figure 10.22. A ring structure for D-glucose.

A ***structural formula*** is needed to identify which $C_6H_{12}O_6$ is being discussed. A structural formula identifies the *kind*, *number* and ***arrangement*** of atoms. Structural formulas can be in a chain (see figure 10.21) or ring form (see figure 10.22).

425   Also called dextrose or grape sugar.

The Hydrocarbon Series

Hydrocarbons are composed of C and H atoms. *Methane* is the simplest molecule in the hydrocarbon series. The *molecular formula* for methane's is CH_4. See figure 10.23. The formula indicates that the methane molecule is made up of C and H atoms. The molecule has one C atom and four H atoms. A structural formula gives the same information but also shows the arrangement of the atoms in the molecule. The hydrocarbon series is built up by adding C atoms to methane to make larger molecules of unlimited length and arrangement. If a second C atom is added to methane, ethane is produced. See figure 10.23 and tables 10.8, 10.9 and 10.10. Most hydrocarbons are found in crude oil.

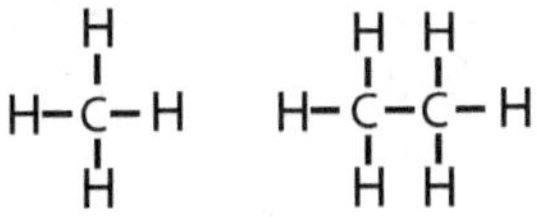

Figure 10.23. Structural formulas for methane (left) and ethane (right).

Table 10.8. A partial list of the alk*ane* hydrocarbon series.

Carbon Atoms in Chain	Alkane	Empirical Formula	State
1	Methane	CH_4	Gas
2	Ethane	C_2H_6	Gas
3	Propane	C_3H_8	Gas
4	Butane	C_4H_{10}	Gas
5	Pentane	C_5H_{12}	Liquid
6	Hexane	C_6H_{14}	Liquid
7	Heptane	C_7H_{16}	Liquid
8	Octane	C_8H_{18}	Liquid

Table 10.9. A partial list of the alk*ene* hydrocarbon series.

2	Ethene (ethylene)	C_2H_4	Gas
3	Propene (propylene)	C_3H_6	Gas
4	Butene (butylene)	C_4H_8	Gas
5	Pentene	C_5H_{10}	Liquid

Table 10.10. A partial list of the alk*yne* hydrocarbon series.

2	Ethyne (acetylene)	C_2H_2	Gas
3	Propyne	C_3H_4	Gas
4	Butyne	C_4H_6	Gas
5	Pentyne	C_5H_8	Gas

Saturated and Unsaturated Hydrocarbons

Carbon combines with itself using single, double and triple bonds.
See figure 10.24. Carbon atoms bond to each other to make
long chains and branching structures to create a great variety
of carbon compounds. Carbon combines with other elements
such as halogens, column 17 on the periodic table, the
nitrogen family, column 15, oxygen and hydrogen. If only a
single bond exists between carbon atoms, the compound is
an *alkane*. *All positions are filled with hydrogen atoms. It is*
saturated. If a compound has at least
one double bond between carbon
atoms, it is an *alkene*. If *triple bonds*
are present, it is an *alkyne*. *Alkenes*
and *alkynes* are *unsaturated* because more hydrogen atoms can be added to these molecules. See
figure 10.25. Hard fats are saturated fats. A hard fat is the fat found on
the outside edge of a steak. Saturated fats are not desirable in the diet.

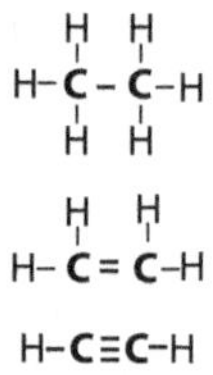

Figure 10.24. carbon to carbon bonds.

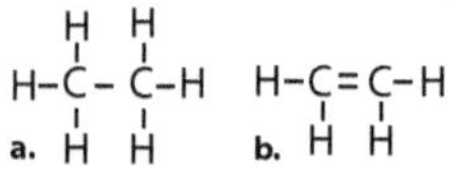

Figure 10.25. Saturated (a) and unsaturated molecules (b).

Functional Groups

Organic molecules may have specific arrangements of atoms called
functional groups (*radicals)* that determine their chemical function.
Each group adds special characteristics and functionality to the molecule.
A part of the molecule that is not changed in the reaction is represented
by *the abbreviation –R*. See table 10.11.

Table 10.11. Selected functional groups.	
Compound	Functional group
Alcohol	**R– OH**
Aldehyde	**R– CHO**
Amine	**R– NH$_2$**
Carboxylic	**R– COOH**
Ether	**R– CH$_2$ – CH$_3$**
Ester	**R– O – R**
Ketone	**R^1 – C = R^2**

Monomers, Dimers and Polymers

A single molecule is termed a *monomer*, "mono" means *one* and
"mer" means molecule. Examples of monomers are simple sugars,
amino acids, fatty acids and nucleotides. Small monomers build larger
molecules called *polymers* ("poly" many). Examples of polymers are:
starches, proteins, fats and nucleic acids. See table 10.12.

174

A *dimer* consists of two molecules. Monomers of glucose and fructose can combine to form the dimer *sucrose* (table sugar). Glucose can form long polymeric chains with *very little branching* to create plant starch (referred to as *starch*). The upper case "G" stands for a glucose molecule in figures 10.26 and 10. 27.

If the glucose molecule is stored as *glycogen* (animal starch), it forms *highly branched* glucose molecules. See figure 10.28. Glycogen is made by the liver and stored in the liver and muscles. If energy is needed fast, glucose can be mobilized quickly from glycogen.

Fats are a secondary source of energy. Fat cells (adipocytes) make up adipose tissue. Each cell contains a single large droplet of lipid composed mainly of *triglycerides*.

Table 10.12. Major biological polymers and their building blocks.		
Polymer	Building Block	Function
Glycogen (animal starch)	Glucose	Stored energy
Starch (plant)	Glucose	Stored energy
Proteins	Amino acids	Muscle, enzymes and hormones
Lipids	Fatty acids and glycerol	Stored energy
Nucleic acids	Nucleotides (DNA and RNA)	Genetic information and protein synthesis

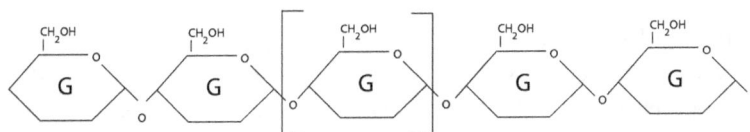

Figure 10.26. The polymer starch found in plants has very little branching.

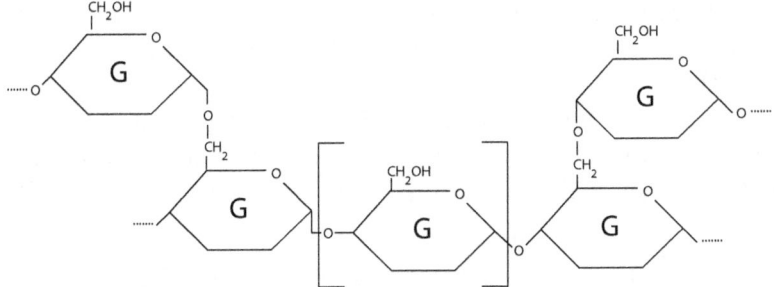

Figure 10.27. Glycogen or animal starch. It is a highly branched polymer that serves as an energy reserve in animals and fungi.

175

Enzymes – The Lock and Key Mechanism

Enzymes are *organic catalysts* that regulate almost all chemical reactions that take place in an organism. The name of an enzyme always ends in "*ase*." Some enzymes build up large molecules from small molecules. *Anabolic* reactions build big molecules from small molecules. Some enzymes break down large molecules into small molecules, these are *catabolic* reactions. In the example to the right, the dimer *sucrose* is the *substrate*. A substrate is the molecule being acted on. The enzyme (**E**) *sucrase* fits *specifically* to the substrate at an *active site*. An active site is the point of attachment of the enzyme to the substrate.

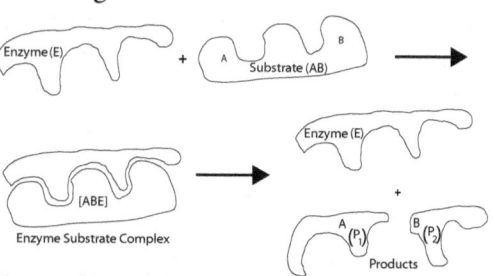

Figure 10.28. The lock and key mechanism of enzyme action.

This is the *lock and key mechanism* of enzyme action. An *enzyme substrate complex* [**ABE**][426] is formed. When sucrose is hydrolyzed by sucrase, the products, glucose (**P_1**) and fructose (**P_2**), are released both monomers. See figure 10.28.

$$E_{(ENZYME)} + AB_{(SUCROSE)} = [ABE] = E + P_{1 (GLUCOSE)} + P_{2 (FRUCTOSE)}$$

Classes of Enzymes

There are two classes of enzymes: *simple* and *conjugated*. Simple enzymes consist of *protein molecules only*. If an enzyme consists of protein and some other part, it is a **conjugated**[427] enzyme. When there are two parts to an enzyme, the protein part of a conjugated enzyme is called an *apoeznyme*. The non-protein part is called a *cofactor*. Cofactors may be *inorganic* metallic ions such as Fe^{++}, Mg^{++}, K^+, Mn^{++} or Cu^+ or *organic* molecules called *coenzymes*. Coenzymes may be vitamins such as B1, B2, B6, B12, niacin or folic acid. Enzymes cannot work if the proper dietary vitamins and minerals are not supplied in the diet. If enzymes do not work, the organism becomes ill and dies. *Metabolic pathways* such as vision, **olfaction**,[428] taste and digestion are mediated by enzymes.

---

426   A molecular complex is a close association between molecules. The complex is not covalently bound as glucose is to fructose forming sucrose.
427   Also referred to as holoenzymes
428   The sense of smell.

176

Metabolism and Metabolic Pathways

Metabolism is the total number of chemical reactions that take place in an organism. Metabolism may be broadly divided into **anabolism** and **catabolism**.

Dehydration Synthesis[429]

Dehydration synthesis is an enzyme-mediated **anabolic biochemical reaction** that makes large molecules from simpler molecules. Anabolic reactions proceed by **dehydration synthesis**. It is building molecules by the removal of water. Small molecules of *glucose, amino acids, fatty acids* and *nucleotides* are built up into *starch, protein, fats* and *nucleic acids*. For example, the enzyme sucrase assists in the **removal of water** to covalently join the monomers glucose and fructose. The result is the double sugar sucrose, a dimer. See figure 10.29. A hydroxyl radical (**–OH**)[430] from glucose (left) unites with a hydrogen (**H**) atom from fructose (right) to join glucose to fructose with the release of a water (**H₂O**) molecule.

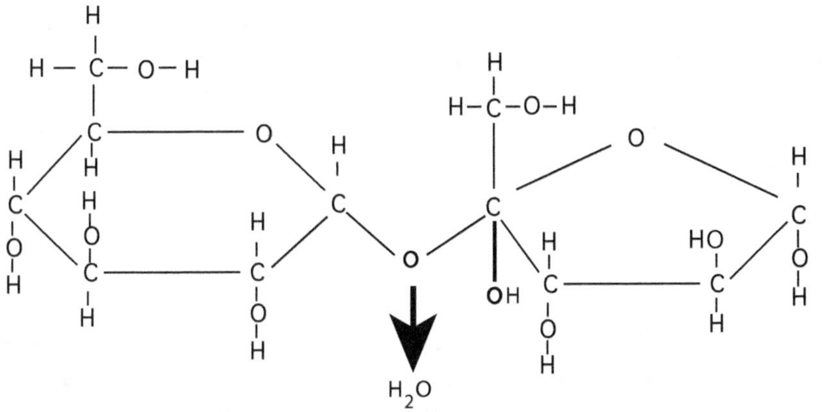

Figure 10.29. Dehydration synthesis. Joining two monomers, glucose (left) and fructose (right) into the dimer sucrose by the withdrawal of water.

Hydrolysis

Hydrolytic reactions are *catabolic*, enzyme-mediated reactions. The **addition of water** splits large molecules into smaller molecules in these reactions. See figure 10.30. A large sucrose molecule is split into small glucose and fructose molecules. The **–OH** of a water molecule is added to the glucose molecule (left) and the remaining **–H** is added to fructose.

429   Dehydration, from the Latin, *de* meaning off and *hydro* from the Greek, meaning water.
430   The hydroxyl radical (–OH) is the electrically neutral functional group of OH¹⁻.

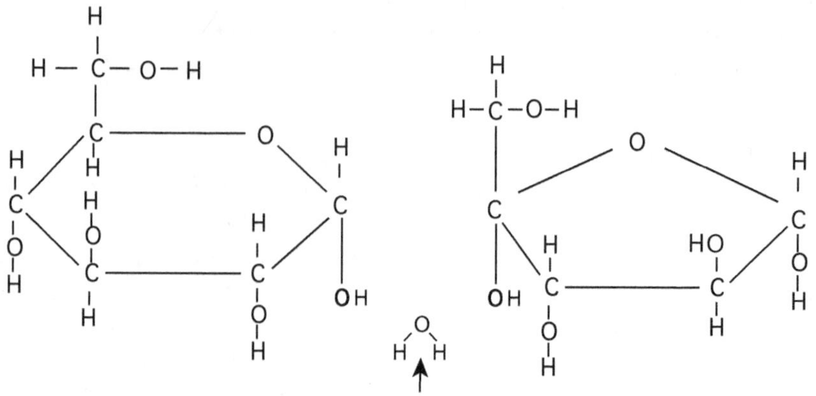

Figure 10.30. The hydrolysis of sucrose into glucose (left) and fructose (right). Water is inserted between the molecules to break the dimer into two monomers.

Energy Producing Metabolic Pathways

Biological systems are *open systems.* Matter and energy *enter* and *leave* the system. Almost all living things directly or indirectly depend on the energy of the Sun to maintain life. *Autotrophs* make the food they need by capturing and **incorporating**[431] the sun's energy into chemical bonds of food molecules they make. *Heterotrophs*, on the other hand, depend on the energy of the Sun *indirectly* because they feed on plants or plant-like organisms that have *captured and stored the sun's energy in the chemical bonds* of large food molecules such as carbohydrates. The digested food molecules are then reassembled into carbohydrate, fat, protein and nucleic acid molecules identical to the consuming organism's carbohydrate, fat, protein and nucleic acid molecules.

After food molecules have been made soluble by digestion, some are stored, and some are metabolized for energy several *metabolic pathways.* *A metabolic pathway is the sequence of enzymatically controlled chemical reactions* that leads from a starting point of initial reactants to an end point of final products. The primary source of energy is glucose. These pathways extract energy trapped in glucose's chemical bonds.

$$C_6H_{12}O_6 + 6O_2 \longrightarrow 6CO_2 + 6H_2O + 36\ ATP$$

Figure 10.31. Overall chemical equation for aerobic cellular respiration.

There may be thousands of intermediate reactions that take place before the final products are arrived at. An overall chemical reaction shows only the reactants and the final products. See figure 10.31.

---

431   From the Latin, *in-* + *corpor-*, body, meaning to unite or to take in.

Aerobic Cell Respiration: Respiration With Molecular Oxygen

Enzymes that mediate aerobic cell respiration are *attached to the membranes of* **mitochondria of eukaryotes**. The enzymes aerobic cell respiration are present in some prokaryotes, but the enzymes are *attached to the prokaryotic* **cell membrane**. Most descriptions of biochemical pathways are shown as an overall chemical reaction. The myriad chemical reactions that occur, are not shown. Figure 10.31 shows an overall chemical equation representing aerobic cellular **respiration**. It shows that glucose and oxygen will ultimately become carbon dioxide, water and energy (ATP). Figure 10.32 shows a more detailed explanation of aerobic cellular respiration.

Heterotrophs need carbon sources for energy. Aerobic cellular respiration consists of enzymatically controlled energy producing pathways that use molecular oxygen (O_2) as their final electron acceptor. Some bacteria, archaea and almost all eukaryotes use aerobic pathways.

Below is a brief summary of the many hundreds of *enzymatically mediated* steps involved in aerobic cellular respiration. Each reaction is made possible by specific enzymes. If an enzyme is missing or not functional, the biochemical reaction will not take place.

(1) A six carbon glucose molecule enters a cell **facilitated**[432] by insulin.

$C_6H_{12}O_6$ + insulin $\longrightarrow$ Cell $\longrightarrow$ Cytoplasm

(2) Glucose is split by the process of glycolysis *in the cytoplasm* into two 3-carbon pyruvate molecules. Two **(2)** ATPs are produced.

$C_6H_{12}O_6 \longrightarrow 2\ CH_3COCOOH + 2\ ATP$

(3) Pyruvate enters the Krebs or **TCA**[433] cycle. Pyruvate is modified, producing CO_2 and electrons (e^-). The electrons are donated to the electron transport chain (ETC). Two **(2)** ATPs are produced.

$CH_3COCOOH + Acetyl\text{-}CoA + NAD^{+}$ [434] $\longrightarrow 2\ CO_2 + 2\ ATP$

(4) Electrons are exchanged between donors and acceptors in the electron transport chain. Thirty four **(34)** ATPs are produced.

$NADH \longrightarrow e^- + H^+$

(5) Oxygen, the final electron acceptor, accepts harmful electrons and hydrogen ions with the formation of harmless water. See figure 10.33.

$e^- + H^+ + O_2 \longrightarrow H_2O$

---

432   To ease, to help something to happen smoothly.
433   Also known as: tricarboxylic acid cycle or the citric acid cycle.
434   The coenzyme **n**icotinamide **a**denine **d**inucleotide accepts electrons to yield NADH. NADH is a coenzyme that is needed to impart biological activity to an enzyme.

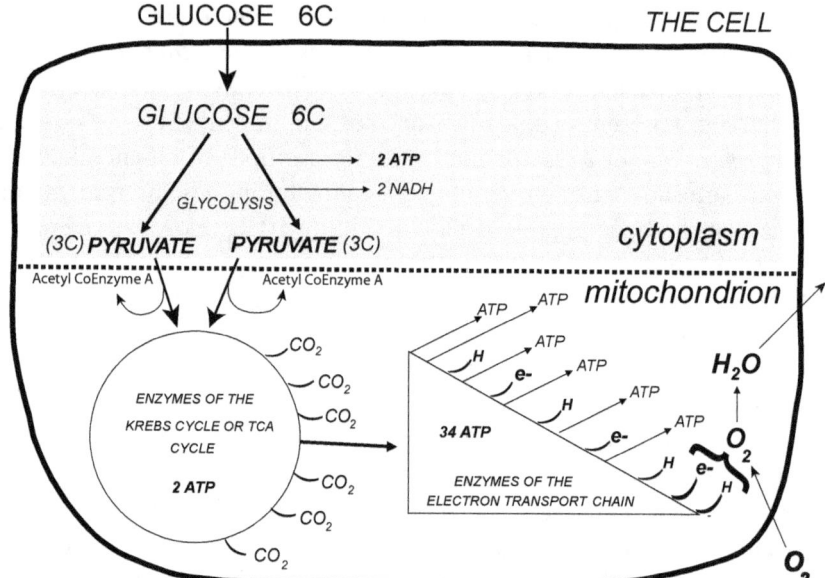

GLUCOSE 6C

THE CELL

Figure 10.32. Illustrating of the conversion of simple sugars into carbon dioxide, water and ATP. Glycolysis produces 2 ATP molecules. The Krebs or tricarboxylic acid cycle (TCA) produces 2 more ATP molecules. The electron transport chain produces approximately 34 more ATPs for a total of 38.

Anaerobic Cell Respiration: Respiration Without Molecular Oxygen

Some bacteria and *some* archaea can extract energy from carbon sources using **anaerobic cellular respiration**. These reactions use the electron transport system *but do not use molecular oxygen as a final electron acceptor*. They use **oxygen containing compounds** such as sulfate (SO_4^{2-}), nitrate (NO_3^{1-}), or carbon dioxide (CO_2). These electron acceptors trap harmful electrons and hydrogen ions to make them harmless.

Fermentation

Fermentation is a second energy producing pathway that does not use molecular oxygen. Fermentation extracts energy from simple sugars by *glycolysis*. *Two (2) ATPs molecules are produced.* The electron acceptors in

$$C_6H_{12}O_6 \xrightarrow[\text{Yeast}]{\text{Enzymes of}} C_2H_5OH + 2\,ATP$$

Glucose $\longrightarrow$ Ethyl Alcohol + ATP

Figure 10.33. Overall chemical equation for the fermentation of glucose into alcohol and carbon dioxide.

fermentation do not come from outside sources such as O_2 or NO_3, rather they come from compounds from within the organism. Simple sugars are converted *anaerobically* to *alcohol* and *carbon dioxide* or to *organic acids*. See figure 10.33. Beer and wine are produced by fermentation.

180

Pyruvate – The Hub of Metabolic Reactions

The metabolic pathways of *aerobic cellular respiration*, *anaerobic respiration* and *fermentation* pass through the molecule *pyruvate*.

$(CH_3COCOOH)$.[435] Pyruvate is an organic *acid and* a *ketone*. Pyruvate has an acid functional group and the ketone functional group. See figure 10.34. Pyruvic acid is made from glucose by *glycolysis*. One glucose molecule yields two pyruvic acid molecules. Glycolysis takes place *in the cytoplasm* of archaea, bacteria and eukaryotes.

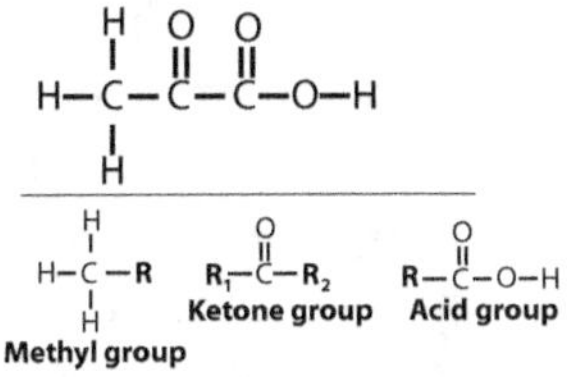

Figure 10.34. The pyruvate molecule showing its functional groups.

In the process of aerobic cell respiration, *pyruvic acid* enters the *mitochondria* of *eukaryotes* and converted to *acetyl coenzyme A* (acetyl – CoA). Acetyl – CoA is processed in the Krebs cycle producing small amounts of ATP and CO_2. From the Krebs cycle, the processed molecules enter the *electron transport chain.* Most of the organism's ATP is produced here. *Harmful free electrons and hydrogen ions are produced during the process of donation and acceptance of electrons in the electron transport chain*. The free electrons (e^-) and hydrogen ions (H^+) are accepted by molecular oxygen (O_2), the *final electron acceptor* to produce harmless water (H_2O). Archaea and bacteria do not have mitochondria. Archaea and bacteria that use respiratory enzymes have them attached to their cell membranes. Organisms that use *anaerobic respiration* use oxygen containing compounds, not molecular oxygen (O_2), as a final electron acceptor. Certain bacteria and yeast[436] use *fermentation*. *They do not use an electron transport system*. In *the fermentation* process, organic molecules *within* yeast cells act as final electron acceptors. They are **endogenous**[437] electron acceptors. In *aerobic* and *anaerobic* respiration the final electron acceptors come from *without* the organism. Molecular oxygen and nitrates come from the organism's environment. They are termed **exogenous**[438] electron acceptors.

Symbiotic Theory

Prokaryotes do not have mitochondria. According to **symbiotic theory**,[439] a prokaryotic cell probably became a *symbiont* within a eukaryotic cell. The prokaryotic cell evolved to become a mitochondrion.

435   Structural formula for pyruvic acid.
436   Yeast can carry on aerobic respiration if enough molecular oxygen is present.
437   From the inside. From the ancient Greek: *endo-* inside + *genēs* born.
438   From the outside. From the Greek: *exo-* outside + *genēs* born.
439   Mitochondria share may characteristics of bacterial cells. Mitochondria have sequences of genes in common with bacteria and they reproduce independently.

Comparison of Metabolic Pathways

Figure 10.35 shows common features in the three major enzymatically mediated metabolic pathways; *aerobic* respiratory pathways, *anaerobic* respiratory pathways and *fermentative respiratory pathways*. All split a six carbon sugar into two pyruvate molecules in their cytoplasm.

Anaerobic cellular respiration is used primarily by **prokaryotes** that live in **oxygen-poor environments**. Anaerobic respiration plays an important role in the nitrogen, sulfur and carbon cycles. These cycles allow for the reuse of elemental nitrogen, sulfur and carbon in the environment. Aerobic and *anaerobic* respiration use electron transport chains. The electron transport chain will work only if electrons flow through the system. As the electrons arrive at the end of the "chain" they must be picked up by exogenous final electron acceptors. Aerobic respiration uses molecular oxygen as its final exogenous electron acceptor.

Anaerobic respiration uses an *oxygen containing compounds* as its final exogenous electron acceptor, *not* molecular oxygen. The fermentation pathway uses final electron acceptors that reside within the cell are endogenous final electron acceptors.

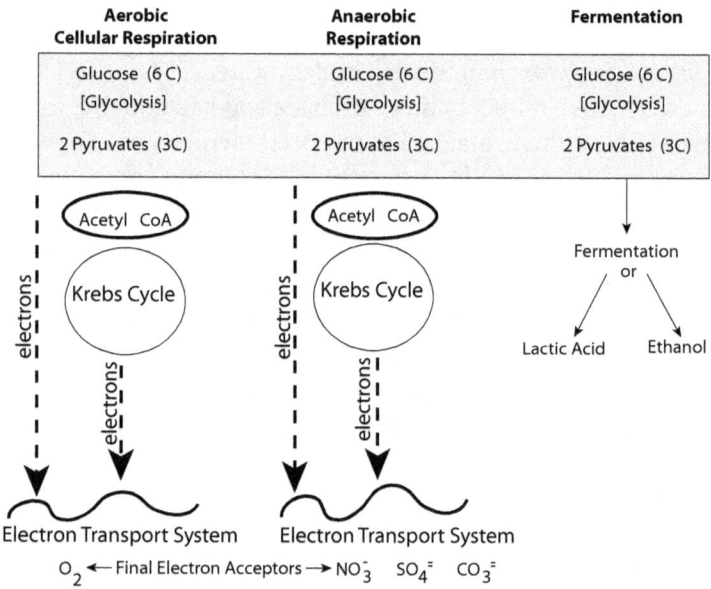

Figure 10.35. Pyruvate is produced in the cytoplasm of cells. See shaded rectangular box. Organisms that use an aerobic respiratory pathway use molecular oxygen (O_2) as their final electron acceptors. Organisms that use anaerobic respiratory pathways use compounds that contain oxygen as their final electron acceptors.

182

Chapter 11 Prokaryotic Cells: Archaea and Bacteria

Bacteria

Bacteria are **ubiquitous**.[440] They are found almost everywhere on, above and below the surface of the **Earth**.[441] Bacteria are in the Earth's air, waters and foods. Bacteria have coevolved with humans. It is estimated that there is a total of 5 nonillion bacteria on Earth (5 x10^{30} or 5 followed by thirty zeros). The human body has as many as 100 trillion cells, but the human gut has ten times that number of bacteria.

Most bacteria are helpful. Only a few cause disease. Yogurt, cheese and antibiotic production are just a few of the products that result from bacterial action. Bacteria are important in the recycling of nutrients and are important commensals in the human gut. As commensals, bacteria help humans, and humans help bacteria by providing them with food and shelter. The gut is host to hundreds of species of bacteria that help *control the growth of harmful bacteria*, *aid in digestive processes*, participate in *immune function* and the *production of* **biotin**[442] and **vitamin K**.[443]

Among the many microorganisms that are helpful to humans are the *Lactobacilli*. There are over 100 species of this genus. *Lactobacillus acidophilus* is a **probiotic**[444] and other species are useful for the production of yogurt, cheese, sauerkraut, pickles, beer and wine.

In the past, archaea and bacteria were placed together in the kingdom monera because the belief was that both were closely related. Molecular biologists discovered bacteria and archaea have distinctly different DNA, RNA and evolutionary history. Presently each is their own domain.

Structure of a Typical Bacterial Cell

Bacteria and archaea *do not* have organelles. They *have structures* that perform specialized functions. See figure 11.1. *Fimbrae* and *pili* are **appendages**[445] on the surface of bacteria. Fimbrae enables bacteria to adhere to each other and to inanimate surfaces. Pili are hollow structures that can penetrate and "inject" genetic material into other bacteria.

The *glycocalyx* is an extracellular coating that lies just above the bacterial cell wall. This structure may be called a *capsule* or an **outer membrane**[446] in some bacteria. The outer membrane should not be confused with the cell membrane. All cells have a bounding cell membrane.

440   Found everywhere.
441   Bacteria have been found 2.8 kilometers below the surface of the Earth.
442   Biotin (B_7), is a water soluble vitamin necessary for cell growth.
443   Necessary for proper clotting of blood and healthy bones by helping with transport of calcium along with vitamin D that plays a major role in adsorption of calcium.
444   Microorganisms that are beneficial to the host they live in or on.
445   Appendage is an outgrowth or a part of the body that is external to the rest.
446   Gram-negative bacteria have an outer membrane that is antigenic.

One strain of *Streptococcus pneumoniae* is surrounded by a ***capsule***. The capsule helps *S. pneumoniae* attach to lung cell membranes. This structure makes *S. pneumoniae* extremely **virulent**.[447] Its colony appears smooth (S form) on nutrient agar. A second strain of *S. pneumoniae* lacks a capsule and its colony appears rough (R form). It is not virulent.

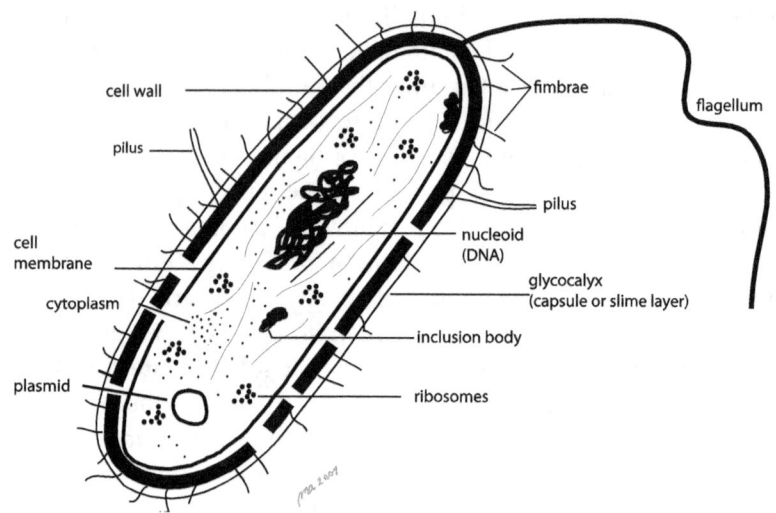

Figure 11.1. Structure of a typical prokaryotic cell.

The Bacterial Cell Wall

GLYCOCALYX LAYER
PEPTIDOGLYCAN LAYER
CELL MEMBRANE

Figure 11.2 **Gram-positive** cell wall.

GLYCOCALYX LAYER
OUTER MEMEBRANE
PEPTIDOGLYCAN LAYER
CELL MEMBRANE

Figure 11.3 **Gram-negative** cell wall.

Bacteria have a rigid ***cell wall*** located above its cell membrane. The cell wall protects the cell and retains its shape. See figure 11.1. Bacterial cell walls are composed mostly of ***peptidoglycan***. Plant cell walls contain cellulose. ***Archaeal*** cell walls contain mainly ***pseudopeptido glycan*** *and* ***glycoprotein***. Glycoproteins are a sugar/protein **moiety**.[448]

The **Gram**[449] stain helps scientists place bacteria into ***Gram-positive*** or ***Gram-negative*** groups depending on the cell's ability to retain a purple dye. ***Gram-positive cell walls*** (see figure 11.2) are ***simple*** and have a ***thicker peptidoglycan*** layer compared to ***Gram-negative cell walls***. See figure 11.3. ***Gram-negative cell walls*** are ***complex*** and have a thinner layer of pepidoglycan.

---

447   Virulence refers to the *pathogenicity* of a disease-causing organism.
448   A part or portion.
449   Hans Gram developed this staining technique in 1884.

Bacterial Chromosomes

Bacteria and archaea have *nucleoids* that float in the cytoplasm. The word nucleoid means "like a nucleus." Nucleoids' chromosomes lack a surrounding membrane. Bacterial chromosomes may be linear or circular. They also have small bits *circular or linear DNA* outside of its nucleoid called *plasmids.* Plasmids are pieces of DNA. Plasmids carry additional genetic information, often related to antibiotic resistance.

The bacterial *cell membrane* is a *phospholipid bilayer that* regulates substances that enter and leave the cell. The cell membrane's fatty acid composition varies in prokaryotes, but not in eukaryotes.

Ribosomes

Proteins are synthesized on *ribosomes*. Bacterial and archael ribosomes float freely in the cytoplasm. Ribosomes consist of a large 70S unit and a smaller 30S unit. A Svedberg (S) is a unit of sedimentation.

Inclusion Bodies

There are different kinds of *inclusion bodies* in prokaryotes. They are structures that store *glycogen* as an energy reserve or contain *protein*. Some are *droplets of lipids* scattered in the cytoplasm.

Bacterial Shapes Aid in Disease Characterization

The three basic shapes of bacteria are: *cocci* (round), *bacilli* (rod) and *spiral*. Spiral bacteria have one to several curves. Spirals can be *vibrios* (comma-shaped), *spirilla* or *spirochetes* (highly coiled like a corkscrew). See page 136.

Many cocci are Gram-positive, non-spore forming and nonmotile. Some bacilli are *motile* and form spores. Vibrios are Gram-negative, non-spore formers. Vibrios are all motile with a single polar flagellum. Vibrios cause food and water borne diseases. *Vibrio cholerae* causes cholera, a serious infection of the small intestine that is contracted by ingesting contaminated food or water.

Most bacteria occur as single cells, but some species have characteristic groupings. Cocci may occur in grape-like clusters called *staphylococci* or "staph" for short. *S. aureus* can cause different kinds of skin infections, such as **folliculitis**,[450] **boils (furuncles)**[451] and **carbuncles**.[452]

Staphylococcus aureus can cause a **septicemia**[453] in women, a condition formerly known as *childbed fever* (*puerperal fever*). Antibiotics and improved hygiene in the delivery room made this a rare infection.

450   Folliculitis can be caused by bacteria, fungi or viruses.
451   An infection, deep in a hair follicle.
452   A collection of pus containing boils that usually drain pus through an opening.
453   Septicemia (bacteremia) refers to having pathogenic bacteria in the bloodstream. This condition leads to sepsis or an inflammation throughout the body.

Cocci also occur as chains called **streptococci**. *Streptococcus pyogenes* causes **streptococcal pharyngitis** ("strep throat"), streptococcal impetigo and necrotizing fasciitis (NF), the so called "flesh eating disease." *Streptococcus pneumoniae*[454] is another dangerous pathogen among the streptococci. It is the cause of bacterial or **lobar pneumonia**.[455] Pneumonia is can also be caused by an infection with *Haemophilus influenzae* and *Moraxella catarrhalis*.

Cell Wall Deficient Bacteria

Emmy Klieneberger-Nobel, a German microbiologist, discovered a unique bacterial cell termed cell wall deficient (CWD) bacteria in 1935. He named them **L-forms**.[456] *Mycoplasma* is a genus of parasitic L-forms. *Mycoplasma pneumoniae* is an important member of this group that causes **primary atypical pneumonia** (PAP) or walking pneumonia. Since *Mycoplasma pneumoniae* do not have cell walls, beta lactam antibiotics are useless. Fortunately, primary atypical pneumonia rarely requires hospitalization.

Antibiotics

Antibiotics[457] are chemical compounds made by bacteria or fungi that kill or inhibit the growth of bacteria. Alexander Fleming discovered the first antibiotic, *penicillin*, in 1941. He noticed that the fungus *Penicillium notatum*, now known as *Penicillium chrysogenum*, inhibited the growth of the bacterial species *Staphylococcus aureus*. He named the substance penicillin. Penicillin (G) is the drug of choice for use against *Gram-positive cocci*. Penicillins have a chemical structure termed *beta-lactams*. Beta lactams interfere with cell wall synthesis. The bacterial cell wall becomes porous, and cytoplasm leaks out of the bacterial cell, thus killing the bacteria.

Aminoglycoside antibiotics are chemically different from beta lactams. Aminoglycosides are used mostly against *Gram-negative bacteria*. These antibiotics interfere with protein synthesis in the bacterial cell. *Streptomycin* was the first aminoglycosides to be discovered.

The activity of antibiotics is classified as a broad spectrum or a narrow spectrum. Broad spectrum antibiotics work against Gram positive and Gram negative bacteria. Narrow spectrum antibiotics act against specific microbes.

454   The old name for the bacterium *Streptococcus pneumonia* is *Diplococcus pneumoniae* because of the pairs of bacteria that occur among the chains in slide preparations.
455   Pneumonia is any inflammation of the lung. It may be caused by bacteria, viruses or inhalation of small physical agents. Lobar pneumonia is a bacterial infections of one or more of the lobes of the lungs.
456   Discovered by in 1935. Named L-forms after Lister Institute in London, UK.
457   Meaning against life.

Chapter 12 Eukaryotic Cell Structure and Function

Eukaryotic Structures and Their Functions

Protists, fungi, animals and *plants are* members of the ***domain Eukarya*** because they possess ***nuclei***. A ***double layered membrane*** surrounds the nucleus and is **contiguous**[458] with the rough endoplasmic reticulum (RER) located in the cytoplasm. ***Chromosomes*** are located in the nucleus and surrounded by a fluid ***nucleoplasm***. Different species have different chromosome numbers. The nuclei of humans contain 46 chromosomes, the dog (*Canis lupis familiaris*) has 78, and a chimpanzee (*Pan troglodytes*) has 48. The ***nucleolus, located in the nucleus***, contains mostly RNA. The nucleolus is the site of ribosome assembly.

Almost all cells (figures 12.1, 12.2 and table 12.1) have an outermost ***glycocalyx***. The *eukaryotic* glycocalyx lies *above the cell membrane*. The *prokaryotic* glycocalyx lies above its cell wall (see figure 11.1). The glycocalyx is composed mainly of glycoproteins that *protect* the plasma membrane, contains *cell recognition molecules* and helps *immune function*. Below the glycocalyx is a ***plasma membrane***. The plasma membrane, like most membrane bound eukaryotic structures, is a ***selectively permeable lipid bilayer*** that regulates movement of material into and out of the cell.

The **fluid mosaic model**[459] is the current model of the plasma membrane. This model proposes that a membrane is a collection of proteins, sterols and phospholipids that behave like a fluid with an oil-like **viscosity**.[460] *The cell membrane separates a cell's external environment from a cell's internal environment.*

Cytoplasm lies between the cell membrane and the nucleus of cells. Cytoplasm is composed of ***cytosol*** and many ***organelles*** that float in it. ***Organelles are subcellular structures that carry out specific functions within a cell***. The non-granulated, outer part of the cytoplasm is termed *ectoplasm*, and a granular central area is termed *endoplasm*. Many chemical reactions take place in the cytoplasm. One reaction, ***glycolysis***, is an important chemical reaction that splits six carbon glucose molecules into two three carbon pyruvate molecules. The small pyruvate molecule can enter mitochondria to participate in the ***Kreb's cycle***. A supporting structure within the cytoplasm of eukaryotes and prokaryotes is called the ***cytoskeleton***. The cytoskeleton helps maintain the cell's shape and consists of ***microfilaments***, ***intermediate filaments*** and ***microtubules***.

---

458   Contiguous means continuous with, bordering on or next to.
459   First proposed by Seymour Singer, an American cell biologist, in the 1970s.
460   A liquid's "flowability." Resistance to flowing is the definition of viscosity. Water's viscosity is 1 centipoise at 20 °C. Other liquids are more or less viscous. Pour one cup of water and one cup honey at the same time. Honey has a resistance to flowing.

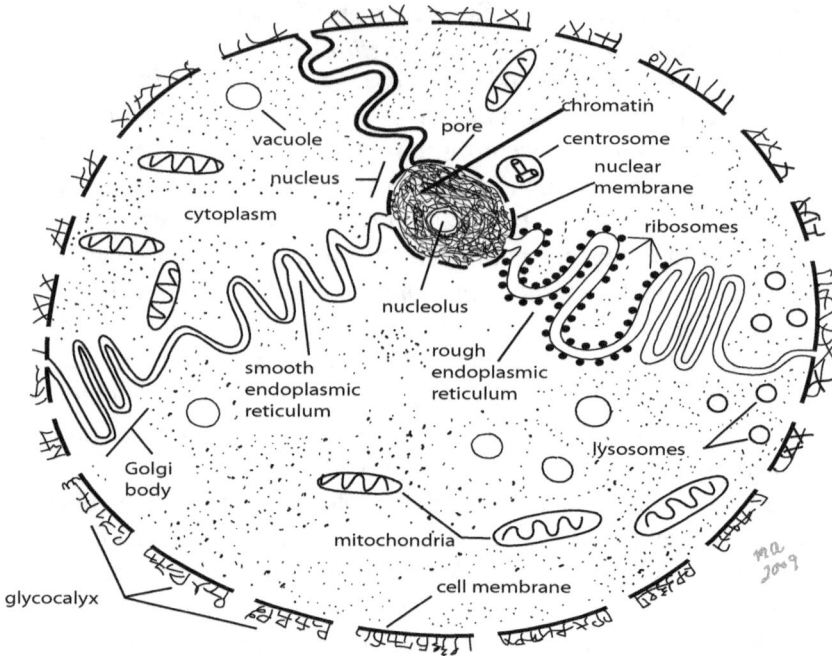

Figure 12.1. A typical generalized eukaryotic animal cell.

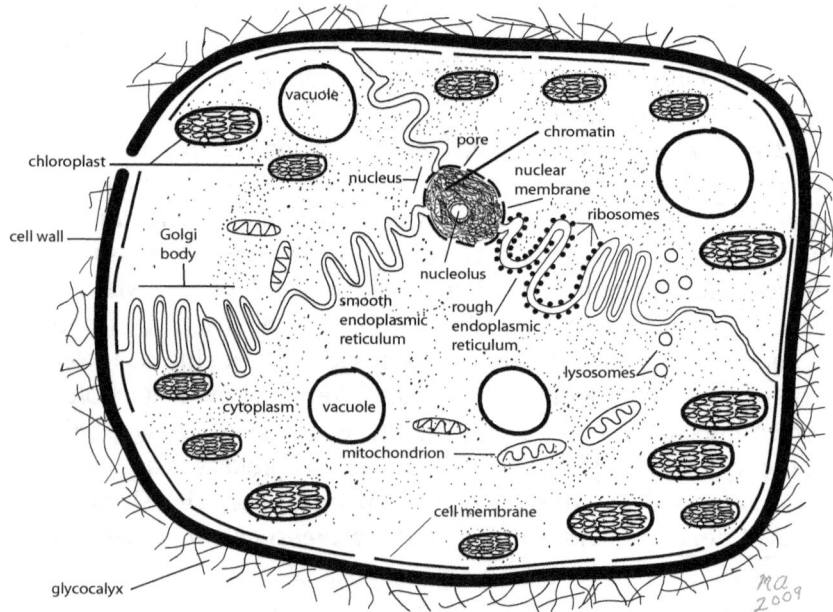

Figure 12.2. A typical generalized eukaryotic plant cell.

Table 12.1 Eukaryotic organelles.		
1	Nucleus	Contains chromosomes that make up the genome of an organism.
2	Nucleolus	A structure found within the cell's nucleus and composed of ribosomal RNA (rRNA). Produces ribosomes. It is not membrane bound.
3	Centrosomes (animals only)	Microtubules are produced here. The tubules form spindle fibers for chromosomes to travel on as they separate at anaphase.
4	Mitochondria	They make most of the cell's ATP. For this reason, they are called "power-houses" of the cell. It is the size of an average bacterial cell.
5	Lysosomes	Contain enzymes that break down waste materials, cellular debris, organelles that are worn out. They can also destroy viruses and bacteria that have entered the cell.
6	Peroxisomes	Contain enzymes that metabolize long-chain fatty acids, branched-chain fatty acids and amino acids.
7	Transport vesicles	Small round structures in the phospholipid bilayer membrane. They are used to move materials around or out of the cell.
8	endoplasmic reticulum (rough) (RER)	A network of interconnected tubules. Its has a rough appearance under the electron microscope because of the many ribosomes.
9	Ribosomes	Site of protein synthesis.
10	Golgi apparatus	Activates, packages and exports materials, such as enzymes from the cell.
11	endoplasmic reticulum (smooth) (SER)	A network of interconnected tubules. The SER is responsible for the synthesis of lipids, steroids and phospholipids and secretory proteins. It does not have ribosomes.

Table 12.1 Eukaryotic organelles.		
12	Cilia	Hair-like structures used for locomotion in protists (paramecia), ciliated epithelia in the respiratory system and fallopian tubes of women.
13	Flagella	Used for locomotion by some protists such as *Giardia lamblia* or *Euglena gracilis*. It is an organ of locomotion in male animal sperm cells.
14	Proteasomes	Located in the nucleus and cytoplasm of all eukaryotes and archaea (some bacteria). They break down damaged protein molecules.
15	Cell wall (plants)	Gives support and shape to plant cells. Composed of cellulose.
16	Tonoplast (plants)	A vacuole that takes up most of the cell's volume. It stores nutrients and water. It is vital to preserve turgor pressure.
17	Chloroplast (plants)	Chloroplasts contains chlorophyll. Site of photosynthesis.
18	Chromoplast (plants)	Plastids that make and store pigments.
19	Amyloplast (plants)	Synthesis and storage of starch.
20	Plasmodesmata (plants)	Channels that pass from one cell wall to another. Allow for transport of material and communication between cells.
Cell structures not considered organelles.		
Cytoskeleton		Made of a protein-like material that helps the cell retain its shape. It functions in movement and cell division.
1	Microfilaments	The thinnest filaments composing the cytoskeleton. Made of actin molecules.
2	Intermediate filaments	Are structural proteins.
3	Microtubules	Functions in structure and transport.

Cell Division and Reproduction of Animals

There are four types of *animal tissue*: (1) *connective*, (2) *muscle*, (3) *nervous* and (4) *epithelial*. Animal tissue is made up of cells that have a limited **proliferative**[461] lifespan. There are about 200 different kinds of cells in the human body. Each cell lives and becomes two at the end of their respective lifespan. Normal epithelial cells undergo cell division about every 20 or 30 minutes. They do not go into a **quiescent**[462] state. A cell's *nucleus* divides (*mitosis*) followed by division of cell's *cytoplasm* (*cytokinesis*). Neurons may never undergo mitosis and may always remain in a quiescent state known as Gap_0 (G_0).

Functions of the Cell Cycle

Interphase and *mitosis* are two stages cell cycle. See figure 12.3. The cell cycle allows for:

(1) an orderly distribution of chromosomes to each new cell.

(2) a proper distributed of chromosomes to each new cell.

(3) one cell to become two.

The Cell Cycle

Part I *Interphase is the first part of the cell cycle. Interphase consists of three phases.*

(1) *Growth Phase One* (Gap 1): *Cell size increases.*

(2) *Synthesis (S) phase:* **Replication**[463] *of chromosomes* *a)* The normal human chromosome number is **46** (occurring as 23 pairs). *b)* Each chromosome doubles, but stays attached at the **centromere**.[464]

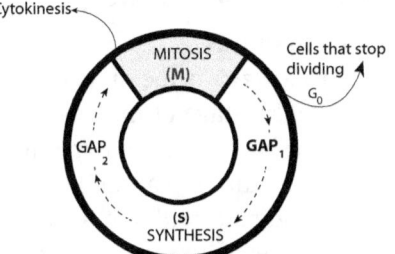

Figure 12.3. Interphase has a Gap_1 where the cell increases in size, S synthesizes its DNA to double the chromosome number and Gap_2 prepares itself for mitosis.

(3) *Growth Phase Two* (Gap 2). *The nucleus is now prepared for mitosis.*

Part II *Mitosis* (**M**)[465] *is the second part of the cell cycle. Mitosis is division of a eukaryotic cell's nucleus.* Mitosis proceeds as a continuous process, but can be broken down into four principle stages. See figure 12.4.

---

461   Proliferate means to make more of or increase in numbers.
462   Quiescent means quiet, silent, resting.
463   To make an exact copy of something.
464   A structure on the chromosome that holds duplicated chromosomes called sister chromatids together.
465   Mitosis is term for division of the cell's nucleus. Karyokinesis is another name.

Principle Stages of Mitosis
Only nuclei of eukaryotic cells undergo mitosis.

I. Prophase: *Chromosomes* become **visible**. The nucleolus is no longer visible. There are now **two identical copies** of the original 46 chromosomes. They doubled during interphase. ***The 46 doubled chromosomes are attached at the centromere.*** *Centrosomes* have doubled and move to opposite sides of the cell. Spindle fibers span from centrosome to centrosome. Centromeres, along with sister chromatids, attach to the spindle fibers.

II. Metaphase: Pairs of **homologous**[466] chromosomes *line up along the middle axis of the cell.* The nuclear membrane is no longer present.

III. Anaphase: Pairs of *homologous chromosomes separate*. The attached centromeres travel to the centrosomes on opposite side of the cell.

IV. Telophase: Pairs of homologous chromosomes are on *opposite sides* of what will be two cells. Two nuclear membranes begin to form.

Cytokinesis
Cytokinesis is the division of the cell's cytoplasm. Cytokinesis follows the last stage of mitosis, but cytokinesis is *not* a stage of mitosis. Cytokinesis in eukaryotic mammalian cells begins with the formation of a contractile ring around the center of the cell resulting in a cleavage furrow and ends with the formation of two identical cells. Cytokinesis in plants begins with the formation of a cell plate between the two daughter cells.

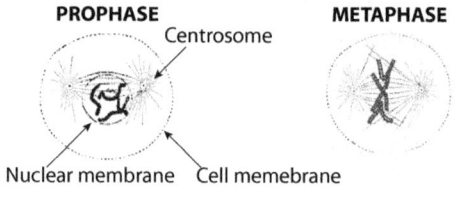

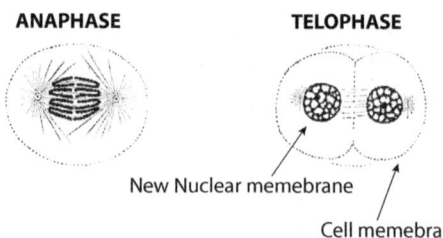

Figure 12.4. Four major stages of mitosis.

466   Pairs of chromosomes that have genes for the same trait opposite each other. One of the pair is from the female and the other form the male.

Meiosis

The normal chromosome number for humans is 46. Forty-six is the *diploid (2n)* number of chromosomes. *Meiosis (*reduction division) *is essential for proper distribution of maternal chromosomes to eggs during* oogenesis.[467] When *gametes* (egg and sperm) form, the chromosome number is halved. *Meiosis* reduces the chromosome number to 23. In the human, egg and sperm are *haploid (n).* Each gamete has 23 chromosomes. Meiosis makes it possible to restore the proper chromosome number to 46 at fertilization. The chromosome number becomes *diploid (2n)* at fertilization. See figure 12.5.

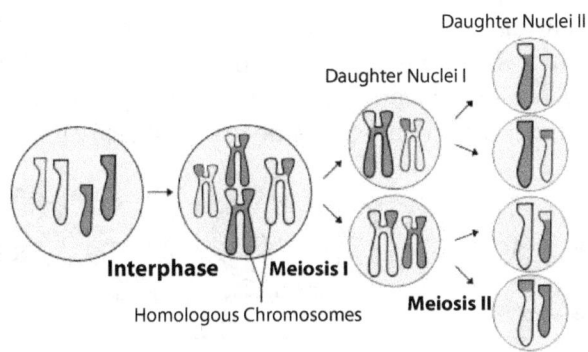

Figure 12.5. Stages of meiosis.

Essential differences between meiosis and mitosis are: (1) meiosis allows for chromosomes to be "mixed up" and recombined. (2) meiosis produces four genetically different gametes. (3) Mitosis produces two identical nuclei containing identical chromosomes.

Fertilization

A sperm cell (spermatozoon) and egg cell (ovum) is *haploid (n).* Each contains 23 chromosomes. At fertilization (see figure 12.6), when a sperm cell penetrates an egg, a *zygote* is formed. A zygote is a fertilized egg that contains the *diploid (2n)* chromosome number (46). The male contributes 23 and the female contributes 23. The zygote will divide many times to develop into a multicellular embryo.

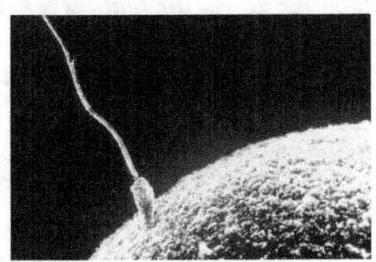

Figure 12.6. Fertilization of an egg by a sperm cell.

467   Egg production.

Development

The **continuum**[468] of human development progresses from fertilization of an egg with a sperm, through *prenatal* development, birth and then growth. The *prenatal period* is from *fertilization to birth* of the fetus.

1. *Embryo* – *Embryology* is the study of growth and development of an *embryo*. Weeks 1-8 is a period of *embryogenesis*. See figure 12.7. The zygote undergoes *cleavage* as it passes down the fallopian tube. Cleavage is a series of cell divisions of the zygote, first into two cells, then 4, 8, 16 and so on until a solid ball of cells called a *morula* results. The morula develops into a fluid-filled ball of cells called a *blastula*. The blastula attaches to the lining of the uterus in most mammals. This attachment initiates the formation of the **placenta**.[469] The blastula is then reorganized into a three-layered structure termed a *gastrula*. The gastrula consists of the three primary **germ layers**,[470] of *ectoderm*, *mesoderm* and *endoderm*.

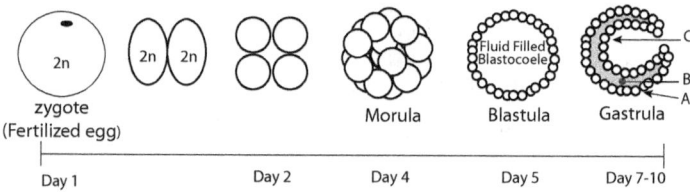

Figure 12.7. Stages of embryogenesis.

In figure 12.7 the ectoderm (A) is the outermost layer, mesoderm is the middle layer (B) and endoderm (C) is the innermost layer. Each layer gives rise to specific tissues and organs. In general, the ectoderm gives rise to the nervous system and skin. The mesoderm gives rise to the circulatory system, connective tissues, muscles and bones. The endoderm gives rise to the digestive system, lungs and the urinary system. See table 12.2. By *the tenth week*, the embryo has developed into a *fetus*.

2. *Fetus* – Human development is defined as *beginning* at the 11th week. By the 11th week the fetus looks like a human being. At *birth* (parturition) and for the first thirty days thereafter, the fetus is called a *neonate* or newborn.

3. *Infant* – From birth up to the twelfth month, the term *infant* or newborn is applied to the fetus.

468   A progression of events where it is not possible tell one event from the next, except by making an arbitrary division.
469   An organ that interfaces with the mother to provide nourishment for the embryo.
470   Animals with radial symmetry produce two germ layers. Those with bilateral symmetry produce three germ layers.

Differentiation

Differentiation means becoming different. *Differentiation* is the development of an *undifferentiated*, *unspecialized* cell, into a differentiated, *specialized* cell. Muscle cells, liver cells (hepatocytes), neurons, adipocytes, epidermal cells and erythrocytes are examples of *terminally differentiated cells*. They cannot go back to the undifferentiated state. There are about 200 terminally differentiated cells in the human body. Differentiation is a *one way* process that takes place at different times as the fertilized egg develops into a multicellular organism.

Adult stem cells (somatic stem cells) are undifferentiated cells. They replace worn out body cells. Adult stem cells can only develop into a *few* cell types. They can only develop into a *few* cell types to aid in a renewal of the organism. Adult stem cells regenerate the cells in the organ they reside. A few types of adult stem cells are hematopoietic, mammary, intestinal, mesenchymal, endothelial, neural, olfactory, neural crest, and testicular.

Embryonic stem cells are *pluripotent*, meaning pluripotent cells can give rise to all three primary germ layers. Below are the three embryonic tissues that give rise to various specialized tissues and structures of the human body.

Table 12.2. Structures derived from ectoderm, mesoderm and endoderm.	
Embryonic Tissue	Derived Tissues
Ectoderm	Nervous system: Spine, peripheral nerves, brain, tooth enamel, epidermis (outer layer of skin), lining of mouth, anus, nostrils, sweat glands, hair and nails.
Mesoderm	Mesenchyme (connective tissue) Mesothelium: Muscles, septa and mesenteries. Lines fluid filled cavities called coeloms: (peritoneal, pleural and pericardial cavities) of mammals.
Endoderm	Epithelial lining of many systems. Alimentary canal except part of the mouth, pharynx and terminal area of rectum). Respiratory tract: (trachea, bronchi and alveoli). Endocrine system: lining of the follicles of the thyroid gland and thymus glands. Auditory system: epithelium of the auditory canal and tympanic cavity. Urinary system: urinary bladder and the urethra.

Chapter 13 Genetics
Mendelian Genetics

Genetics[471] is the science of *inheritance*. Inheritance is the passing on of observable characteristics or *traits* from one generation to the next. Sometime after the domestication of **animals**[472] for meat and milk production, farmers noticed that some animals possessed superior traits, such as better milk production or bigger and more muscular bodies. They learned to breed animals with superior traits to produce better livestock without understanding how inheritance worked.

Gregor Mendel (1822-1884), the "Father of Genetics," began his systematic studies of inheritance by experimenting with animals first and later with **pea plants**.[473] His studies uncovered the mechanisms of inheritance. Born in what is today the Czech Republic, Mendel grew up on a farm and developed an interest in animal breeding. He became a Catholic priest and continued his interest in animal breeding to understand *how* traits were passed from one generation to another. People believed traits seen in offspring were a *blend* of parental traits. Mendel did not.

Mendel's superior encouraged him experiment with organism other than animals. Fortunately, Mendel selected the common garden pea plant, *Pisum sativum*. This model organism enabled him to discover the mechanism of inheritance because *one gene* controls *one trait* in the pea plant. See table 13.1. What are call **genes**[474] today, Mendel called "factors." *He hypothesized that these factors were discrete particles that controlled the expression of traits.*

Mendel published his discoveries in 1866. They were soon forgotten, but his findings were independently rediscovered in 1900 by Hugo de Vries, Carl Correns and Erich von Tschermak. Mendel received credit for his discoveries by each of the three scientists even though Mendel died many years earlier.

Dominant and Recessive Genes

Mendel discovered traits in pea plants to be either *dominant* or *recessive*. He found that one trait, he called the dominant trait, masked the expression of another gene, he called the "recessive" trait. His experiments with pea plants showed that recessive traits did not disappear. They were hidden by the dominant gene and *not expressed* (not visible).

471   Coined by an English embryologist William Bateson (1861-1926). Bateson was greatly influenced by Mendel's "rediscovered" paper on plant hybridization in 1900.
472   Selecting animals (or plants) that have superior traits for breeding to produce better organisms for humans.
473   *Pisum sativum* in the binomial system.
474   Wilhelm Johannsen, a Danish botanist, used "gen." From the Danish and German to denote the fundamental structural and functional unit of inheritance in 1909.

Mendel also discovered that *genes* in pea plants occur in *pairs*. One gene of the pair is an *allele*. An allele is *an opposing or variant gene of a pair of genes*. For example, if one gene controls tallness (T) in a pea plant, then the opposite gene, its allele, controls shortness (t). The plant will have two genes, T and t for height. If one gene controls for the production of yellow pods (Y) and one controls for the production of green pods (y), then Y and y are alleles of each other. Notice an upper-case letter represents a dominant gene, and a lowercase letter represents a recessive gene. There are many example of dominant and recessive traits in the pea plant. See table 13.1 for a partial list of genes in pea plants.

Table 13.1. Selected phenotypes and genotypes in pea plants		
Phenotype	Genotype	Symbols
Plant height	Tall or short	T or t
Seed shape	Round or wrinkled	R or r
Pod color	Yellow or green	Y or y
Flower color	Purple or white	P or p

The observable traits of an organism such as height, seed pod color, flower color and shape of peas are termed *phenotypes*. Phenotypes are traits that *can be seen*. The *genotype* refers to the genes possessed by an organism on located on its chromosomes. Genotypes *cannot be seen*.

Obtaining Pure Tall (TT) and Pure Short (tt) Patents

Since Mendel observed some traits to be dominant and some to be recessive or "hidden" by the **dominant gene**,[475] he knew he must start his experiments with plants of a known genotype. To do this, he crossed tall plants with other tall plants until the plants were "true breeding" tall plants. He did the same thing with short plants.

Mendel began his experiments with "pure" tall plants and "pure" short plants. *If* he was correct, and only two genes controlled tallness, *then* pure tall plants would be **TT**. A pure short phenotype pea plant would be **tt**. Plants that only produce tall plants, are genetically *homozygous tall* and plants that produced only *pure short* plants are genetically *homozygous short*. *Homozygous tall and homozygous short were his starting points*. Homozygous refers to an organism having two identical genes on separate chromosomes as in **TT**. *Heterozygous* refers to having two different genes on separate chromosomes as in **Tt**. He correctly hypothesized that factors (genes) controlling tallness come in pairs and genes controlling shortness also come in pairs.

---

475   One masks another gene called the "recessive" gene.

First Generation (F₁)

First Generation (F$_1$)

Mendel crossed pure tall plants, homozygous tall (TT), with pure short, homozygous short plants (tt), to yield the *first generation* (F$_1$) of pea plants. The F$_1$ offspring were *all tall*. Their *phenotype* was 100% tall, and their genotype was 100% Tt. See figure 13.1. The trait for shortness seemed to *disappear*, but it really did not disappear.

Second Generation (F₂)

Second Generation (F$_2$)

Mendel then crossed F$_1$ generation tall plants (Tt) with other F$_1$ generation tall plants (Tt) to yield the *second generation* (F$_2$). He ob-

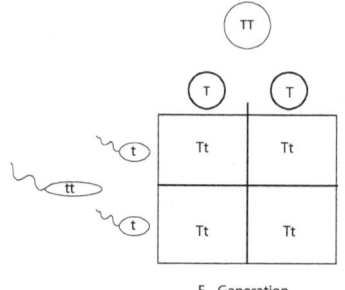

F$_1$ Generation
Phenotype 100% Tall
Genotype 100% Hybrid Tall

Figure 13.1 The first generation (first filial generation (F$_1$)). Pure tall (TT) is crossed with pure short (tt) pea plants.

served tall and short plants. *The trait for shortness reappeared* in the F$_2$ generation. He concluded that the gene for short did not go away, but was masked by the dominant gene for tall in the F$_1$ generation. The *phenotypes* in the F$_2$ generation were: 75% tall and 25% short, a ratio of 3:1. The *genotype* was: 25% homozygous tall, 25% homozygous short and 50% heterozygous tall, a ratio of 1:2:1. *One gene located on one chromosome was responsible for "tallness," and one gene on the opposing chromosome was responsible for "shortness."* See figure 13.2.

Mendel's Laws of Heredity

I. The *Law of Segregation:* Every organism has *two genes for a trait,* and each parent contributes only one copy *randomly* to its offspring.

II. The *Law of Independent Assortment:* Individual genes for individual traits pass from parent to offspring *independently,* if genes are on different chromosomes.

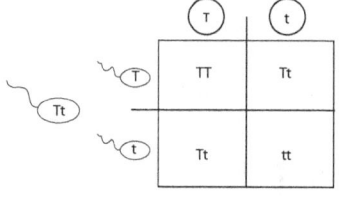

F$_2$ Generation
Phenotype 75% Tall
Genotype 25% Pure Tall
50% Hybrid Tall
25% Pure Short

Figure 13.2. The second generation or the second filial generation (F$_2$). Hybrid tall (Tt) is crossed with hybrid tall (Tt) pea plants.

Patterns of inheritance that follow these laws are referred to as *Mendelian genetics.* Mendel's laws work if alleles are on opposite chromosomes. When chromosomes are *segregated* or distributed to egg and sperm, each parent gives one copy to the offspring.

Human Diseases That Follow the Laws of Mendelian Genetics

Some human diseases are controlled by a single gene. They follow Mendelian patterns of inheritance. The diseases that follow Mendelian patterns of inheritance may be **autosomal**[476] **dominant**, *autosomal recessive*, and *sex linked, either X or Y*.

Autosomal Dominant Genes

If a child inherits *one copy* of an ***autosomal dominant*** gene, the child will get the disease. Autosomal dominant diseases occur where only one copy of a gene is necessary to cause a disease condition. See table 13.2. The dominant gene will have to be introduced by one parent. For example, with Huntington's disease, if one parent carries a copy of the gene called **Huntingtin**,[477] then there is a 50% chance that the child will get one copy of the gene. If a child inherits one copy of the *Huntingtin gene*, then the disease will appear between the ages of 35 to 45 years of age.

Table 13.2. Selected autosomal **dominant** inherited diseases.

Name of disease	Description of disease
Brachydactyly	short fingers
Achondroplasia	Dwarfism
Huntington's disease	Neurodegenerative disease, cognitive decline and psychiatric problems.
Marfan syndrome	Disorder of connective tissue
Neurofibromatosis -1	Unusually tall, long limbs and long, thin fingers.
Familial hypercholesterolemia	High cholesterol levels, very *high* low-density lipoprotein, early cardiovascular disease.

Autosomal Recessive Genes

Of 2000 recessive genes that exist in humans, ***none*** will be expressed *if* the "wild type" (normal gene) is present. If *two* recessive genes are inherited, then the disease trait will appear. See table 13.3.

Both parents ***must donate a recessive gene*** for the disease to appear. *Sickle-cell trait* is an example. In order for a child to develop sickle cell anemia, both parents must contribute one copy of an autosomal recessive gene. If each parent has one copy of the gene, neither is affected. A parent will not know if they have a copy of a recessive gene unless ***genetic counseling*** is sought, or if a child shows symptoms of the disease. *Sickle cell disease* is the most common inherited, life shortening, childhood inherited disorder in the United States. Cystic fibrosis is second.

---

476   Autosomal refers to all human chromosomes other than X or Y chromosomes.
477   Huntingtin refers to the gene.

Table 13.3. Selected **autosomal recessive** inherited disease conditions.	
Name of disease	Description of disease
Sickle cell anemia "HbSS" A single nucleotide mutation on chromosome 11. (A to T)	Abnormally crescent or sickle shaped red blood cells, anemia, abnormal cell shape blocks blood flow causing pain and organ damage.
Cystic fibrosis	Abnormally thick mucus leads to airway obstruction, recurrent and progressive pulmonary infections.
Albinism	Lack of pigmentation in skin, eyes and hair.
Gaucher disease	Bone pain, fractures, cognitive impairment, easy bruising, splenomegaly, hepatomegaly, heart valve problems, lung disease and seizures.
Phenylketonuria (PKU)	Brain damage due to accumulation of phenylalanine.
Familial Mediterranean fever	Recurrent fever without other symptoms, hereditary autoinflammatory disease, abdominal pain, appendicitis, joint inflammation, pleuritis, pericarditis.
Tay-Sachs disease	Lipid storage disorder, build up in tissues and nerve cells in the brain.

Sex-Linked Genes

Sex-linked traits are inherited by an *X or Y chromosome*. Some traits can be linked to the X chromosome, other traits can be linked to the Y chromosome. Traits can be either *dominant or recessive*. See table 13.4 and 13.5. The sex determination system *governs* sexual traits.

Sexually reproducing organisms have a *sex determination system* that causes an organism to be a male or a female. Mammals have one pair of sex chromosomes. One pair may be XX and the other may be XY. *One chromosome of the pair is inherited from each parent.* If an organism inherits two X (**XX**) chromosomes, it will be a female. If an organism inherits an X and a Y chromosome (**XY**), it will be a male. Sex-linked *characteristics* are traits controlled by genes on an X or Y chromosome. The human X chromosome carries many genes. The Y chromosome has fewer than 50 genes.

Several traits, including red-green color blindness, are inherited from genes located on the X chromosome. A male, a Y carrier, has no opposite-acting genes on the Y chromosome that could influence or suppress the action of the genes on the X chromosome. Hemophilia (bleeder's disease) is another trait controlled by genes located on the X chromosome. Thus, the disease is transmitted to males *only* through the female line. Webbed toes and several other minor traits are determined by genes on the Y chromosome and thus are carried and transmitted only by males.

Table 13.4. Common X-linked **dominant** disease conditions.

Name of disease	Description of disease
Vitamin D resistant rickets	Osteomalacia, bone deformity, short stature and bow legs.
Klinefelter's Syndrome (XXY syndrome or XXY trisomy)	Hypogonadism and reduced fertility. Possession of an extra X chromosome. **It is a result of an error in cell division. It is not inherited**.
Fragile X syndrome (Martin – Bell syndrome, or Escalante's syndrome)	Most common form of autism caused by this genetic mutation on the X chromosome.

Table 13.5. Common X-linked **recessive** disease conditions.

Name of disease	Description of disease
Hemophilia A	Bleeding disorder caused by a lack of blood clotting factor VIII.
Hemophilia B	Bleeding disorder due to a deficiency of clotting factor IX.
Duchenne muscular dystrophy	Muscle degeneration, difficulty walking, breathing and death.
Color blindness	Inability to perceive red, green or both colors.
Glucose-6-phosphate dehydrogenase deficiency	Low levels of enzyme important in pentose phosphate pathway for red blood cell metabolism. This is the most common human enzyme defect. It does confer protection to malaria.
Hypoxanthine-guanine phosphoribosyltransferase (HGPRT)	The HGPRT gene is carried by the mother and passed to the son. Present in boys at birth. Severe gout, mental and physical problems result.

Linked Genes

If genes are located on the same chromosome, they may be located close together and therefore, inherited together at meiosis. It is at meiosis that one chromosome is donated by the mother (m) and one by the father (f) to the offspring. See figure 13.3 and table 13.6.

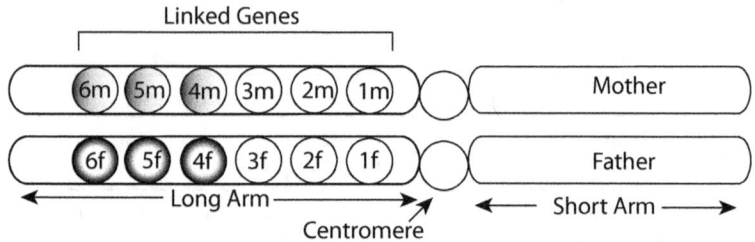

Figure 13.3. Diagram representing maternal and paternal chromosomes donated to an offspring illustrating linked genes 4-6.

Table 13.6. Linked traits.		
A	B	Dominant or Recessive
Blond hair	Blue eyes	both are recessive
Flexibility	Anxiety disorder	A is recessive B is dominant
Large ears	Broad nose	Both are dominant
Red hair	Freckles	A is recessive B is dominant

Codominance

Codominance is equal expression of alleles. See figure 13.4. ***One allele does not mask the other.*** One gene of a pair of genes is an **allele** or ***alternate form*** of the other gene. The A, B, and O blood groups in humans are an example of codominance. Blood group A has antigen A on its red blood surface, Blood group B has antigen B, blood group AB has antigens A and B and blood group O has no antigens on its surface. Although there are multiple alleles available in a population, offspring can only inherit two alleles.

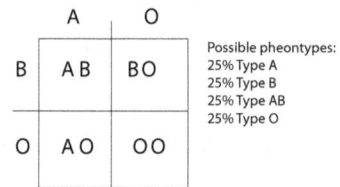

Figure 13.4. Codominance displayed by inheritance of the three alleles A, B, O that determine blood groups in humans.

The possibilities for an offspring can be AA (type A), OO (Type O), AB (Type AB), BO (Type B). In the A, B and O blood groups the alleles A and B are coequal. Gene A and gene B will both produce antigens.

Multiple Alleles

Multiple alleles refers to having three or more genes for a single trait. Offspring cannot inherit three copies for the single trait. Only two are possible. The others are just there. Since there are two parents, offspring can only inherit one copy from one parent and one copy from another parent The codominance of blood groups, figure 14.4, is an example of multiple alleles.

Polygenic[478] Inheritance

Traits such as human skin, hair and eye color, are polygenic. **These are traits controlled by more than one gene.** Most human inheritance is polygenic. The genetics of hair color is not fully understood.

Two genetically controlled pigments, *eumelanin* and *pheomelanin*, play a major role in hair color. Eumelanin can be brown eumelanin or black eumelanin. *Low* concentrations of brown eumelanin produces **blond hair** and *high* amounts of brown eumelanin produces **brown hair**. Black eumelanin produces **black hair**. High concentrations of pheomelanin produces **red hair**. The genes for the two substances do not account for the many different shades of hair color.

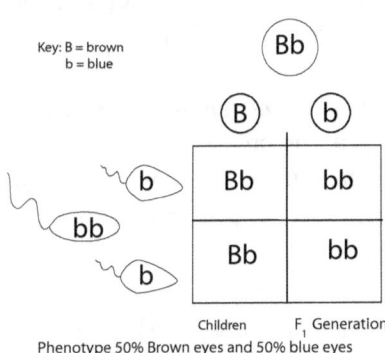

Key: B = brown
b = blue

Children   F₁ Generation
Phenotype 50% Brown eyes and 50% blue eyes
Genotype 50% Hybrid brown and 50% pure blue eyes

Figure 13.5. Inheritance of eye color.

Eye color is another trait that is controlled by many genes. The genes will control the expression of eye color. The expression of genes that control eye color results in different pigments in the iris of the human eye. The most common eye color is brown. **Gray[479]** is the most rare. In general, eye color ranges from dark brown to light blue. Eye color is controlled by green, blue and brown alleles. Green is dominant over blue, and brown is dominant over green and blue. See figure 13.5.

If a brown-eyed mother and a blue-eyed father have four children and half are blue-eyed and half are brown-eyed, then the only way to explain having blue-eyed children is that the mother is a carrier for the gene for blue eyes. This means that she is *not homozygous* for brown eyes, BB, rather *she is heterozygous* for brown eyes, Bb.

---

478   Many genes.
479   Gray eyes may indicate uveitis, an inflammation of the uvea or middle layer of the eye (iris, ciliary body and choroid). This is a vascular and pigmented area. Uveitis presents a risk of blindness.

Intermediate Expression[480]

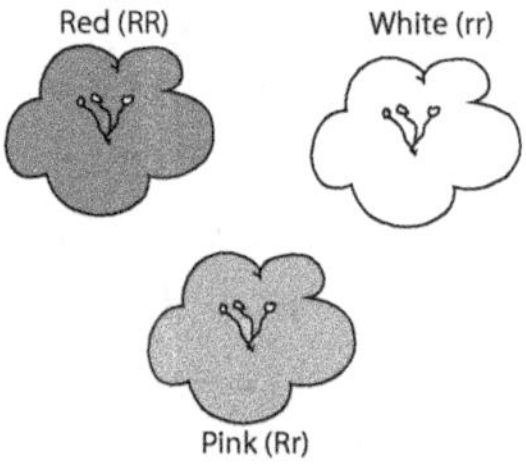

Red (RR)   White (rr)

Pink (Rr)

One allele is **not dominant** over another with some heterozygous traits. An **apparent blending** occurs. The "four-o'clock," is a plant that has red or white flowers. Red (RR) and white (rr) are both *homozygous*. Mating a homozygous red plant with a homozygous white plant produces a plant with pink petals. See figure 13.6. This outcome is an exception to **Mendels Law of Dominance**. Both alleles (R and r) are inherited in the first generation (F_1). Neither is dominant. Pink is heterozygous. Its **genotype** is Rr. The gene that codes for "red" pigment *do not code for enough red pigmen*t to produce red petals. The second generation (F_2) will be 25% RR, 50% Rr and 25% rr.

Figure 13.6. Intermediate expression in the four-o'clock (*Mirabilis jalapa*).

Incomplete Penetrance

Incomplete penetrance results if certain environmental factors are present. Some individuals have a mutation in the breast cancer genes BRCA1 or BRCA2, but *not all* who carry the mutated gene will develop breast cancer. Penetrance refers to the proportion of people that will show clinical symptoms of a disease.

Sex-Limited Genes

*Men and women inherit sex-limited genes, but only the male **or** only the female* expresses the trait. *Sex-limited genes* are limited to a male or to a female. An example is facial hair in males. Females have facial hair also, but it is not generally visible because the gene is turned off.

Sex-Determined Genes

Sex-controlled genes may be expressed in males and females, but inherited differently. Gout is a painful inflammatory disease that can affect many joints of the body, but more commonly the big toe. If the gene is present in a brother and sister, the male is eight times more likely to express the disease. The female sibling will not show symptoms.

Pleiotropy

Pleiotropy is the effect that one gene has on several traits. **Phenylketonuria** (PKU) in humans is one example. The enzyme **phenylalanine hydroxylase** is essential to convert the amino acid **phenylalanine** to tyrosine. A mutation in the gene that codes for phenylalanine hydroxylase causes **phenylalanine** accumulate in the body. This can cause mental retardation and abnormal skin and hair pigmentation. Eating less phenylalanine rich foods can make up for the deficiency of the enzyme.

480   Also called incomplete dominance.

The History of DNA as Genetic Material

Deoxyribonucleic acid (DNA) is a simple substance that makes up genes. DNA is a *polymer* composed of four bases: *adenine (A)*, *thymine(T)*, *guanine(G)* and *cytosine (C)*. Deoxyribonucleic acid (DNA) is one of the four major **biological polymers**[481] that are necessary for living things. The total DNA in a cell's nucleus is the genetic information an organism inherits, its genome.

Johann Miescher (1844-1895), a Swiss physician, identified nucleic acid in the nuclei of white blood cells in 1869. He named it "nuclein."

Robert Feulgen (1884-1955), a German chemist, developed a specific staining method for DNA called the Feulgen reaction. This staining technique proved DNA exists in plant and animal cells. He hypothesized DNA was made of four **nucleotides**.[482]

Phoebus Levene (1869-1940), a Russian American physician, characterized *two forms* of nucleic acids, DNA and RNA. He found that DNA was made up of four bases: adenine, thymine, guanine and cytosine. Adenine and guanine are nitrogenous bases called *purines*. Thymine and cytosine are nitrogenous bases called *pyrimidines*. He named these building blocks of DNA, *nucleotides*. See figure 13.7. Levene also discovered that the DNA polymer was composed of repeating units of sulfur, phosphorus and a base.

Erwin Chargaff (1905-2002), an Ukranian American biochemist, discovered that the amount of *purine bases is equal to the amount of pyrimidine bases in DNA*. Therefore, the *pyrimidine* bases *cytosine* (C) and *thymine* (T) and the *purine* bases, *adenine* (A) and *guanine* (G) *are always equal in quantity*. He also found that an "**A**" *always pairs with* a "**T**" and a "**G**" *always pairs with* a "**C**." Every middle school student now learns the rule that A attaches to T and G attaches to C in the DNA molecule.

Nitrogenous Bases found in DNA.	
Purines	Pyrimdines
Adenine	Thymine
Guanine	Cytosine

Figure 13.7. The four bases that make up DNA.

Frederick Griffith (c. 1879-1941), a British bacteriologist, experimented with *Streptococcus pneumoniae*, the causative agent of bacterial (lobar) pneumonia. Pneumonia was the leading cause of death at this time. He knew of two forms of *S. pneumoniae*, a smooth *(S) form* that is *virulent* that causes pneumonia, and a *rough (R)* form which is **avirulent**.[483]

---

481   Other polymers include starches, fats and proteins.
482   Nucleotides are composed of a base, either adenine, thymine, guanine or cytosine, a sugar (deoxyribonucleic acid) and a phosphate. Four nucleotides make up an oligonucleotide.
483   Avirulent: does not cause a pathology, not disease causing or infective..

The S-form of *S. pneumonia* has a *polysaccharide* capsule that makes its colony look smooth. The *S. pneumonia* strain that does not have a capsule is designated the rough (R) form because its colony does not appear smooth.

Griffith did an ***in vivo***[484] experiment. He killed the S-form and injected it into a live test animal along with live R-form. The animal died. Only live S-forms of *Streptococcus pneumonia* were recovered from the dead animal. ***The R-avirulent forms were transformed into the S-virulent form***. This is the first documented case of ***bacterial transformation***. Griffith never followed up his finding.

Oswald Avery (1877-1955), a leading American researcher on *Streptococcus pneumoniae,* heard of Griffith's work and decided to replicate bacterial transformation *in vitro*. In 1944, Oswald Avery, Maclyn MacCarty and Colin McLeod published their findings demonstrating bacterial transformation *in vitro* and concluded that DNA was the substance that caused the transformation of avirulent R-forms of *S. pneumonia* into virulent S-forms by using DNA from the S-form. By the early 1950s, most scientists believed that DNA is the chemical substance that makes up genes.

In 1952, ***Alfred Hershey*** (1908-1997) and ***Martha Chase*** (1927-2003) carried out a series of experiments that *confirmed* DNA is genetic material, not protein, as a few still hypothesized. Hershey and Chase's clearly confirmed Avery, MacCarty and McLeod's 1944 paper demonstrating DNA as the chemical substances of genes.

Once DNA was proven to be the chemical substance that made up genes, an American, ***James Watson*** (1928-) and a British scientist, ***Francis Crick*** (1916-2004), along with ***Rosalind Franklin*** (1920 -1958), a British biophysicist, set out to discover the structure of DNA. In 1953, Watson and Crick used Franklin's X-ray diffraction data to figure out DNA was a double helix. *Watson and Crick published the model of the double-helical structure of DNA.* See figure 13.8.

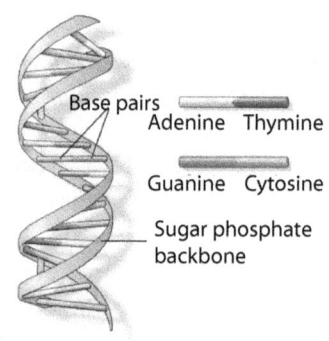

Figure 13.8. The DNA double helix and base pairing.

---

484   From the Latin, meaning in a living thing.

DNA Contains the Code of Life

DNA *is a polymer that contains the instructions for all the inheritable physical traits of an organism.* The instructions are in the form of **genes**. Genes are sequences of the bases A, T, G and C that occur in pairs. The base pairs are held together by a "backbone" of the sugar, **deoxyribose**, and phosphate molecules to form the **double helix**. See figure 13.8.

Base pairs are the "code of life." **A group of three bases is a code for an amino acid.** Each amino acid is directed *to a precise location in the* **primary structure**[485] *of a protein.*

The smallest genome is that of a bacterium and includes about 160,000 base pairs. The **human genome** has about three billion base pairs that code for about 25,000 genes.

Mutations

A **mutation** is a change in a gene. For example, each group of three bases such as TAG, codes for an amino acid. See figure 13.9. If the first A becomes a T, then it becomes GAG instead of TAG. This will cause an amino acid in a protein to be missing or misshaped. The protein may not work. If the protein is a critical enzyme, such as **phenylalanine hydroxylase,** then a critical biochemical step will *not* occur and a disease condition results called **phenylketonuria**. Many things cause mutations. Ultraviolet light, certain chemical agents, mistakes in DNA replication are all possible causes of mutations.

CCG-CTT-**TAG**-GCG-TAA   Normal Sequence

CCG-CTT-**GAG**-GCG-TAA   Mutation in Sequence

Figure 13.9. Mutated DNA sequence

Living things inherit biochemical traits as well as phenotypic traits (appearance). For example, **primary lactase deficiency,**[486] is the most common *genetic cause* of lactose intolerance in adults. Primary lactase deficiency is the inability to digest foods containing the sugar *lactose*. If the code for the *lactose persistence gene* is not inherited, then there is an inability to make the enzyme *lactase*, usually after the age of two. If the code for the lactose persistence gene is inherited, then lactase is made in response to consuming lactose containing foods such as milk.

---

485   Primary structure is like a chain of amino acids that resemble beads on a string.
486   In addition, there is a secondary deficiency that is caused by some injury to the small intestine by Crohn's disease, chemotherapy, intestinal parasites or some other physical cause.

DNA Triplet Code

A group of three bases code for one amino acid. For example, UUU codes for phenylalanine (Phe) and GGU codes for glycine (Gly). Twenty amino acids makes about 100,000 protein molecules for the human body. Proteins are enzymes, receptors, antibodies, hormones like insulin and structural proteins of muscle. Of the 20 amino acids, nine are ***essential amino acids*** (indispensable amino acids). They have to be provided in the diet. Humans cannot synthesize them. The nine essential amino acids are, phenylalanine, valine, threonine, tryptophan, methionine, leucine, isoleucine, lysine, and histidine.

Phe
Val
Gly   Asn
Ile   Gln
Val   His
Glu....Leu
Cys------- Cys
Cys   Gly
Thr   Asp
Ser   His
Ile   Leu
Cys   Val
Ser   Glu
Leu   Ala
Tyr   Leu
Gln   Tyr
Leu   Leu
Glu   Val
Asn   Cys
Tyr   Gly
Cys   Glu
Asn   Arg
Gly
Glu
Arg
Gly
Phe

Figure 13.10. Linear structure for insulin molecule. Dashed lines represent S-S bonds.

Protein Synthesis

Transcription of RNA

Transcription is the process of copying a gene. If the body needs a substance, cells that make the substance receive a ***signal*** to start its production. When the enzyme *lactase* is needed, cells in the intestinal lining are stimulated to produce ***lactase.*** The DNA in their nuclei unwinds to expose the ***lactase gene***. The gene is copied or ***transcribed*** onto a small messenger molecule called ***mRNA***. The ***mRNA*** contains the ***codon***. The codon is RNA copy of the DNA code. The codon for lactase is *translated* on ***ribosomes*** of intestinal cells where lactase is synthesized. Only these cells will make lactase.

Translation of RNA

Translation is the process of making a protein molecule on a ribosome. When a messenger molecule (mRNA) attach to a ***ribosome***, individual amino acids are brought to the ribosome by ***transfer RNA*** (tRNA) ***molecules***. The amino acids arrive at the proper location on the mRNA molecule because tRNA molecules have an ***anticodon*** that pairs with the codon on the mRNA. See figure 13.12. Amino acids are connected to each other like beads on a string. This is a primary protein structure. The process can be summarized as:

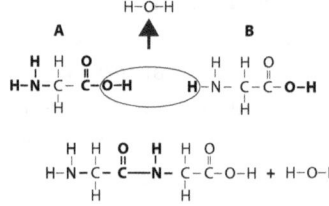

Figure 13.11. A peptide bond.

DNA → RNA → Protein. The DNA (*code*) directs the synthesis of RNA (*messenger*). The RNA messenger directs the synthesis of lactase molecules (***protein***). Amino acids are held together by peptide bonds produced by ***dehydration synthesis.*** See figure 13.11.

In figure 13.11 the hydroxyl group of amino acid "A" is joined to the hydrogen atom of amino acid "B" by the removal of a molecule of water. A covalent bond forms between the carbon of amino acid "A" and the nitrogen of amino acid "B."

Activation of an Enzyme

After an enzyme has been made, it must be folded correctly to work. Correct folding produces a three dimensional *functional* shape. The correct three dimensional shape determines if the enzyme will fit on a *substrate* according to the lock and key mechanism of enzyme action. This process is called *conformation* and is completed in the *Golgi body*.

Primary structures are like beads on a string. The primary structure is an unfolded chain if amino acids produced by translation on a ribosome. Primary protein structures are complete molecules, but not active. A *secondary folding* of the primary structure is the alpha helix (∂) and the beta (ß) sheet (ß-pleated sheet). The *tertiary structure* is a folding of the ∂ and ß secondary structure to form a tertiary structure. The last folding of the molecule is the *quaternary structure*. The quaternary structure is a collection of several folded subunits. The final shape gives activity to the molecule. The last folding of the protein is referred to as the *native state*. Hemoglobin is an example of quaternary structure.

The Golgi body sorts, folds, packages and exports active proteins. If enzymes are exported from the cell, they are termed *exoenzymes*. Examples of exoenzymes are *carbohydrases, proteinases, lipases, DNAases* and *RNAases*. All of the above are digestive enzymes exported from digestive organs to the lumen or space within the small intestine and the mouth where digestion of food molecules take place. Enzymes activated in the Golgi body and *retained* within the cell are *endoenzymes*. Endoenzymes are used within the cell for metabolic processes such as cell respiration and other metabolic processes required for use within the cell.

Biotechnology

Biotechnology is the use of living things *to make useful products*. Humans have been using yeast to make beer, wine and bread since ancient times. Fungi and bacteria help in the manufacturing of cheeses. These biotechnological processes have been used for thousands of years, but not well understood until relatively recently.

Genetic Engineering

Genetic engineering is the process of *modifying the genome* of an organism that allows the organism to produce a product that it normally does not produce. Genes from one species may be removed (knocked

209

out) through use of an enzyme. The knocked out gene is then inserted into the genome of another organism. Genetically engineered bacteria were first produced in 1972.

Genetic engineering began by altering the genome of *E. Coli.* The gene responsible for producing human insulin was inserted into the *E. coli* genome. In doing so, the genome of the normal *E. coli* genome now contains a gene that it did not contain before. Every time an *E. coli* cell reproduces, it also reproduces the insulin producing gene located in its genome. Trillions of insulin producing bacterial cells now make identical human insulin. The bacteria have become human insulin factories. Human insulin has been synthesized commercially using *E. Coli* since 1982.

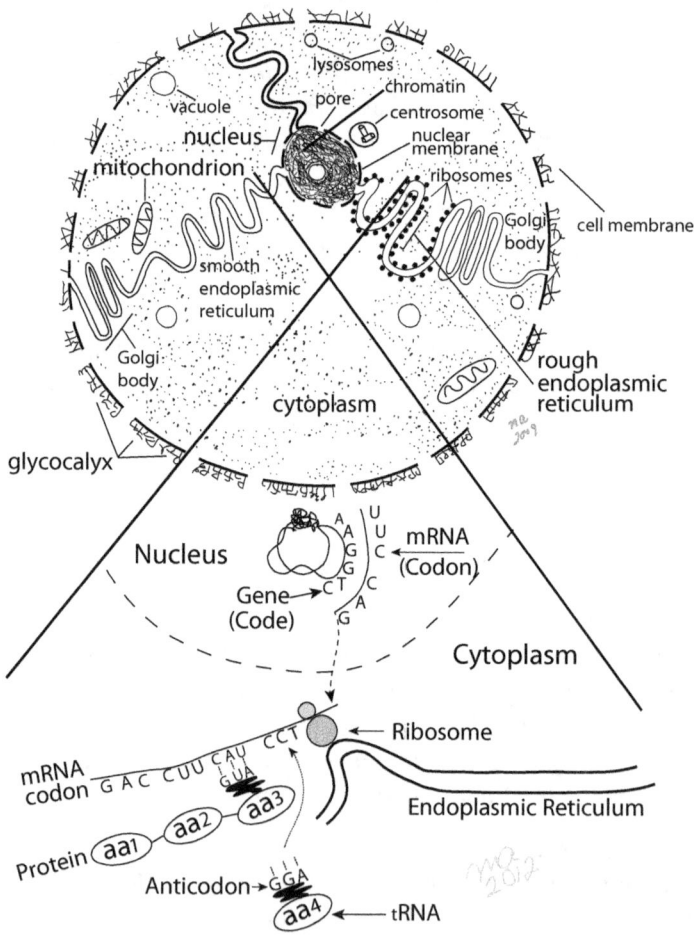

Figure 13.12. Diagram showing the major steps in protein synthesis.

Chapter 14 Interrelationships Among Living Things
Ecology

Ecology studies interactions among and between living things and their physical environment. Ecological interactions take place in a *defined area*. The defined area in which a species or group of species lives and interact is a *habitat*. For example, a single lion species, *Panthera leo*, inhabits an area of about 100 square miles. The 100 square miles is its habitat. The lion interacts with other lions, other animal species and the lion's physical surroundings. *Ecosystems* are environments that are made up of *living things* and *nonliving things*. A pond, wetland or the forest floor are examples of ecosystems. A pond has many different organisms living in, below and at the edges of the pond. Living things may be algae, protozoa, frogs and snails. Water, soil, rocks and dissolved gases such as oxygen are some of the nonliving components. Living things that interact with each other within an ecosystem are termed a *community*. All ecosystems taken together is termed the *biosphere*. The biosphere is the total of all living things on Earth.

Producers and Consumers

Autotrophs are *producer* organisms. Producer organisms are multicellular green plants, single celled algae and cyanobacteria. Producers use the photochemical process of *photosynthesis* to make *organic food substances*. *Autotrophs use carbon dioxide* from the air, *water from the soil and the energy of sunlight to produce and release oxygen into the atmosphere.* **The organic catalyst chlorophyll traps the energy of sunlight and transfers it to the chemical bonds of organic food molecules.**

Autotrophs also use *aerobic cell respiration* to metabolize some of the food they make for their energy needs. **Heterotrophs** are **consumers**. Heterotrophs *ingest* food molecules in many different ways and metabolize them for energy. See figure 14.1.

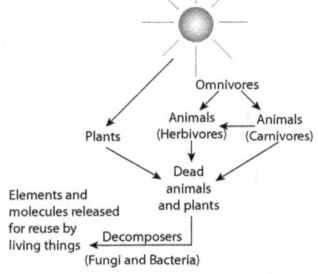

Food Webs

A *food web* is a diagrammatic way of showing how different species of living things are linked together by their feeding

Figure 14.1. A simplified food web.

habits. *Autotrophs* and *heterotrophs* **participate** in a food web. Autotrophs are self-feeders that are **primary producers**. Heterotrophs feed on autotrophs for their life sustaining energy needs. Some heterotrophs feed on autotrophs directly by eating plants or indirectly by eating animals that feed on plants. See figure 14.1.

Feeding Methods

Four major *modes of nutrition* are *autotrophic*, *heterotrophic*, *saprophytic* and *parasitic*. **Troph-**[487] refers to feeding in ancient Greek. Algae are one celled *autotrophs* that live in the oceans and fresh waters of the Earth. Multicellular green plants, photosynthetic bacteria and algae are autotrophs. Algae serve as food for filter feeding organisms like whales. Algae in the oceans help keep the Earth's oxygen levels in balance. *Saprobes* are primarily fungi and many bacteria that live on dead or the products of dead organisms. Saprobes send digestive enzymes into the surrounding environment and digest a food source extracellularly. Food webs show how various organisms depend on each other and how each obtains their food. See figure 14.1.

Different Ways Organisms Obtain Food

A community consists primarily of *producers*, *consumers* and *decomposers*. *Consumers* are *heterotrophs that* feed on other living things. Consumers can be either *herbivores*, *carnivores* or *omnivores*. *Herbivores* eat plants, c*arnivores* eat animals and *omnivores* eat plants and animals. *Decomposers,* bacteria and fungi, also called *saprophytes*, live on dead organisms or waste products of living things. *Decomposers break down and* recycle elements and compounds in living thing that have died. If decomposers did not exist, the elements necessary for life would be locked up in the cells and tissues of dead animals and plants forever.

Relationships Among Organisms in an Ecosystem

Different types of relationships among living things exist in an ecosystem. Organisms live together in *symbiotic* relationships. *Symbiosis* is a general term that includes three kinds of interactions displayed by organisms: *mutualism*, *commensalism* and *parasitism*.

Mutualism is a symbiotic relationship between two species where each organism benefits from the relationship. **Lichens** are composed of a

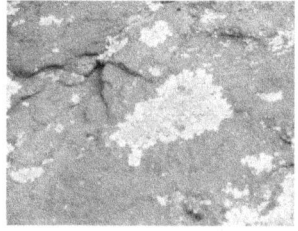

fungi and *algae*. Fungi provide shelter for the algae, and the algae provide food for the fungi. Lichens can often be seen growing on the north side of trees and rocks as a gray-green patch. See figure 14.2.

Termites feed on dead organic matter such as wood and leaves. The termite can not digest cellulose, but termites shelter certain protozoa in its gut that can, thus providing a place for protozoa to live. The protozoa, in turn, digest

Figure 14.2 Lichen growing on the north side of a rock.

the cellulose for the termite.

487   From the Greek: to feed.

212

Commensalism *is a symbiotic relationship whereby one organism is* *helped*, *and the other organism is* *unaffected*. A barnacle, in the larval stage, will settle on a whale and become sessile. The barnacle feeds on material that flows by the whale as barnacle remains attached to the whales back. The barnacle is helped. The whale is unaffected.

Parasitism *is a symbiotic relationship whereby one organism is* *helped*, *and the other is* *hurt* or *killed*. A parasite-host relationship exists when one organism lives at the expense of the other. Parasites such as tapeworms, protozoa such as the plasmodia species that cause malaria, viruses and some bacterial parasites cause harm or may kill the host.

Cycling Nature's Elements

Biological systems are **open systems**.[488] Open systems are those in which matter and energy enter and leave the system. Matter and energy have to be recycled. Many substances necessary for life have to be reused in nature. Elements move from the biological world into the nonliving physical world. Elements or compounds that accumulate in one place is called a *reservoir*. For example, huge amounts of carbon in is stored in the Earth's oceans, crust and biosphere. A less than 1% is found in the atmosphere in the form of CO_2. Listed below are several of the most significant cycles in nature.

The Oxygen Cycle

The *oxygen cycle* describes how molecular oxygen moves from one reservoir to another. The main reservoirs are the atmosphere, the **biosphere**[489] and the **lithosphere**[490] or crust of the Earth. Occasionally, a local failure in the hydrosphere such as a lake, river or bay and water becomes depleted of dissolved oxygen. Living things that inhabit those waters die. This phenomenon is called *hypoxia* or low oxygen. See figure 14.3.

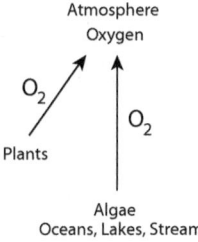

Figure 14.3. The oxygen cycle.

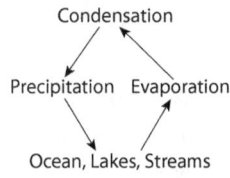

Figure 14.4. The water cycle.

The Water Cycle

The **water cycle** *describes how water moves* *into and out of its major reservoirs, oceans, lakes,* *streams and the atmosphere*. The physical changes that allow for this interchange are *precipitation,* *evaporation* and *condensation*. Water evaporates from the oceans and returns it to the atmosphere.

488   Earth is an open system. Matter and energy flows in and out of the system.
489   All living things on Earth.
490   The rocks of the Earth.

Salt and other solids stay behind in the oceans. The content of salt in the ocean is about 3.5%. The ocean's salt content comes from salts in the soil. When rain falls on the soil, water runs across the land, dissolves salts and brings the salts to the oceans. Humans brought their saltwater environment with them, the blood. Homeostatic mechanisms keep human salt content at about 0.9%. See figure 14.4.

The Nitrogen Cycle

The **nitrogen cycle** *describes how N_2 passes from one form to another by undergoing a chemical change.* Seventy-nine percent of the atmosphere is molecular nitrogen (N_2). Bacteria "**fix**" atmospheric nitrogen into forms plants can use. Nitrogen fixation refers to an organism's ability to naturally **convert** atmospheric nitrogen (N_2) into ammonium ions (NH_4^+), nitrites (NO_2^-) and nitrates (NO_3). Plants have the ability to take up and utilize ammonium, nitrites and nitrates.

Bacteria of decay break down dead organisms and release nitrates (NO_3) into the soil. See figure 14.5.

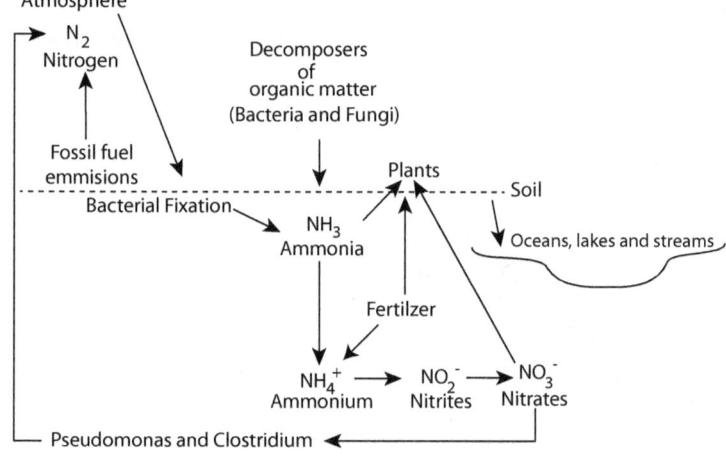

Figure 14.5. The nitrogen cycle.

The Carbon Cycle

The **carbon cycle** *is the movement of carbon between different reservoirs.* Important reservoirs for carbon are the atmosphere, the biosphere, the land, the oceans and fossil fuels. The release of CO_2 and taking in of CO_2 by different reservoirs has been in balance for thousands of years – **until recently**. See figure 14.8.

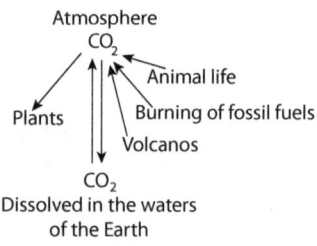

Figure 14.6. The carbon cycle.

214

Autotrophs, such as multicellular green plants, algae and photosynthetic bacteria produce food they need by photosynthesis. *Autotrophs use CO_2 from the air H_2O from the soil and the energy of sunlight to make food. They release O_2 into the atmosphere. CO_2 is removed from the atmosphere, and the Earth's O_2 reservoir is replenished.* See figure 14.7.

$$6\,CO_2 + 6\,H_2O \longrightarrow C_6H_{12}O_6 + 6\,O_2$$

(6) Carbon Dioxide + (6) Water $\xrightarrow[\text{CHLOROPHYLL}]{\text{ENERGY OF SUNLIGHT}}$ (1) Glucose + (6) Oxygen

Figure 14.7. The overall chemical equation for photosynthesis.

The use of CO_2 by autotrophs is not the only way CO_2 is removed from the atmosphere. Most of the Earth's carbon is stored in **reservoirs**. Coal, oil, and organic matter such as trees and other vegetation lock up excess carbon. See figure 14.6.

Huge amounts of ancient carbon previously locked up for millions of years has been released into the atmosphere since the discovery of fossil fuels such as coal, natural gas and oil. This extra carbon released into the atmosphere has resulted in a warming the Earth, termed the "**greenhouse effect.**"

Impact of Humans on an Ecosystem

The amount of carbon dioxide in the atmosphere has increased dramatically since the beginning of the industrial revolution (approximately from 1750 to 1850). Coal was used in increasing amounts as a fuel for running factories, heating homes and transportation. In the 20th century, the burning of more coal along with oil and natural gas added to carbon dioxide production even more. Because of the increased amount of burning fossil fuels, the levels of carbon dioxide, carbon monoxide and other "greenhouse" gases has increased dramatically in the atmosphere. See figure 14.8.

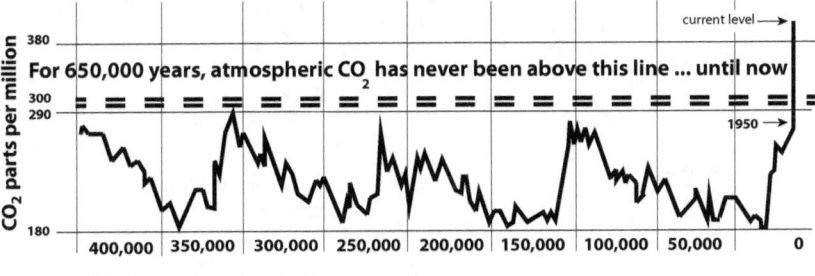

Figure 14.8 This graph, based on the comparison of atmospheric samples contained in ice cores and more recent direct measurements, provides evidence that atmospheric carbon CO_2 has increased since the industrial revolution.

The Greenhouse Effect

A rise in the Earth's average temperature has been accompanied by an unprecedented increase in the Earth's average carbon dioxide levels since the 1950s. See figure 14.8. "Greenhouse" gases are carbon dioxide, water vapor, methane, nitrous oxide and ozone. *Greenhouse gases capture and re-radiate the sun's thermal energy back to Earth much like a greenhouse traps the sun's energy within its walls.* The result has been an increase in the Earth's average temperature over the last 200 years. This phenomenon came to be known as the *"Greenhouse Effect."*

A gradual warming trend has been taking place in the world. This trend is clearly illustrated in figure 14.9 for the United States from 1895 to 2009. Scientific evidence indicates that the Earth as a whole is getting warmer. Population growth has led to an increased consumption of resources with the industrialized nations consuming at the expense of less-developed nations. The depletion of vital resources such as fresh drinking water has already occurred in many areas of the world. Severe droughts have resulted in famines causing the death of millions. There have been five mass extinctions on Earth. Humans may cause the next mass extinction unless they change how they live on planet Earth.

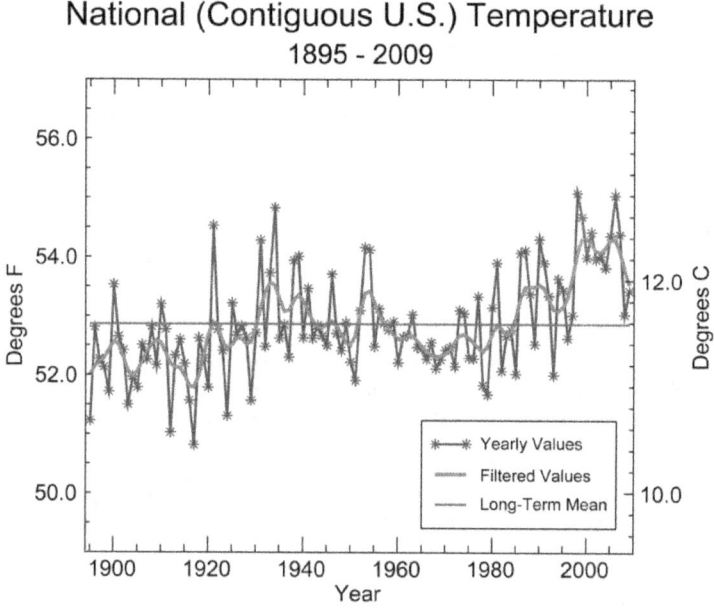

Figure 14.9. United States temperature ranges from 1895 through 2009.

Chapter 15 Introduction to Human Anatomy and Physiology
General Terminology

The sciences of **anatomy**[491] and **physiology**[492] are two divisions of the very broad study of biology. Anatomy studies the *structure* of living things. Physiology studies the *functions* of living things. Anatomy is divided into, *gross anatomy*, *microscopic anatomy and comparative anatomy*. Gross anatomy studies features that can be observed with the unaided eye, such as surface of the body and internal organs. Microscopic anatomy studies structures to small to be seen with the unaided eye. Comparative anatomy studies the similarities and differences between structures of different species.

Dissection

Dissection is an ancient practice. Human curiosity or accident allowed people to see inside dead animals or humans. **Vivisection**[493] was practiced at various periods in human history. It was rare in the past and is even more rare today. If it is performed, it is under heavy sedation.

Historical Figures in the Sciences of Anatomy and Physiology

Galen, the Roman physician and anatomist, was the authority on anatomical knowledge for over 1,000 years. His writings on the human body had significant errors that were corrected over time. A few of these errors were: the interventricular septum of the heart was porous, the mandible was two bones and the sternum was made of **seven parts**.[494]

Mondino de'Luzzi and *Leonardo da Vinci* paved the way for the Dutch anatomist *Andreas Vesalius* (1514-1564), the "Father of Modern Human Anatomy." Vesalius did dissections as his students observed. He believed *only direct observation could yield reliable information about the body*. Vesalius corrected many of Galen's errors in his great work *De Humani Corporis Fabrica* (*On the Fabric of the Human Body*).

Physiology, like anatomy, has its roots in ancient times. Hippocrates, Aristotle, and Galen tried to explain the *function* of living things. In the past, most people believed in the theory of the four humors: black bile, yellow bile, phlegm, and blood. If the humors were in balance, the body was healthy, if the humors were out of balance disease resulted. Bloodletting, to cure disease, was based on this theory of humors. A vestige of this theory is evident in the concept of *humoral* immune **response**.[495]

---

491   From the Greek, to cut open.
492   From the Greek, *physis* nature and *logia* the study of.
493   The dissection of live animals.
494   It is true for primates. Galen erroniously believed it was true for humans also.
495   Humoral immunity is regulated and mediated by antibodies in the plasma.

The *modern* concept physiology began with ***Claude Bernard***, a Roman Catholic priest (1813-1878) and French physician. He introduced the modern concept of ***homeostasis***, *the regulation of body functions such as temperature, glucose levels and pH*. When homeostatic mechanisms fail, disease conditions occur and manifest as ***symptoms*** and ***signs***.

Symptoms and Signs

Symptoms are subjective. Symptoms are ***not*** easily ***measured***. For example, a patient may complain of having pain, a headache, being nauseous or feel fatigued. Symptoms like these are *not measurable*.

Signs, on the other hand, are most often *objective* and *measurable*. Blood pressure, blood glucose levels or temperature can be measured. Normal blood pressure is 120 over 80 and blood glucose should be about 100 mg/dL. Signs may also be a *characteristic* observed by a physician, such as a skin rash that is associated with a specific condition. The most commonly observed sign is **dermatitis**.[496] Knowing the kind of dermatitis is crucial to prescribing the proper topical steroid. Psoriasis is a less frequently encountered sign. The hallmark of psoriasis is red patches with silvery flaking skin. Signs like these are not quantifiable.

Anatomical Position

Standing erect and *facing forward* with the ***palms facing forward*** is the anatomical position. This position is universally used to reference structures on or in the human body. See figure 15.1. ***Planes*** superimposed on the anatomical position are helpful to reference sections of the body. *Planes are imaginary lines connecting two points.* Commonly used planes are ***coronal*** (frontal), ***mid-sagittal*** (medial), ***sagittal, parasagittal*** and ***transverse*** (horizontal, axial or transaxial). See figure 15.2.

Directional Terms

Navigation around the body requires knowledge of the ***anatomical position***, ***planes*** and ***directional terms***. ***External*** refers to the *outside* and ***internal*** refers to the *inside* of the body. ***Deep*** refers to *away from the surface* and ***superficial*** means *close to the surface*. ***Anterior*** is *toward the front*. ***Posterior*** refers *to the back*. ***Superior*** is *above* and ***inferior*** is *below*. Superior is closer to the head and inferior is closer to the feet. ***Proximal*** is *close to the point of attachment of a muscle (origin)* and ***distal*** is *further away from the point of attachment*. ***Medial*** indicates *toward the midline* of the body and ***lateral*** *away from the midline*. ***Intermediate*** means *between two structures*. ***Supine*** means *lying on the back* with palms face up. ***Prone*** is lying on the belly (ventral) portion of the body with palms facing down.

---

496   Inflammation of the skin caused by **contact**, **atopic** (hyperallergic), **stasis** (flow of a body fluid blocked), or **seborrheic** (excess secretion of sebum) .

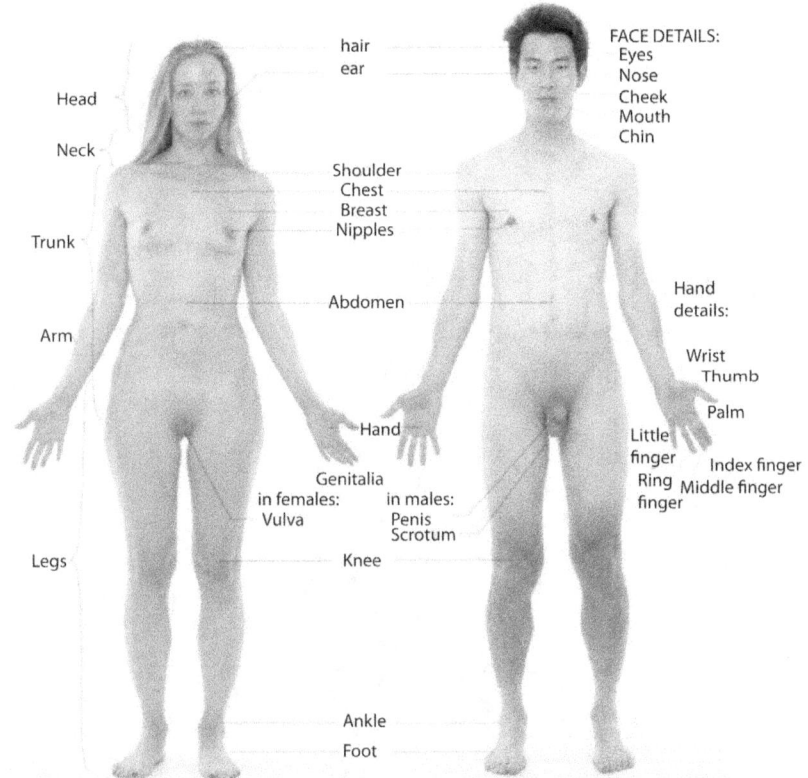

Figure 15.1. The anatomical position and surface gross anatomy.

Anatomical Planes of the Human Body

Planes are *imaginary lines connecting two points.* See figure 15.2. Planes of the human body are:

1. **Coronal** (frontal) plane: *divides the body into a ventral (front) and dorsal (back) sections.*

2. **Mid-sagittal** (medial) plan is *a vertical plane passing through the midline of the body divides the body into equal right and left sides.*

3. **Sagittal plane** is *a vertical plane parallel to the median plane dividing the body or organs into a right and left portion.*

4. **Parasagittal plane** is *a plane that is parallel to a sagittal plane but does not include the median plane. It divides the body into unequal right and left sections.*

5. **Transverse** (horizontal, axial or transaxial) plane *divides the body into a superior and inferior section.*

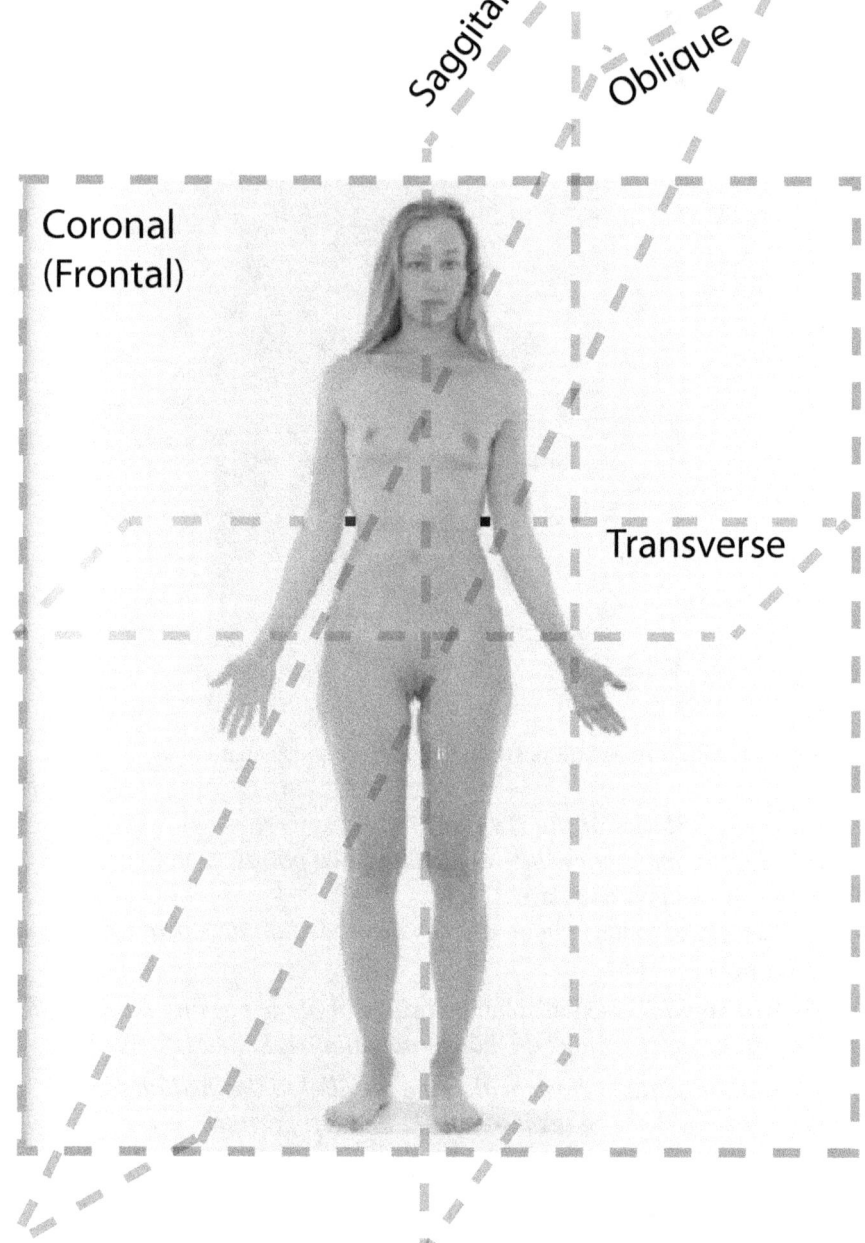

Figure 15.2. Anatomical planes of the human body.

Abdominopelvic Quadrants of the Human Body

In addition to planes of the body, it is helpful to physicians and nurses to divide the body into *four quadrants*. The quadrants are right upper (RU), left upper (LU), right lower (RL) and left lower (LL). Located in the *right upper quadrant* is most of the liver, the gallbladder, the right kidney. In the *right lower quadrant*, appendix, caecum, right ovary (female), right Fallopian tube (female) and right ureter. In the *left upper quadrant,* the stomach, spleen, pancreas and left kidney are found. Some major organs found in the *left lower quadrant* are: the sigmoid colon, left ovary (female), left Fallopian tube (female) and left ureter. Clinicians are more likely to use the four quadrants in the abdominopelvic cavity rather than the nine regions. The *nine regions* are more frequently used by anatomists. See figure 15.3 and table 15.1.

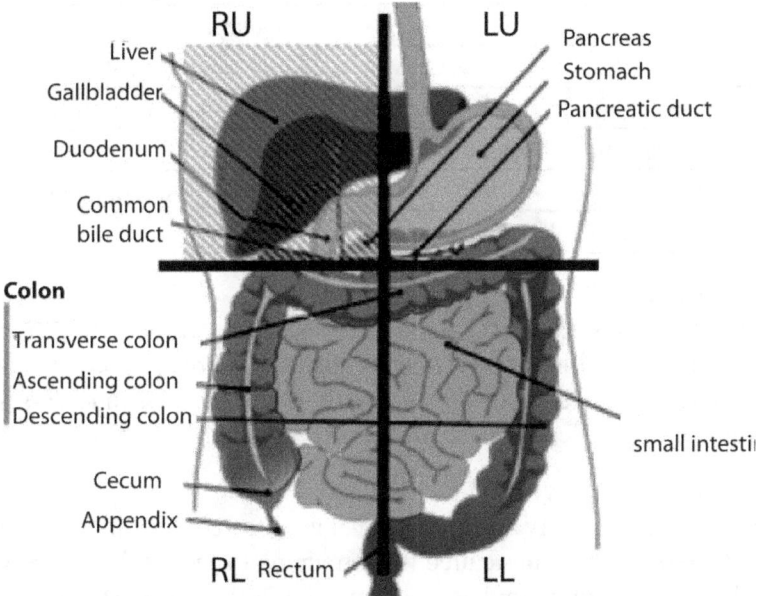

Figure 15.3. Quadrants of the body and the organs located in each quadrant. RU, LU, RL and LL are right upper, left upper, right lower and left lower respectively.

Abdominopelvic Regions of the Human Body

The *nine regions* used to locate structures within the abdomino-pelvic cavity. The nine regions are (from the upper right and ending in the lower left):

1. Right hypochondriac 2. Epigastric 3. Left hypochondriac.

4. Right lumbar 5. Umbilical 6. Left lumbar.

7. Right inguinal 8. Hypogastric (pubic) 9. Left inguinal.

Table 15.1. Structures located in each of the four quadrants.	
Right upper quadrant	**Left upper quadrant**
Most of the liver	Stomach
Gallbladder	Spleen
Right kidney with adrenal gland	Left kidney with adrenal gland
Duodenum	Most of pancreas
Part of the transverse colon	Part of the transverse colon
Part of the ascending colon	Part of descending colon
Part of the pancreas	
Right lower quadrant	**Left lower quadrant**
Appendix	Sigmoid colon
Caecum	Left ovary (female)
Right ovary (female)	Left Fallopian tube (female)
Right Fallopian tube (female)	Left ureter
Right ureter	Most of descending colon
Most of ascending colon	

Body Cavities

Body cavities are spaces within the body that **contain** and **protect** organs. Body cavities are *not open to the exterior of the body*. See figure 15.4. These cavities are located ventrally (see table 15.2) and dorsally in the human body. See table 15.3.

The Dorsal Body Cavity

The cranial cavity and spinal canal are located dorsally. The **cranial**[497] **cavity**, sometimes called the intracranial space is the space within the **skull**.[498] The cranium has a volume of 1,200-1,700 cm^3. The cavity contains and protects the brain, eyes and ears. The skull is lined with membranes called the **meninges**.[499] The space between the brain and the meninges is filled with cerebrospinal fluid (CSF). CSF is made in the ventricles of the brain. CSF helps provide a cushion against shock to the brain. CSF has immunological functions as well. Intracranial pressure (ICP) is kept constant by its production and absorption.

---

497   Pertaining to the space inside the skull.
498   The skull is made up of eight fused bones; they are: frontal, occipital, sphenoid, ethmoid, two parietal and two temporal bones.
499   From the Greek, "membrane." Meninges surround nerves of the central nervous system and consist of the *dura mater*, the *arachnoid mater* and the *pia mater*.

A *vertebral column* (the backbone or the spine) is found in vertebrates. The vertebral column is made up of individual bones called vertebrae. Vertebrae inclose a vertebral canal. The **vertebral canal**[500] contains the spinal cord. In reality, the cranial cavity and the spinal canal are one continuous fluid-filled space, enveloping the brain and spinal cord.

The **Ventral**[501]*Body Cavities*

The thoracic cavity contains two **pleural cavities**[502] and a *pericardial* cavity. Each pleural cavity contains one lung surrounded by a double layered pleural membrane that contains pleural fluid.

The heart is located in the *pericardial cavity*. The *pericardium* is a double layered membrane that forms a fluid filled sac that surrounds and protects the heart. The **mediastinum**[503] is a region located between the pleural cavities.

The Abdominopelvic Cavity

Inferior to the **thoracic**[504] cavity is the *abdominopelvic cavity*. The abdominopelvic cavity is divided into a superior **abdominal**[505] **cavity**, and a **pelvic**[506] **cavity** located inferior to the abdominal cavity. Both cavities are separated from each other by the **diaphragm**.[507] See figure 15.4.

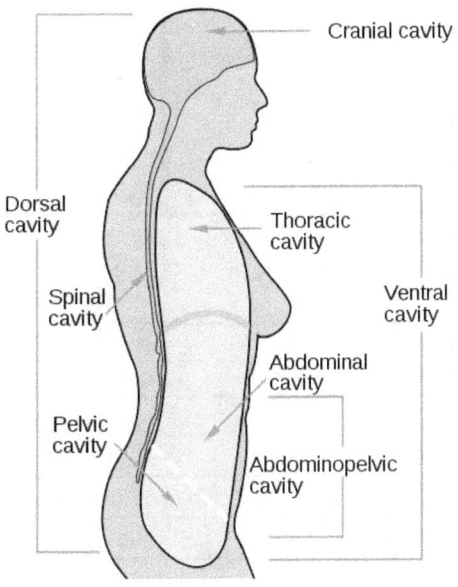

Figure 15.4. Body cavities.

---

500   A opening (foramen) in the vertebrae through which the spinal cord passes.

501   From Latin, *venter* meaning "belly."

502   The space between pleural membranes, the visceral and the parietal. The parietal pleural membrane covers the chest wall. The visceral pleural membrane covers the lung.

503   The space between the right and left pleura from the sternum to the vertebral column.

504   Chest.

505   Commonly called the belly. Located from the diaphragm to the pelvic brim.

506   Enclosed by the bones of the pelvis. Upper boundary is an oblique line along the pelvic brim to the pelvic opening to the muscles of the pelvic floor (*levator ani* and the *coccygeus*) and connective issue.

507   Diaphragm mean partition in Greek.

Table 15.2. **Ventral** body cavities and organs they contain.		
Cavity	Structures contained	Lining membrane

Thoracic (chest) cavity	Heart
	Great vessels
	    Thoracic aorta
	    Pulmonary artery
	    Superior vena cava
	    Inferior vena cava
	    Pulmonary veins
		Pleural

	Lungs
	    Trachea
	    Bronchi
	Esophagus
		Pericardial

Abdominopelvic cavity Abdominal	Digestive organs, spleen and kidneys	Peritoneum
Pelvic	Bladder, reproductive organs	Peritoneum

Table 15.3. **Dorsal** body cavities and organs they contain.		
Cavity	Contents	Lining membrane
Cranial	Brain, eyes and ears	Meninges
Vertebral	Spinal cord	Meninges

Chapter 16 Tissues

Histology[508] is the study of tissues. *Tissues* are groups of similar cells that have a ***common origin.*** Tissues carry out ***specialized functions*** and consist of cells and matrix.[509] *There are four kinds of tissues in the human body:* **1.** *epithelial*, **2.** *connective*, **3.** *muscular* and **4.** *nervous.*

1. Epithelial Tissue

Epithelial tissue *covers*, *protects*, *lines* body cavities and hollow organs. In addition, epithelial tissue functions in the *exchange of materials*, *secretion*, *absorption* and *sensation*. *Glands* are formed from epithelial tissue. Epidermal cells of the skin are tightly packed together. The upper part of an epithelial cell is the ***apical end***. The ***basal surface*** is the bottom of the cell that is in contact with the **basement membrane**.[510] See figure 16.1.

One side of epithelial tissue faces the external environment. The skin and a hollow portion of a gland or duct face the environment. The other side is attached to loose connective tissue by the basement membrane. The basement membrane anchors the epidermis to the dermis. See figure 16.1.

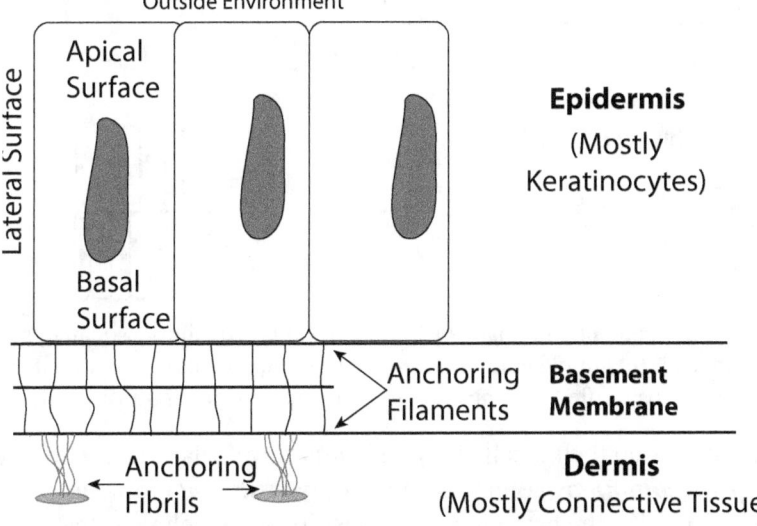

Figure 16.1. Generalized epithelial cells. The diagram shows the basal end of simple epithelia attached to the upper part of the basement membrane.

---

508   *Histo-* from the Greek, tissue and *logos* meaning knowledge.
509   Matrix includes ground substance. It is made of sugar, protein and fibers like collagen. Ground substance surrounds and supports cells. It is gel-like and has no specific shape.
510   A thin sheet of membrane that lies between the skin and the underlying connective tissue.

Arrangement and Shape of Epithelial Cells

Cell shape and number of layers of cells determines how epithelial tissue is classified. *Epithelial cells are the building blocks of epithelia.* *Simple epithelia* are a single layer of epithelial cells. *Stratified epithelia* consist more than one layer of epithelial cells. *Pseudostratified* epithelia are, in fact, simple epithelia. *Pseudostratified epithelia* appear to be stratified, but upon close inspection, it is observed to be made up of a single layer of epithelial cells. The basal portion of each pseudostratified epithelial cell is in contact with the basement membrane. The arrangement of the nuclei in the single layer causes pseudostratified epithelia to appear to be stratified. True stratified epithelia have its *basal* end in contact with the *apical* end of another cell. See figure 16.2 *f*.

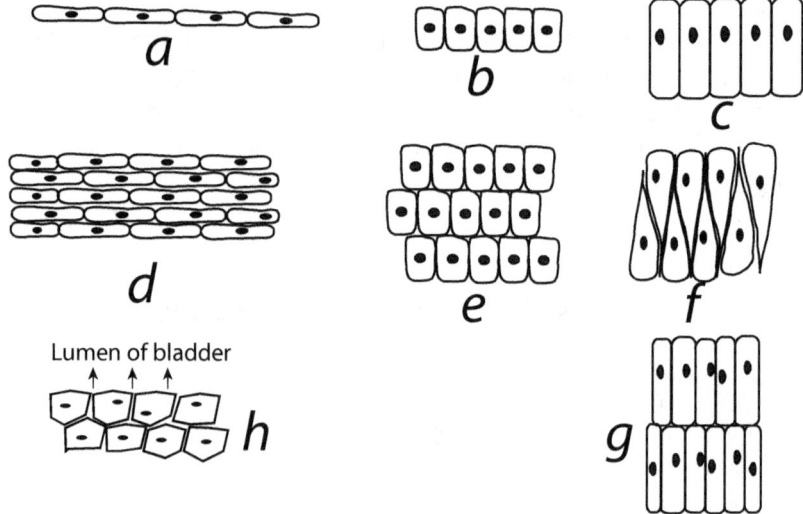

Figure 16.2. Arrangement and Shape of Epithelial Cells. (*a*) simple squamous, (*b*)simple cuboidal, (*c*) simple columnar, (*d*) stratified squamous, (*e*) stratified cuboidal, (*f*) pseudostratified columnar (*g*) stratified columnar (*h*) transitional.

The shapes of epithelial cells may be *squamous*[511] (flat), *cuboidal* (cube-shaped), *columnar* (column-shaped) or *transitional* (variable shapes) epithelia. As the name implies, transitional epithelia change shape. Transitional epithelium consists of several layers of epithelial cells that expand and contract depending on changes in volume of liquids within structures. Transitional epithelia are found in the urethra, ureters and urinary bladder and gland ducts associated with the prostate in males. See figure 16.2.

---

511   From the Latin, *squama* meaning "scale." Scales are flat and arranged in layers.

2. Connective Tissue

Connective tissue is the most widely distributed type of tissue in the human body. *Connective tissue connects, supports, separates structures and stores energy reserves in the form of fat.* In addition, connective tissue *manufactures blood cells.*

Connective tissue consists of **cells, fibers** *and* **ground substance** (extrafibrillar matrix). There are few cells in ground substance. Fibroblasts secrete ground substance, collagen, reticular and elastic fibers. There are several different kinds of connective tissue. See table 16.1.

Table 16.1. Types of connective tissue.	
Mature Connective Tissue	
1. Loose Connective Tissue	
a. Areolar	Supports almost every structure in the body. Areolar connective tissue attaches the skin to underlying tissues.
b. Adipose	Found in the subcutaneous layers of skin, around heart, kidneys, yellow bone marrow and around joints.
c. Reticular	Supporting framework of liver, spleen, lymph nodes, red bone marrow; wraps around muscles and blood vessels.
2. Dense Connective Tissue	
a. Dense *Regular*	Tendons, ligaments and aponeuroses.
b. Dense *Irregular*	Fascia beneath skin, periosteum of bone, joint capsules.
c. Elastic	Lung, walls of arteries, trachea and bronchial tubes.
3. Cartilage	
a. Hyaline	Ends of long bones, anterior ends of ribs, nose, embryonic and fetal skeleton.
b. Fibrocartilage	Pubic symphysis, interverterbral discs, cartilage pads of knee.
c. Elastic cartilage	Epiglottis, auricle of ear.
4. Bone	
a. Compact	Skeleton: Diaphysis of long bones.
b. Spongy (Cancellous)	Skeleton: Ends of long bones.
5. Blood (Liquid Connective Tissue)	Contained within the circulatory system.

3. Muscle Tissue

Muscle tissue functions in voluntary and involuntary movements. There are three types of muscle cells, *cardiac*, *smooth* and *skeletal*. See figure 16.3. *Muscle cells (muscle fibers) are the structural and functional units of contraction.* Skeletal muscles are under conscious control. *Skeletal muscle cells* are voluntary, multinucleated and **striated**.[512] These muscles can move at will.

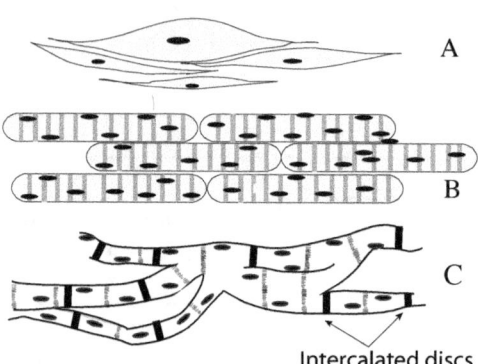

Figure 16.3. Three basic types of muscle cells. (A) smooth, (B) skeletal or involuntary and (C) cardiac.

Smooth muscle cells lack striations and have a single nucleus. These cells are responsible for many involuntary functions, one of which is *peristalsis*. Peristalsis is the involuntary contraction of the gastrointestinal wall that moves food along from the mouth to the anus.

Cardiac muscle tissue is found only in the heart. Cardiac muscle tissue is not under conscious control. The microscopic examination of cardiac muscle cells show that they are (1) *multinucleated*, (2) have *striations* and (3) are *branched*. The branches of cardiac cells connect one cell to another by structures called *intercalated discs*. These structures allow the heart to contract as one unit.

Over six hundred skeletal muscles work in a coordinated manner with the nervous system. The muscular system and the bones of the skeleton form a system of *levers* that perform movement.

The Motor Unit

A single *motor neuron* innervates ten to one thousand skeletal muscle fibers (cells). One neuron branches off to several muscle fibers. The neuron and the muscle fibers it innervates is called a *motor unit*. See figure 16.4.

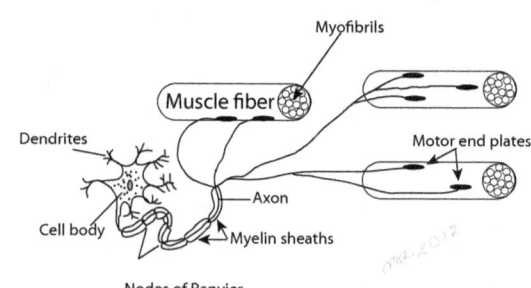

Figure 16.4. A simplified motor unit.

---

512   Striated means appearing to have ridges or stripes.

Mechanism of Muscle Contraction

Muscle contraction begins when Ca⁺⁺ ions flow into neurons. The flow of Ca⁺⁺ ions into a neuron causes the neuron to release the neurotransmitter acetylcholine (Ach) from the synaptic bulbs into the synaptic cleft of a neuromuscular junction. See figure 16.7. Binding of Ach to its receptors allows Na⁺ to flow into a muscle fiber and K⁺ to flow out; resulting in the contraction of the muscle fiber. See figure 16.8.

Hierarchy of Skeletal Muscle Structure

Each *muscle* is an organ. Muscles are composed of smaller units called *fascicles*. Fascicles are bundles of muscle fibers (muscle cells). Muscle fibers are composed of *myofibrils*. Myofibrils are the contractile elements within muscle cells. Myofibrils are composed of *thin and thick filaments* arranged in units called *sarcomeres*. *Sarcomeres are the functional units of contraction.* *Thin* filaments are made of *actin*. *Thick* filaments are made of *myosin*. The thick myosin filaments "crawl" over the stationary thin actin filaments causing contraction takes place. See figure 16.5.

Organizational Levels of Muscle Coverings

Muscle, fascicles and muscle fibers have coverings. The outermost covering of a muscle is the **epimysium**.[513] A fascicle is covered with a **perimysium**.[514] The **endomysium**[515] covers a muscle fiber. See figure 16.5 a, b and c and table 16.2.

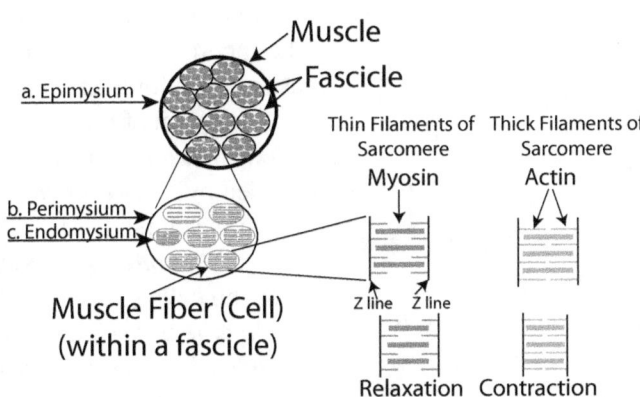

Figure 16.5. Muscle organization. (A) Muscle cross section, (B) Fascicles (bundles of nerve cells), (C) Nerve cell (muscle fibers) and thick and thin filaments organized into sarcomeres (D) and (E) Sarcomeres: (D) relaxed and (E) contracted

---

513	*Epi* - above.
514	*Peri* - around.
515	*Endo* - inside.

Table 16.2. Muscle organization.	
Structure(s)	Covering
Muscle, the organ	Epimysium
Fascicle	Perimysium
Muscle Cell (Muscle Fiber)	Endomysium
Myofibrils	No covering

Filaments
    Thin (protein actin)
    Thick (protein myosin)
	No covering
	Arranged into sarcomeres

4. Nervous Tissue

The basic unit of structure and function of the nervous system is the neuron (nerve cell). The nervous system functions in rapid communication for the body and perception. Perception is the gathering of sensory information by detecting light, sound, odor, touch and taste. The nervous system works with the **endocrine system**[516] to regulate homeostatic regulatory mechanisms such as blood pH, body temperature and blood pressure. The nervous system also produces **emotions**.[517]

The Neuron

A neuron is an electrically excitable cell that transmits information electrically and chemically. The *cell body* of a neuron has *dendrites* that

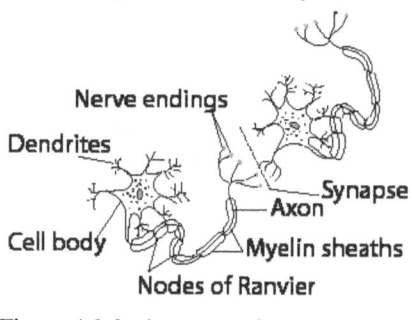

receive electrical information from other neurons. Electrical stimuli pass along a long, thin protoplasmic extension of the cell body called an *axon*. An axon is surrounded by *Schwann cells* arranged in segments separated by gaps called *nodes of Ranvier*. Schwann cells supply *myelin* for the *sheaths* of *peripheral nerves*. *Oligodendrocytes* make *myelin*

Figure 16.6. A synapse between two neurons.

sheaths for *nerves of the central nervous system. Myelin insulates* axons from each other, *anchor* them in place and *transport* nutrients and oxygen. See figure 16.6.

---

516   The endocrine system consists of many glands that produce hormones (chemical messengers). Each hormone produces different effects in the body. The "fight or flight" hormone epinephrine (adrenalin) is well known. Hormones play a major role in homeostasis as well.

517   Physical and psychological expressive and arousal reactions to stimuli.

Each nerve ending of an axon terminates in a *synaptic bulb*. The synaptic bulb forms a junction called a *synapse* with other *neurons*, *glands* or *muscle cells*. See figure 16.8. The *neuromuscular junction* is a place where a nerve ending comes close to a muscle cell, but does not touch it. The *neurotransmitter acetylcholine* (ACh) is released from the synaptic bulb and muscle contraction begins. See figure 16.7.

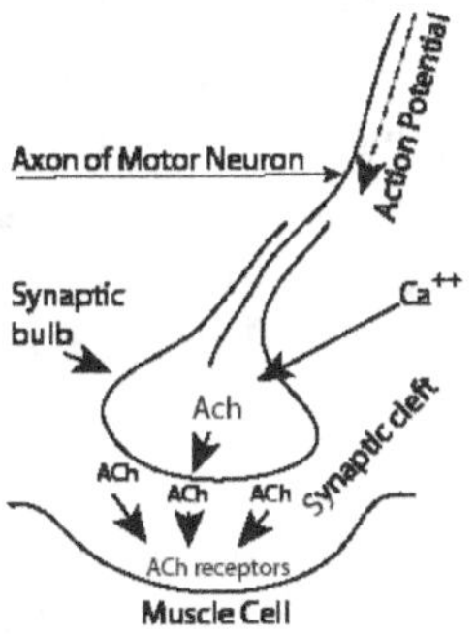

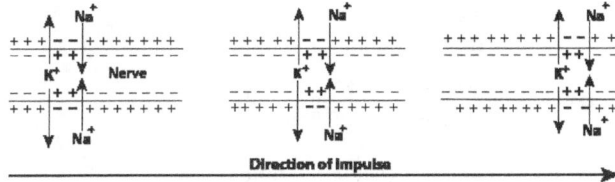

Figure 16.7. A synapse between a neuron and a muscle cell.

Conduction Along a Nerve

Figure 16.8 shows sodium ions (**Na⁺**) *on the outside* of the nerve cell membrane and potassium ions (**K⁺**) *on the inside* during the resting phase. When a neuron is stimulated, an *action potential* is generated in the axon. During an action potential, Na⁺ moves into the axon, depolarizing the membrane. Potassium (K⁺) then moves out of the axon repolarizing the membrane. The ion movements continue along the length of the axon causing the release of a chemical signal called a neurotransmitter. Resting takes place, and *repolarization* occurs as K⁺ ions *move to the outside*. Sodium ions remain inside the nerve cell membrane. A refractory period puts everything back to normal. Potassium returns to the inside and sodium returns outside.

Figure 16.8. Conduction along a nerve.

Reflexes

A *reflex* is a rapid, involuntary action that occurs as a result of a stimulus. A *stimulus* is something in the internal or external environment that causes a response in an organism. Examples of reflexes are the startle reflex, spinal reflex (patellar reflex or knee jerk reflex), **Babinski reflex**,[518] blushing, **palmar grasp reflex**,[519] **plantar reflex**,[520] **Moro reflex**[521] (infantile reflex) and the **pupillary light reflex**.[522]

518   Normal infants up to one year of age spread their toes in response to stroking the side of the foot.
519   Grasping an object placed in their hand of infants of (up to six months of age).
520   Stroking the sole with a blunt object produces a downward flexion of the big toe (flexor plantar reflex).
521   If an infant senses loss of balance it will cling to the mother.
522   Decrease of pupil size in response to light.

The Reflex Arc

Sensory neurons synapse in the spinal cord. A synapse is a gap between two neurons that allows for the passage of electrical impulses from one neuron to another or from a neuron to a muscle. A *reflex arc* is a series of neurons that control an involuntary action called a reflex. The neurons, through which a nervous impulse passes is called a reflex arc. Some reflex arcs may have *two neurons* and one synapse between them. The arrangement is known as a *monosynaptic reflex arc.* A monosynaptic reflex arc is responsible for a simple reflex. A *polysynaptic reflex arc* consists of at least *three neurons.* Polysynaptic reflex arcs have one or more *interneurons* that connect sensory neurons with motor neurons.

A *simple reflex* consists of two neurons, a *sensory neuron* and a *motor neuron.* A *sensory neuron* is an *afferent* neuron. *Afferent* neurons send impulses to the spinal cord. A *motor neuron* (effector neuron) is an *efferent neuron.* Efferent *neurons* send impulses to a muscle or gland (effectors). Efferent impulses are going *away* from the spinal cord.

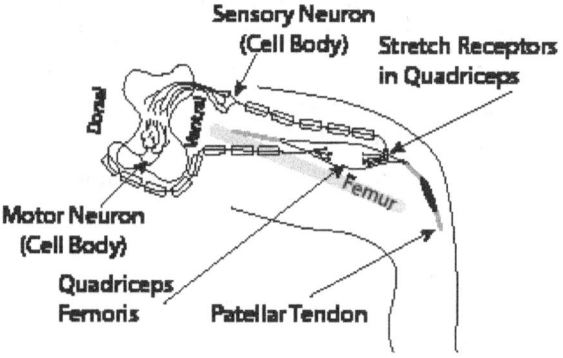

Figure 16.9. Patellar (knee jerk) reflex.

The **patellar[523] reflex**, commonly known as the knee-jerk reflex, is an example of a simple reflex. It is a deep tendon reflex that involves only the spinal cord. The brain is *not* involved. If the patellar tendon is struck just below the patella, the tendon stretches. This action stretches the **quadriceps[524]** muscle, causing an impulse (*afferent nerve*) to be sent to the spinal cord. The impulse passes to a motor nerve (*efferent nerve*) causing the quadriceps to contract. The contraction of the quadriceps causes the lower leg to rise. The entire process occurs in about 50 millionths of a second. See figure 16.9. Absence of patellar reflex helps to determine damage to the central nervous system, assist in the diagnosis of disease conditions and prevents muscles from overstretching.

523   The kneecap. It protects the anterior surface of the knee joint.
524   A group of four muscles located anterior and medial part of the thigh.

Chapter 17 Body Systems
An Overview of Human Body Systems

The human body has twelve **systems**[525] that work together to maintain homeostasis. Each body system is composed of organs. Organs, in turn, are made up of tissues and tissues are composed of cells. The following pages present a brief description of the twelve systems of the human body.

1. The Nervous System

The human nervous system (see figures 17.2 and 17.3) consists of the *central nervous system, peripheral nervous system* and *sense organs*.

Sense organs gather information from the environment. Peripheral nerves transmit sense information to the brain for processing. For example, the human eye detects different frequencies of light. Each frequency is a color. See page 130. The color red is visualized in the brain, not in the eye. Similarly, feeling resides in the brain, not the skin. Nerve endings in the skin detect various stimuli. The brain receives the stimuli where sensations are interpreted as pressure, touch, heat, pain.

The nervous system is a *rapid messaging system*. The speed of an impulse along a nerve can be as fast as 120 meters per second.

Organization of the Human Nervous System
I. The Central Nervous System

The central nervous system consists of the *brain* (figure 17.1) and *spinal cord*. See figure 17.3. The spinal cord is connected to the brain at the level of the **occipital bone**[526] of the **skull**.[527] The occipital bone has a

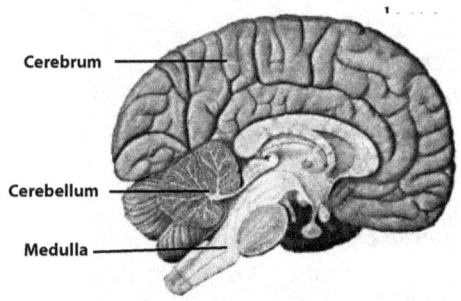

Cerebrum

Cerebellum

Medulla

Figure 17.1. A cross-section of the human brain.

opening called the *foramen magnum* through which the spinal cord passes. The spinal cord terminates at the **medulla oblongata**[528] (medulla) of the brain.

The skull protects the brain. The brain is the central control center of the body. Vertebrae of the spinal column protect he spinal cord. The brain stem includes the *medulla oblongata, pons* and *midbrain*.

525   Some authors include the lymphatic system with the circulatory system while others may combine the urinary system with the genital system (the urogenital system).
526   The occipital bone forms the rear lower part of the skull.
527   The human skull is composed of fourteen facial bones and eight cranial bones.
528   The medulla oblongata is an extension of the spinal cord and is the lower part of the brainstem.

Human Nervous System (NS)

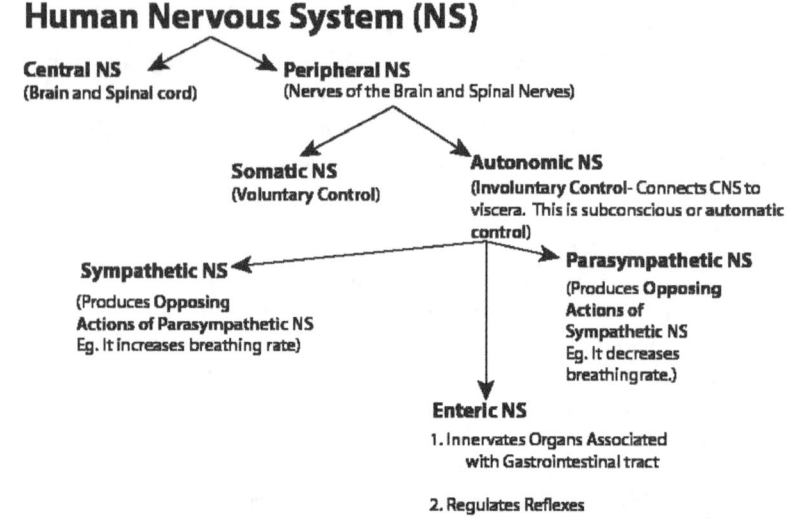

Central NS
(Brain and Spinal cord)

Peripheral NS
(Nerves of the Brain and Spinal Nerves)

Somatic NS
(Voluntary Control)

Autonomic NS
(Involuntary Control- Connects CNS to viscera. This is subconscious or automatic control)

Sympathetic NS

(Produces Opposing Actions of Parasympathetic NS Eg. It increases breathing rate)

Parasympathetic NS

(Produces Opposing Actions of Sympathetic NS Eg. It decreases breathing rate.)

Enteric NS

1. Innervates Organs Associated with Gastrointestinal tract

2. Regulates Reflexes

3. Acts Independently of CNS and PNS

Figure 17.2. Flow chart of subdivisions of human nervous system.

II. *The Peripheral Nervous System*

The peripheral nervous system (see figure 17.2) is not protected by bone, as is the central nervous system. The peripheral nervous system consists of 12 pairs of *cranial nerves*[529] that **emanate**[530] from the brain along with thirty-one pairs of *spinal nerves* that emanate from the spinal cord.

The thirty-one pairs of *spinal nerves* exit the spinal cord are designated: *cervical* (**C1-C8**),[531] *thoracic* (**T1-T12**), *lumbar* (**L1-L5**) and *sacral* (**S1-S5**).[532] In addition to cranial nerves and spinal nerves, *nerve plexuses* and *ganglia* that lie outside of the brain and spinal cord are part of the peripheral nervous system.

A plexus is the *intersection of nerve cells*. It is like a crossroad. Five are listed here: the *cervical plexus, brachial plexus, lumbar plexus, sacral plexus* and the *celiac plexus* (solar plexus). See figure 17.3.

Ganglia are **relay**[533] points between different structures of the nervous system. Ganglia are bundles of *nerve cell bodies* that *relay* information between the peripheral and central nervous systems.

---

529   Only the first two pairs of the twelve pairs of nerves originate from the cerebrum. The next ten pairs emerge from the brain stem.
530   Emanate means to flow out of, originate or origin.
531   There are seven cervical vertebrae, but there are eight cervical nerves.
532   The sacrum consists of 5 fused vertebrae.
533   A place where something can be switched or have its direction changed.

Divisions of the Peripheral Nervous System

The **peripheral** nervous system (PNS) is made up of nerves and ganglia that are not part of the brain and spinal cord. The PNS connects the brain and spinal cord to the rest of the body. The two major divisions of the peripheral nervous system are:

(A) Somatic nervous system (SNS)

(B) Autonomic nervous system (ANS)

The Somatic Nervous System

The **somatic**[534] nervous system (SNS) is made up of sensory neurons and motor neurons. The SNS is responsible for *voluntary* skeletal movement. Sensory neurons carry information *from* the somatic receptors in the body wall, eye, ear, nose and tongue *to* the CNS. Motor neurons carry impulses *from the* CNS *to voluntary muscles.*

The Autonomic Nervous System

The **autonomic nervous system** (ANS), also called the *visceral nervous system,* regulates heart rate, respiratory rate, urination and sexual arousal. Some of these functions can be controlled by the conscious mind. Some functions cannot be controlled by the conscious mind because they are automatic actions. The autonomic nervous system functions *without conscious control.*

Divisions of the Autonomic Nervous System

Three major divisions of the ***autonomic nervous system are the* sympathetic nervous system (SNS), parasympathic nervous system and the enteric nervous system (ENS).**

1. Sympathetic Nervous System

The *sympathetic nervous system* innervates smooth musculature, glands, the heart and other organs. The sympathetic nervous system ***opposes actions of the parasympathetic nervous system.*** The s*ympathetic nervous* system speeds up or slows down homeostatic mechanisms. The *sympathetic nervous system* speeds up the heart and respiration rate in response to danger, and it slows down peristalsis. See table 17.1.

2. Parasympathic Nervous System

The *parasympathetic nervous system* will slow down the heart and respiration rate when it needs to be slowed down. These are automatic, unconscious actions, but are partially controllable by conscious control.

3. Enteric[535] Nervous System

The Enteric nervous system is the third division of the autonomic nervous system. This system is also involuntary. The enteric nervous system is embedded in the walls of the gastrointestinal (GI) tract.

534   From the Greek, *soma* meaning the body.
535   Sometimes called the intrinsic nervous system (INS).

The major functions of the ENS are to:
1) Detect any changes in gut chemistry.
2) Detect stretching of the gut wall.
3) Regulate smooth muscle contraction of the GI tract peristalsis.
4) Regulate hormonal secretions of endocrine cells of the GI tract.
5) Regulate the acid secretion in the stomach.

Table 17.1. Actions of the sympathetic and parasympathetic nervous systems.		
Structures Acted On	Sympathetic NS	Parasympathetic NS
Heart rate	Speeds up	Slows down
Arteries	Constricts most arteries. Dilates arteries leading to the heart and skeletal muscles	Dilates
Peristalsis of intestines	Slows down	Speeds up
Urinary bladder	Relaxes	Constricts
Muscles within bronchi	Dilates bronchi	Constricts bronchi
Pupil	Dilates	Constricts
Hair Smooth muscle fiber	Erection of hair	Relaxation of hair
Sweat gland secretion	Increases	Decreases

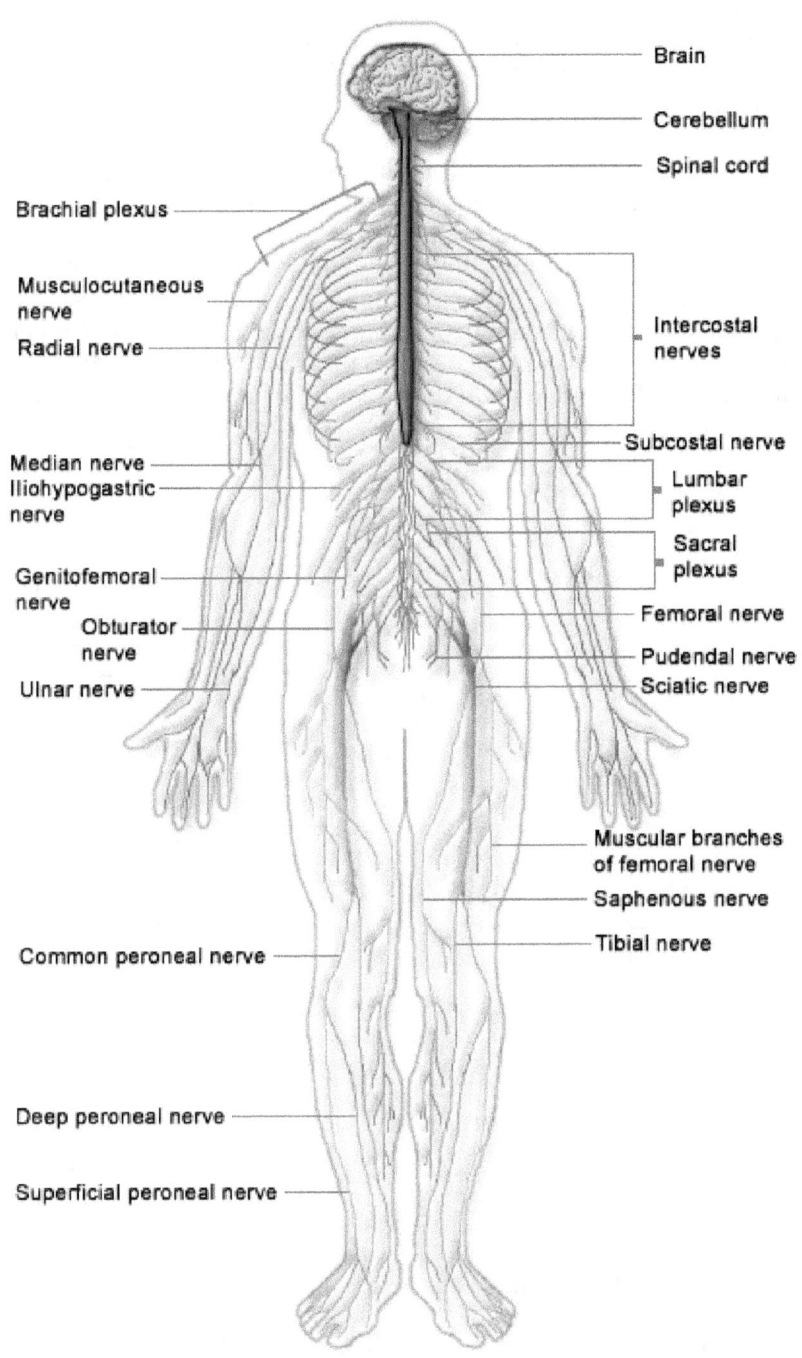

Brain

Cerebellum

Spinal cord

Brachial plexus

Musculocutaneous nerve

Radial nerve

Intercostal nerves

Subcostal nerve

Median nerve

Iliohypogastric nerve

Lumbar plexus

Sacral plexus

Genitofemoral nerve

Obturator nerve

Femoral nerve

Pudendal nerve

Sciatic nerve

Ulnar nerve

Muscular branches of femoral nerve

Saphenous nerve

Tibial nerve

Common peroneal nerve

Deep peroneal nerve

Superficial peroneal nerve

Figure 17.3. The human nervous system.

2. The Skeletal System

An infant has 270 bones. The bones of a fetus are composed of flexible cartilage. The process of **ossification**[536] of cartilage into bone in the infant occurs by week six of development. Some bones fuse together leaving 206 bones in the adult. See figure 17.5. The human skeleton is a framework within the body. It is an **endoskeleton**.[537] The skeletal system provides *support and protection* for structures within the body and allows for *movement* with the help of muscles. Bone produces *blood cells*, *stores minerals*, mainly *calcium* and *phosphate ions*. The skeletal system also performs *endocrine functions* that help regulate blood sugar.

The human skeleton consists of an *axial skeleton* and an *appendicular skeleton*. The axial skeleton includes the *skull, vertebral column, the rib cage, the hyoid bone and the bones of the middle ear* (*malleus–hammer*, *incus–anvil* and *stapes–stirrup*). See figure 17.4.

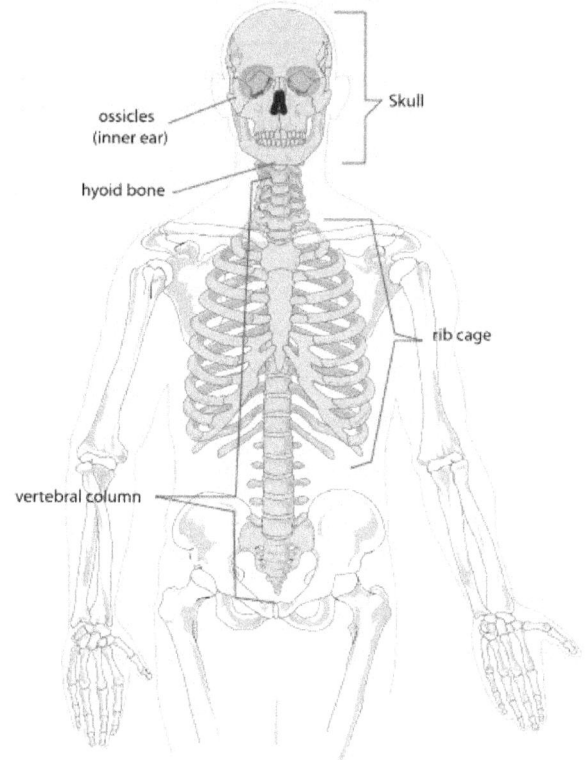

ossicles (inner ear)

Skull

hyoid bone

rib cage

vertebral column

Figure 17.4. The human axial skeleton.

---

536   The laying down of new bone by cells called osteoblasts.

537   Endoskeletons are internal supporting frameworks of an organism and exoskeletons are external skeletons. Grasshoppers, crabs and lobsters have exoskeletons. Snails and clams are organisms that possess a shell. Shells are also exoskeletons.

The appendicular skeleton consists of the ***upper and lower limbs***. The ***pectoral girdle*** connects the upper limbs (arms) to the axial skeleton. The pectoral girdle is made up of two bones, the ***clavicle*** and the ***scapula***.

The ***pelvic girdle*** connects the lower limbs to the axial skeleton. See figure 17.5. The pelvic girdle is composed of a *sacrum* and *coccyx* posteriorly and *two hip bones* (coxal bones). The hip bones consist of three fused bones, the ***ilium***, ***ischium*** and the ***pubis***.

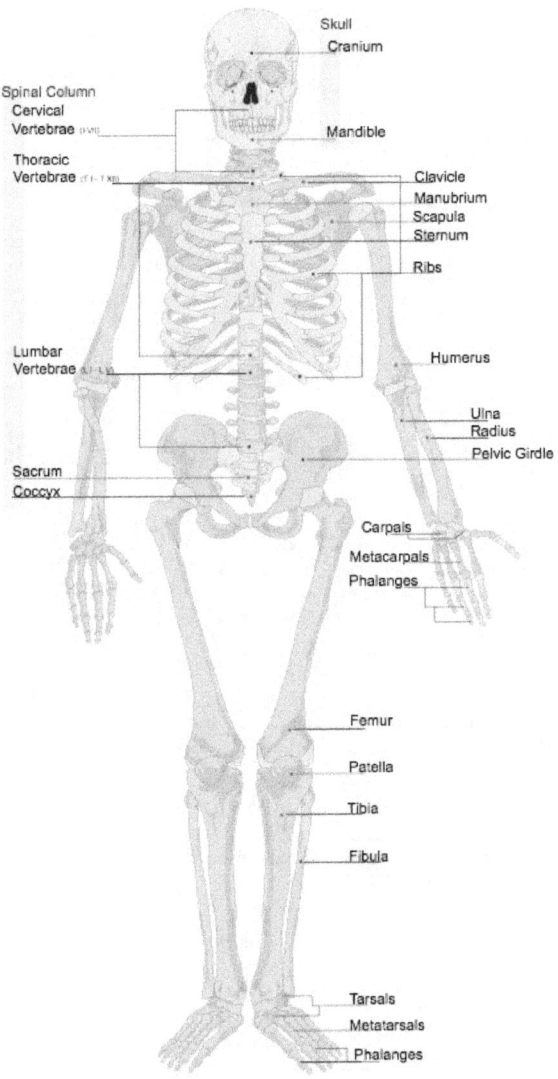

Figure 17.5. Major bones of the human skeleton. Grayed area denotes the axial skeleton.

Normal Curvatures of the Human Spine

The human spine is made up of 26 articulated vertebrae, including the sacrum and coccyx. Vertebrae provide protection for the spinal cord and nerve roots. In addition, the spine provides support for the body, points of attachment for the pectoral and pelvic girdles and muscles.

The curvatures of the spinal column are designated as *cervical (neck), thoracic (chest/trunk), lumbar (lower back) and sacral (pelvic)*. See figure 17.6.

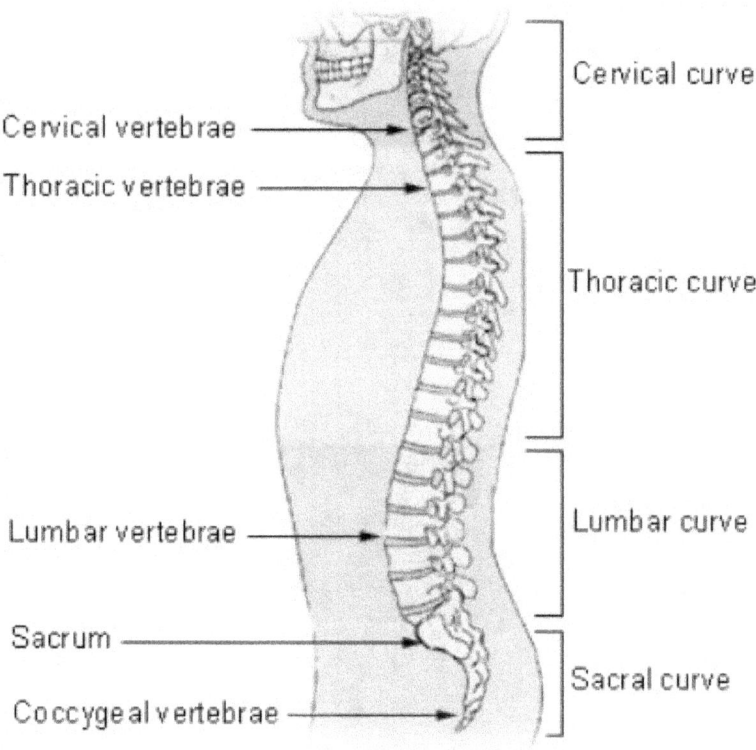

Figure 17.6. Curves and regions of the human spine.

3. The Circulatory System

The main function of the circulatory system is to ***deliver oxygen and nutrients to cells*** of the body. At the same time, carbon dioxide and other waste products of metabolism are picked up by the blood and delivered to organs of excretion. The heart is located in the middle of the thorax in the space between the lungs called the ***mediastinum.*** This area contains the heart, trachea, esophagus, thymus and the ***great vessels.***

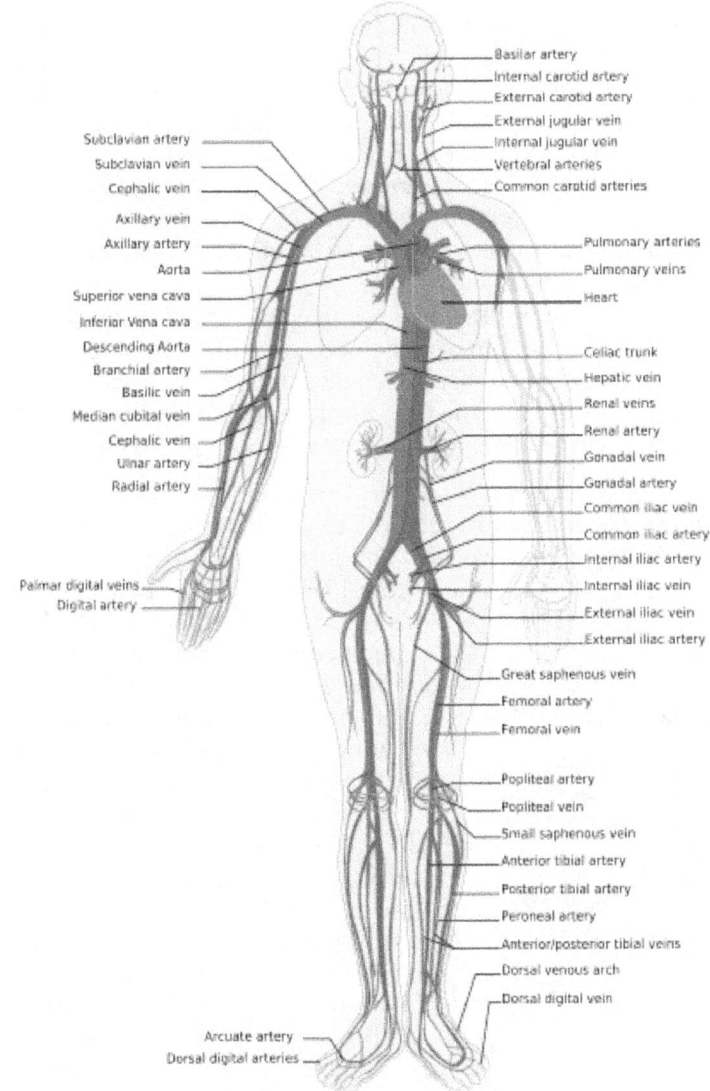

Figure 17.7. The human circulatory system. Arteries are depicted in red and veins are depicted in blue.

The circulatory system consists of the *heart*, *blood vessels* (see figure 17.7) *and blood*. *Arteries*, *veins* and *capillaries* are three kinds of blood vessels that transport blood around the body. Arteries carry blood away from the heart, and veins carry blood back to the heart. Microscopic *capillaries connect arteries to veins*. Capillaries permit the exchange of materials between the blood and body cells.

The Human Heart

The heart has *four chambers*. The two upper, thin walled *receiving* chambers of the heart are called *atria*. The two lower thick walled chambers are *pumping* chambers called *ventricles*. The left ventricle has the thickest walls of the four chambers because it must to pump blood into the aorta and then to the entire body. See figure 17.8

The right atrium receives *deoxygenated* blood from body cells and pumps it to the right ventricle. Deoxygenated blood is blood whose hemoglobin within the cytoplasm of red blood cells has given up most of its oxygen to cells of the body.

The right ventricle pumps the deoxygenated blood to the lungs where oxygen binds to the hemoglobin of red blood cells. The left atrium receives the *oxygenated* blood that has come from the lungs and pumps the blood to the left ventricle. The left ventricle pumps the oxygenated blood to body cells where hemoglobin gives up its oxygen to body cells.

Circulation of Blood Through the Heart

The *right atrium* of the heart receives blood from the systemic circulation via the *superior vena cava* and *inferior vena cava.* See figure 17.7 and 17.8. The blood received by the right atrium has *low oxygen concentrations* and *high carbon dioxide concentrations.* The right atrium pumps this blood to the *right ventricle*, which in turn, pumps this oxygen poor blood to the lungs where oxygen is picked up by red blood cells in the capillary beds surrounding the alveoli. At the same time oxygen diffuses out of the capillary beds to the blood, carbon dioxide diffuses into the alveoli of the lungs, is expelled by *expiration*.

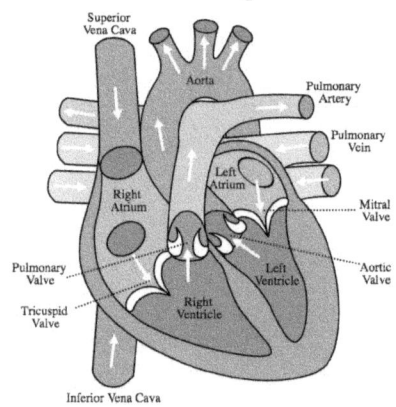

Figure 17.8. Structure of the human heart showing the great vessels and blood flow.

Environmental oxygen enters the lungs by the process of *inspiration*. Oxygen is picked up by hemoglobin within red blood. Oxygen rich blood is pumped to the *left atrium*, which in turn pumps oxygen rich blood to the *left ventricle* where it is pumped through the *aorta* for delivery to body cells. See 17.8 and 17.9.

In the *systemic circulation*, *arteries* bring *oxygenated blood* away from the heart and *veins* return *deoxygenated blood* to the heart. In the *pulmonary circulation*, veins and arteries are exceptions to this rule. In contrast to all other arteries, pulmonary arteries carry deoxygenated blood away from the heart to the lungs, and pulmonary veins carry oxygenated blood from the lungs back to the heart. See figure 17.9.

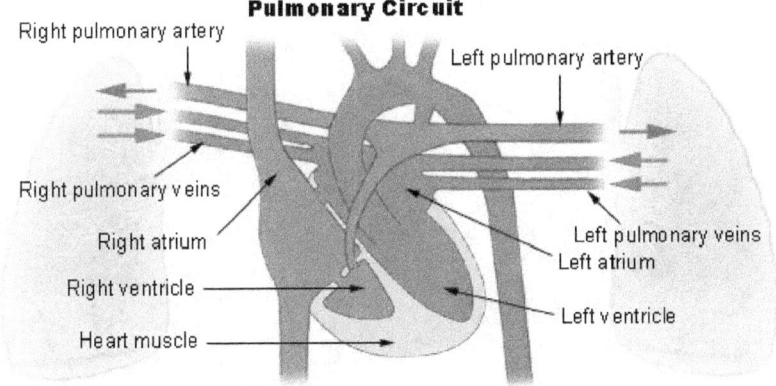

Figure 17.9. The pulmonary circulation. Oxygenated blood is in red and deoxygenated blood is in blue. The great vessels in the mediastinum are seen as the *superior vena cava, inferior vena cava, pulmonary artery, pulmonary veins and the aorta*.

Coronary Circulation

The heart muscle (myocardium) is supplied with blood by relatively narrow *coronary arteries* and drained by coronary veins. The left and right cardiac arteries are *epicardial*; that is, they are observable on the surface of the heart. See figure 17.10. These arteries are subject to hardening and restriction by atherosclerosis. Atherosclerosis is frequently the cause of heart attacks.

Other coronary arteries run deep within the myocardium and are termed *subendocardial*. Coronary arteries can regulate the flow of blood to the myocardium (heart muscle) in a healthy heart.

The right and left coronary arteries emerge from the base of the aorta as the aorta exits the left ventricle. The left coronary artery supplies blood to the left atrium and the left ventricle. The right coronary artery supplies blood to the right atrium and a part of the right ventricle and part of the left ventricle.

In some cases, the right and left coronary arteries join in what is termed anastomoses. In the area of anastomoses, the heart muscle receives blood supply from both blood vessels. If one coronary blood vessel becomes blocked, a supply of blood is still available to the affected region of the heart.

243

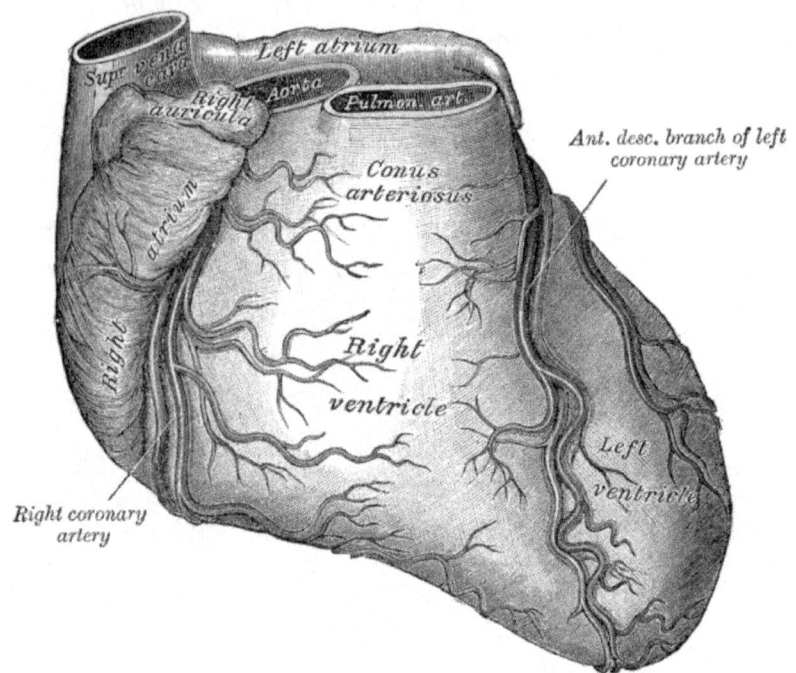

Figure 17.10.

4. The Respiratory System

The respiratory system consists of the ***nose, pharynx, larynx, trachea, bronchi*** and ***lungs.*** See figure 17.11. The chest cavity is a closed cavity that contains lungs that can expand and collapse with changes in air pressure within the cavity. If the ***diaphragm*** (figure 17.11) at the bottom of the cavity moves downward, air pressure is lessened, causing air to flow into the lungs. Oxygen molecules diffuse into the capillaries surrounding the ***alveoli*** (figure 17.11, inset) and it attaches to hemoglobin in the red blood cells. Red blood cells are transported to the left atrium and thence to the left ventricle for distribution to all the cells of the body.

When molecular oxygen reaches a cell, it diffuses into the cytoplasm and then into the mitochondria. Oxygen molecules take its place at the end of the electron transport system where it functions as a final electron acceptor. Electrons and hydrogen ions, both harmful to the cell, combine with oxygen to form harmless water.

Carbon dioxide produced by the Krebs cycle makes its way to the lungs where it diffuses from the blood into the alveoli. When the diaphragm moves upward, air pressure increases in the chest cavity and air is ***exhaled*** from the lungs. Together, inhalation and exhalation is referred to as ***breathing***. See pages 35-36.

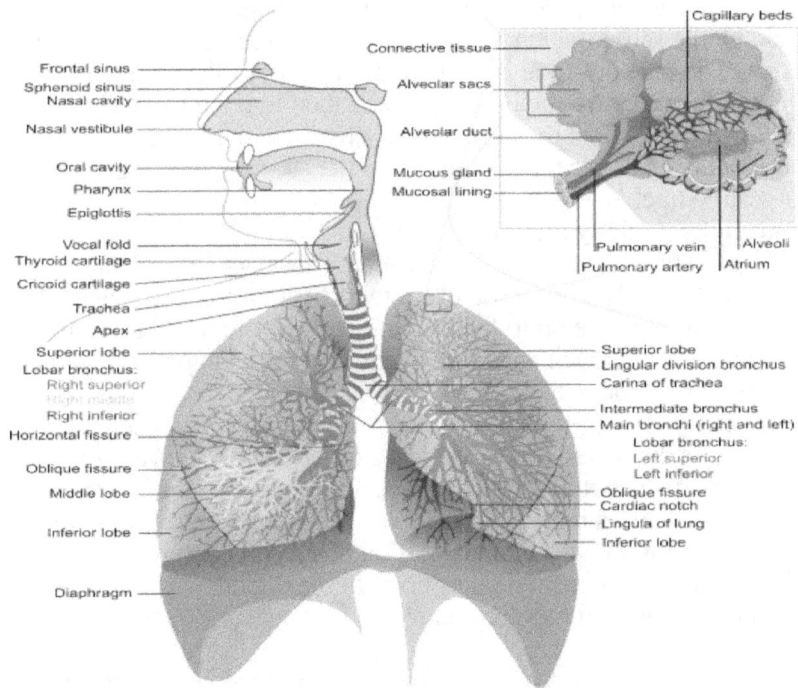

Figure 17.11. The human respiratory system.

5. The Digestive System

The *digestive system* is made up of: the *mouth, teeth, tongue, salivary glands, esophagus, stomach, small intestines, large intestine, rectum* and *anus*. The *liver, gallbladder* and *pancreas* are also digestive structures. See figure 17.12.

Digestion is the process of breaking down large *insoluble food molecules* into small *water soluble molecules*. Large water insoluble food molecules are *starch, protein, fat* and *nucleic acids*. Large molecules are polymers that are too big to pass through the cell's membrane. Digestion makes these polymers into the small monomers of *simple sugars, amino acids, fatty acids* and *nucleotides*. Monomers are small enough to diffuse through the cell membrane and into the cell's cytoplasm where metabolic processes can extract energy from the chemical bonds of glucose and provide molecules for life processes. In the case of the small glucose molecule, a facilitator such as the hormone *insulin*, made in the beta cells of the pancreas, is needed to get a glucose molecule into a cell. The pancreas also produces the digestive enzymes *lipase, protease* and *amylase*. Lipase digests lipids, proteases digest proteins and amylase digests starch.

Digestion starts in the *mouth*. Starch is broken down into simple sugars by the enzyme *ptyalin* (salivary amylase/alpha amylase). Ptyalin is made by the salivary glands of the mouth and breaks down starch into maltose. The bolus of food is pushed to the rear of the mouth and passes into the pharynx, down the esophagus and then into the stomach. No digestion takes place in the esophagus. Rhythmic contractions and relaxations called peristalsis moves food from the esophagus along the rest of the digestive tract.

When the partially digested starches in the bolus reaches the stomach, the digestion of starch is halted because stomach acids inhibit the action of ptyalin. The digestion of proteins begins in the stomach when pepsinogen is converted into the enzyme pepsin with the help of hydrochloric acid. Pepsinogen is an *inactive* form of pepsin that is stored in the lining of the stomach.

The first part of the small intestine is the *duodenum*. There are four regions of the duodenum: the superior, a descending, an inferior and an ascending part. The digestion of fats begin in the duodenum. The common bile duct carries bile, and digestive enzymes are transported by the pancreatic duct to the descending portion of the duodenum.

Most chemical digestion takes place in the duodenum. The digestion of sugars, proteins and fats are completed in the rest of the small intestine.

Digested food passes on to the second part of the small intestine called the *jejunum*. Here digested food is absorbed into the blood by microscopic finger shaped *villi*. Food then passes on to the third part of the small intestine, the **ileum**.[538] The ileum absorbs vitamin B_{12} and bile salts. The villi and *microvilli* of the ileum absorb the products of digestion that did not get absorbed in the jejunum. The large intestine reabsorbs water from the solid waste (feces) products of digestion. The waste is then excreted through the anal canal.

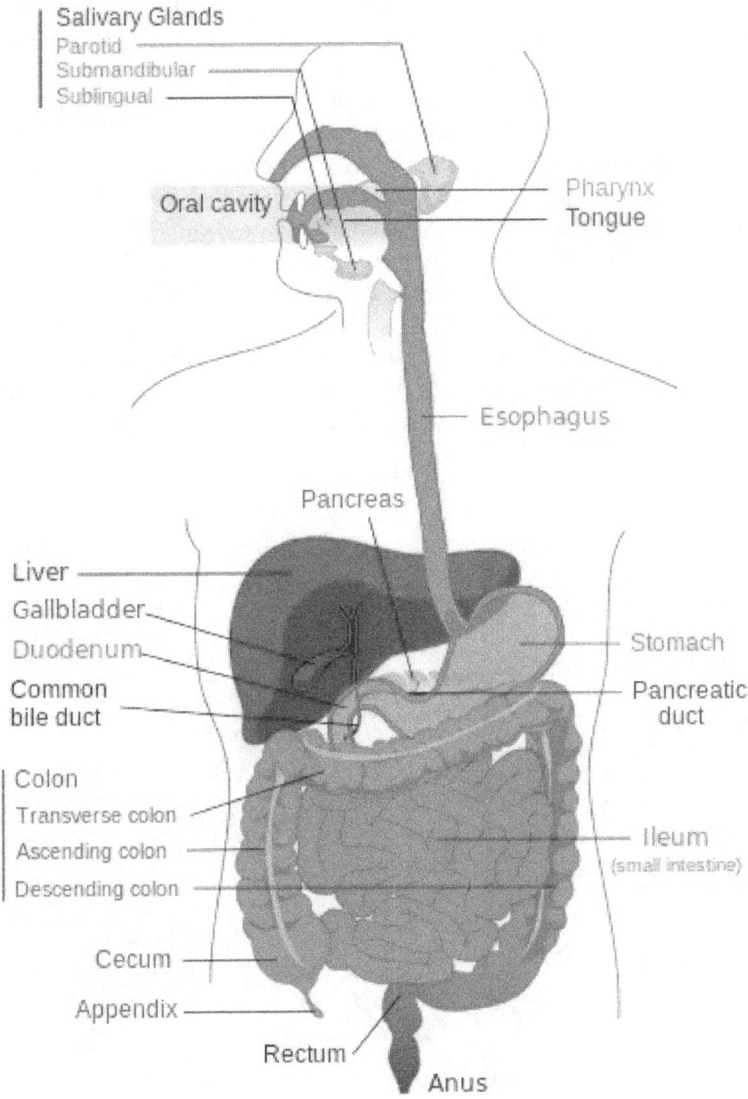

Figure 17.12. The human digestive system.

---

538   Do not confuse ileum, part of the small intestine, with ilium, one of the three bones that make up the hip bone (coxal bone).

6. The Integumentary System

The integumentary system is composed of *skin*, *hair nails* and *glands* of the skin. The skin (integument) is the largest organ of the human body. The integument protects the body from water loss, protects against infection, helps synthesize vitamin D and it excretes wastes. The skin also helps in the regulation of body temperature.

The skin contains sensory receptors. Receptors are specialized cells or nerve endings. Some examples are Ruffini's end organ (sustained pressure), Meissner's corpuscle (texture and vibration), Pacinian corpuscle (pressure, vibration) and Merkel's disc (touch and pressure).

The skin consists of two main layers, the epidermis and the dermis. Below the dermis is the subcutaneous (subQ) layer or hypodermis. The hypodermis consists areolar and adipose tissue. See figure 17.13. The epidermis is the outermost layer and consists of about 90% keratinocytes. Keratinocytes are made up of the protein *keratin*. These keratin-bearing cells form a barrier against microbes, ultraviolet radiation and water loss.

Formation of keratinocyte takes place in the mitotic layer, the *stratum basale (germinativum)*. Keratinocytes undergo *terminal differentiation* as keratinocytes move up through the epidermal layers: *stratum spinosum*, *stratum granulosum* and the *stratum corneum*.

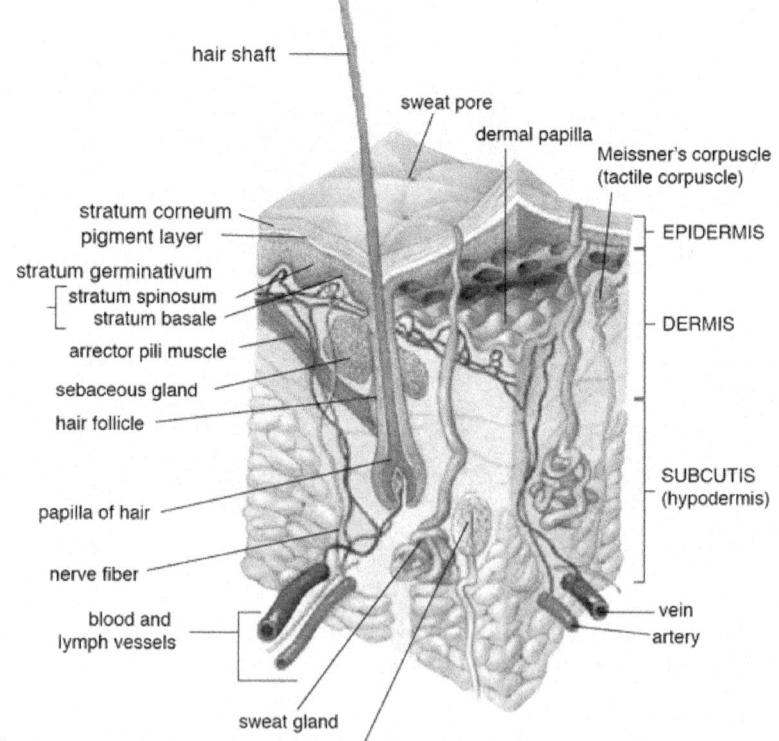

Figure 17.13. A cross section of human skin.

7. Muscular System

The muscular system enables the body to move. *Movement* can be of many kinds. *Locomotion*, *peristalsis* and *beating of the heart* are a few. There are three types of muscle cells in the body, *skeletal* (striated or voluntary), *smooth* (involuntary) and *cardiac* (see figure 16.3 on page 228). Skeletal (striated muscle) enables locomotion to take place. Smooth muscle cells produce a rhythmic contraction of the walls of the intestines that moves food along the alimentary canal by peristalsis. Cardiac muscle (cardiomyocytes) cells enable the heart to beat.

Skeletal muscle, along with the bones, allow for movement of the limbs. Skeletal muscles also produce heat to help maintain body temperature under adverse conditions. The number of skeletal muscles in the human body varies, but the best estimate is 640. See figures 17.16 and 17.17.

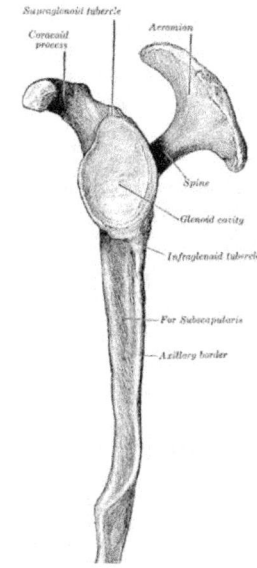

Muscles work in opposable pairs. If one muscle contracts, the other relaxes. For example, the *biceps brachii* is a muscle with two heads. A *head* is the *origin*. See figure 17.16. The *triceps brachii* is a muscle with three heads, (see figure 17.17). This pair of muscles works in opposition to each other. When the *biceps brachii* contracts, the *triceps brachii relaxes*. Conversely, when the *biceps brachii* relaxes, the *triceps brachii* contracts. The muscle that causes movement is called a *prime mover* or *agonist*. When the forearm is raised, the biceps is the cause of the movement and, therefore, is an agonist. As the *triceps* relaxes, it becomes the *antagonist*. Agonists and antagonist are reversed if the motion is reversed.

Origin and *insertions* are points of attachment for muscles. The *origin* of a muscle is in the bone that does not move. The *insertion* of a muscle is in the bone that moves. The biceps brachii has a short head and long head. The short head of the biceps brachii is attached to the end of the *coracoid process* of the scapula. The long head is attached to the *supraglenoid tubercle* of scapula. See figure 17.14. The insertion of the *biceps brachii* is on the *radial tuberosity* of the radius. See figure 17.15.

Figure 17.14. The scapula. Lateral view.

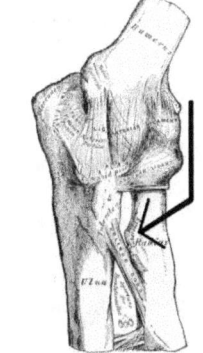

Figure 17.15. Point of insertion of *biceps brachii* at the radial tuberosity of the radius.

Synergists (fixators) help stabilize joints. A synergist works with a prime mover. For example, The *brachioradialis* works the *biceps brachii* to stabilize the elbow joint when the forearm is raised.

Smooth muscle tissue (involuntary or nonstriated) produce involuntary movements. Smooth muscles produce **peristalsis**. Peristalsis is an involuntary movement of the digestive tract that moves food through the digestive tract. Smooth muscle tissue is also found in glands, the walls of blood vessels, lymphatic vessels, urinary bladder, the uterus, the respiratory tract, **ciliary muscle**[539] and **arrector pili**[540] in the skin.

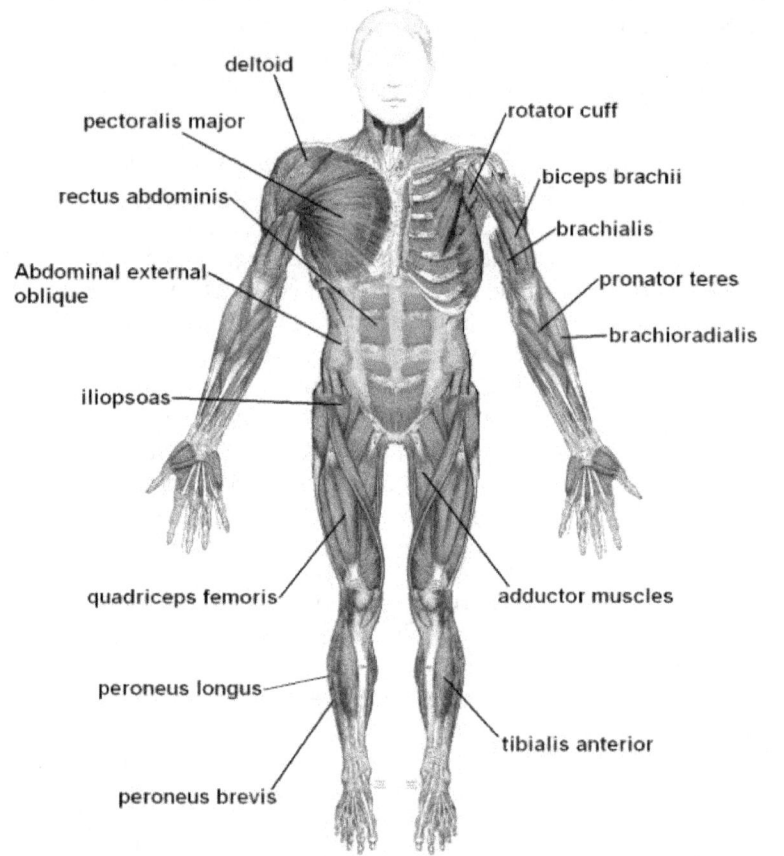

Figure 17.16. Anterior view of the human muscular system.

---

539   The ciliary muscle controls the eye's ability to view objects at varied distances.
540   These muscles are attached to hair follicles in the skin of mammals. When these muscles contract, the hairs stand up. This is known having "goose bumps."

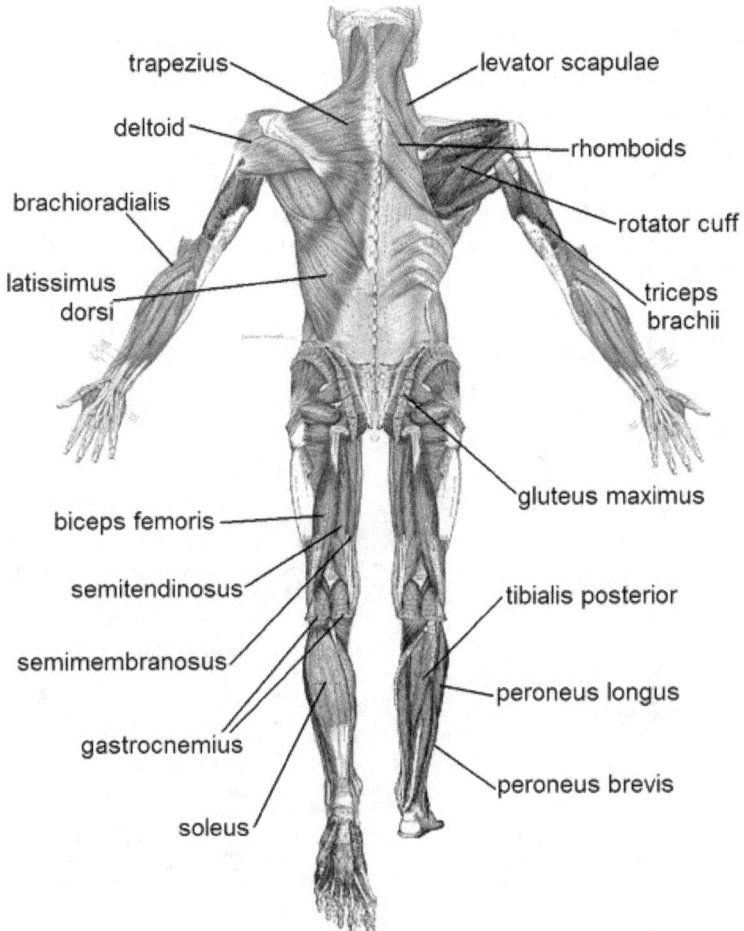

Figure 17.17. Posterior view of the human muscular system.

Cardiac muscle tissue is found only in the heart.

Cardiac muscle tissue is responsible for contraction and relaxation of the heart. Cardiac muscle tissue has striations, but unlike voluntary muscle cells, cardiac muscle cells (cardiomyocytes) are involuntary. Cardiac muscle tissue is a network of connected cardiomyocytes. *Intercalated discs* connect cardiomyocytes to each other. Since all of the cells are functionally connected to each other, it is *syncytium*. The syncytium allows the heart to beat as a unit. The intercalated discs allow for a synchronized of contraction and relaxation of the heart. The muscle tissue of the heart is referred to as *myocardium*. The chambers of the heart are lined with the endocardium. The outer layer of the myocardium is surrounded by the *pericardium*.

8. The Reproductive System

The reproductive systems of humans possess different organs. The purpose of the male reproductive organs is to provide *sperm* for the fertilization of an *ovum* (egg) produced by the female of the species.

The male reproductive system

Some organs of the male reproductive system are located *exterior* to the body; they are the *penis*, *testes* and *scrotum*. Male reproductive structures found *in* the pelvic region are the *epididymis*, *vas deferens*, *urethra*, *seminal vesicles*, *bulbourethral (Cowper's) glands* and the *prostate*. See figure 17.18.

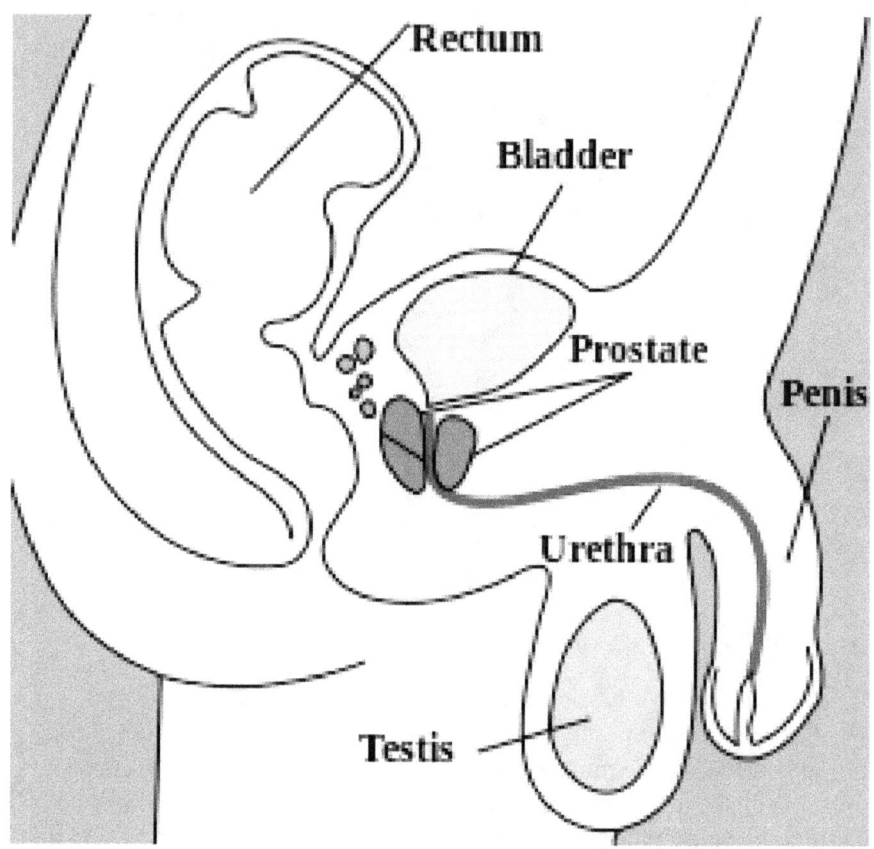

Figure 17.18. Diagram of the male reproductive system.

The female reproductive system
   The *external* organs of the *female reproductive system* are the *mons pubis*, *labia majora*, *labia minora*, *Bartholin's glands* and the *clitoris*. Female reproductive structures located *within* the pelvic region are the *vagina*, *cervix*, *uterus*, *Fallopian tubes* and *ovaries*. See figure 17.19. The ureter is shorter in the female, compared to the male.

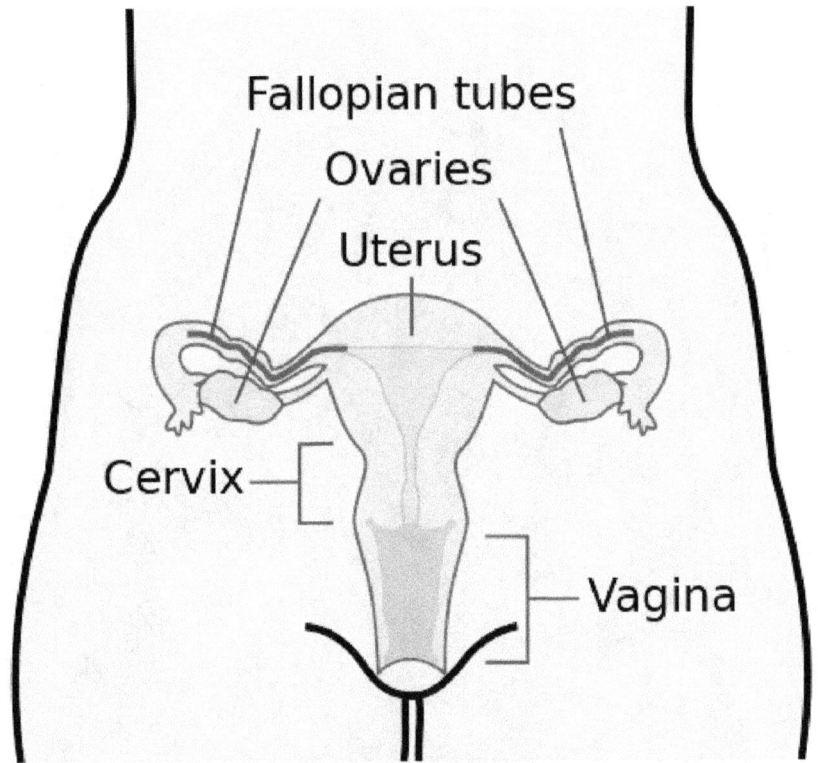

Figure 17.19. Diagram of the female reproductive system.

9. *Urinary System* (renal system)

The urinary system consists of two **kidneys**, two **ureters**, the **urinary bladder** and the **urethra**. See figure 17.20.

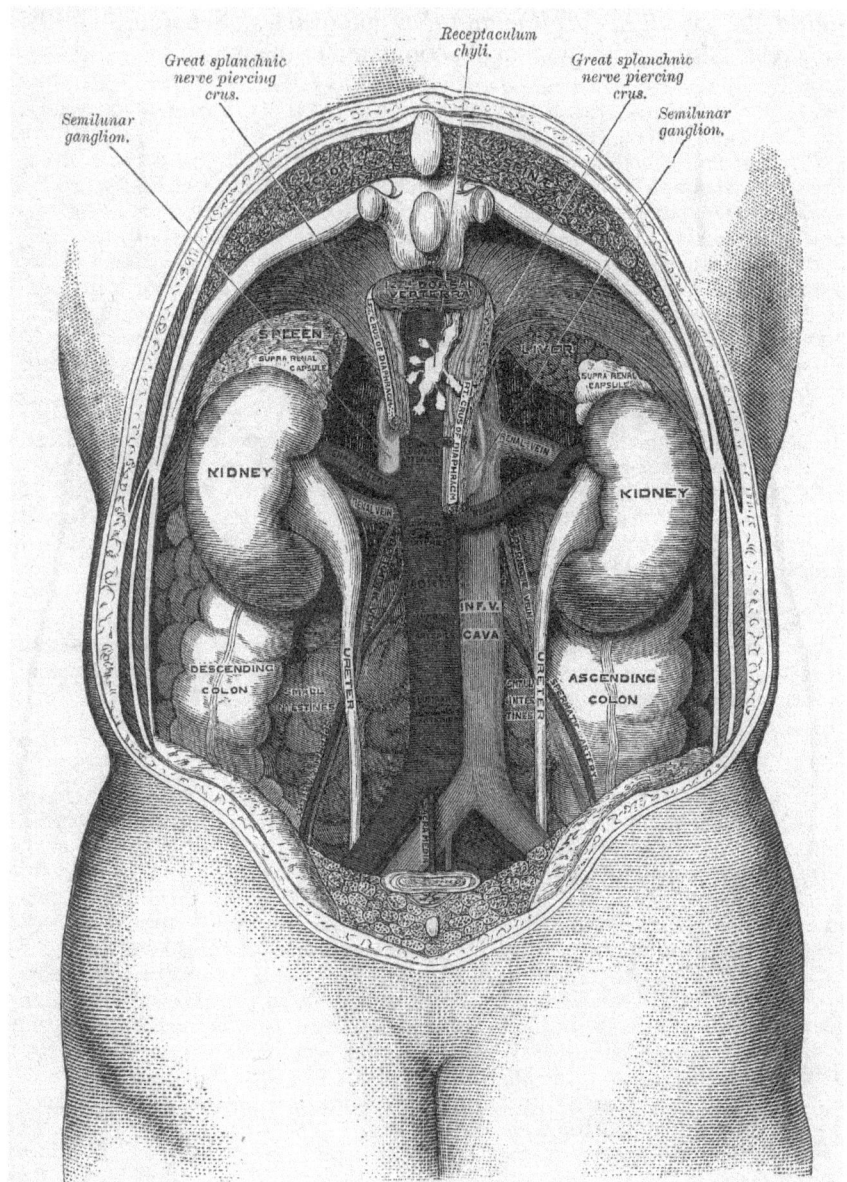

Figure 17.20. The human kidneys. Posterior view.

The *nephron* (see figure 17.21) is the unit of structure and function of the kidney. A healthy kidney contains about 800,000 to 1.5 million nephrons. The nephron excretes materials that are in excess the body and reabsorbs materials that are needed by the body. The nephron *removes urea and uric acid* from the blood. Urea is a nitrogen containing waste product of metabolism. Uric acid is a product of the metabolism of nucleotide *purine* bases.

The nephron also *maintains the electrolyte balance* between sodium, potassium and calcium. It also *regulates acid-base balance*, *controls blood volume* and *maintain normal blood pressure*.

A nephron consists of several substructures. The *glomerulus* is a network of capillaries enclosed in a capsular structure called *Bowman's capsule*. Bowman's capsule receives materials that are filtered out of the blood. The capillaries that exit the glomerulus enter an **efferent**[541] *arteriole*.

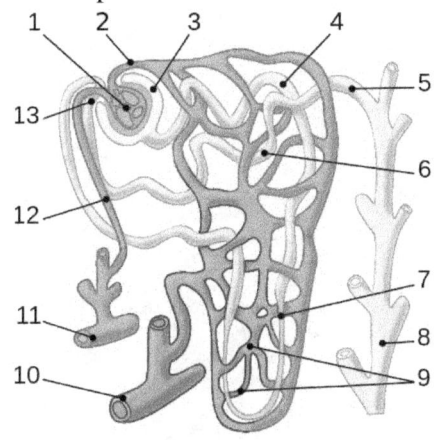

Nephron of the kidney. The labelled parts are 1. Glomerulus, 2. Efferent arteriole, 3. Bowman's capsule, 4. Proximal convoluted tubule, 5. Cortical collecting duct, 6. Distal convoluted tubule, 7. Loop of Henle, 8. Duct of Bellini, 9. Peritubular capillaries, 10. Arcuate vein, 11. Arcuate artery, 12. Afferent arteriole, 13. Juxtaglomerular apparatus.

Figure 17.21. Diagrammatic representation of a nephron.

The filtrate from Bowman's capsule enters a structure called the *proximal convoluted tubule*. The pH of the filtrate is regulated here, and sodium is reabsorbed. The *cortical collecting duct* connects nephrons to the ureter. The *distal convoluted tubule* helps regulate potassium, sodium, calcium and pH. A solution of NaCl and urine passes from the proximal convoluted tubule to the U-shaped *loop of Henle*. The loop of Henle recovers water and sodium chloride from urine. This stage concentrates the urine and allows for a lessened intake of water for the organism.

Urine is primarily a mixture of water and urea. Urine is collected and stored in the urinary bladder and passes to the urethra. The urethra passes through the penis of males and out of the body through the *urinary meatus*. The urinary meatus is the opening in the male penis through which urine and sperm exit the body. In the female, urine also leaves through the *urinary meatus* (external urethral orifice).

Urine gets its yellow color from a pigment called *urochrome*.

541   To carry away.

10. Endocrine System

The *endocrine system* refers to the glands that secrete *hormones* directly into the bloodstream. Hormones are chemical messengers that are carried by the circulatory system to a target organ. Hormones do not travel as fast as nerve impulses, but a hormone's effect is longer lasting than a nervous impulse. Glands of the endocrine system are classified as *endocrine* or *ductless glands*. The major endocrine glands located in the head and neck are the *pituitary gland*, *hypothalamus*, *pineal gland*, *thyroid gland* and the *parathyroid gland*. See figure 17.22.

Glands that secrete substances into a duct are called *exocrine* glands. Salivary glands, sweat glands and glands associated with the gastrointestinal tract are exocrine glands.

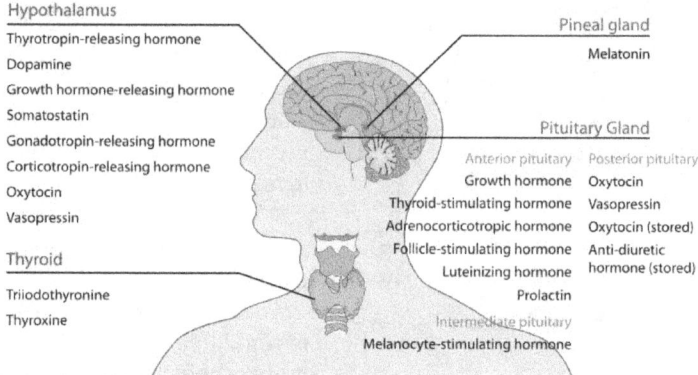

Figure 17.22. Endocrine glands of the head and neck.

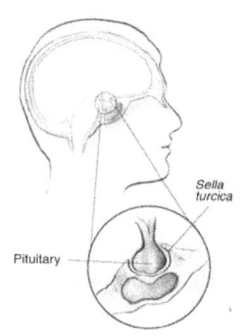

Figure 17.23. Diagrammatic representation showing the location of the pituitary gland in the *sella turcica*.

The *pituitary gland* is often called the "master gland" of the body. The gland is about the size of a pea that sits in a structure called the *sella turcica*. The *sella turcica* (see figure 17.23) located in the butterfly shaped *sphenoid bone* in the middle of the skull. The pituitary extends from the hypothalamus by a small stalk called the pituitary stalk. The pituitary gland consists of three areas: *anterior*, *intermediate* and *posterior lobes*.

The secretions of the *anterior lobe* are controlled by the hypothalamus gland. The anterior lobe secretes *human growth hormone* (HGH, HG, somatotropin), *thyroid-stimulating hormone* (TSH), *adrenocorticotropic hormone* (ACTH), *beta endorphins*, *lactotrophins*, *gonadotropins*.

256

The intermediate lobe of the pituitary gland makes ***melanocyte-stim-ulating hormone*** (MSH). MSH is responsible for the production and release of melanin within melanocytes in the skin and hair. MSH also stimulates appetite and sexual arousal.

The posterior lobe does not synthesize; rather it stores the ***antidi-uretic hormone*** (ADH), also known as *vasopressin* and *oxytocin*.

The ***hypothalamus connects the endocrine system to the nervous system*** by way of the pituitary gland. ***Body temperature, hunger, par-enting and attachment behaviors, thirst, fatigue and sleep*** are controlled by the hypothalamus.

The ***pineal gland*** is a located between the two hemispheres near the center of the brain. The pineal gland produces ***melatonin***, a hormone that regulates sleep patterns. Melatonin is synthesized from the amino acid *tryptophan*.

The ***thyroid gland*** is located in the neck inferior to the thyroid carti-lage or "Adam's apple." The butterfly shaped thyroid gland consists of a right and left "wing" connected by an **isthmus**.[542] The ***thyroid gland regulates*** heart rate, blood pressure, body temperature, and metabolism. Metabolic rate is controlled by the thyroid hormones ***triiodothyronine* (T3)** *and* **thyroxine (tetraiodothyronine -T4)**.

The ***parathyroid gland*** is four glands located on the dorsal side of the thyroid gland. These glands cannot be felt during a physical ex-amination. The parathyroid produces ***parathyroid hormone*** (PTH) and ***calcitonin***. These hormones act in opposition to each other. ***Parathyroid hormone increases*** blood calcium levels, and ***calcitonin decreases*** calci-um blood levels. PTH stimulates the release of calcium and phosphorus from bone tissue. The kidneys convert precursor molecules to vitamin D. Vitamin D increases the absorption of calcium in the small intestine.

Endocrine Functions of Other Structures

The ***liver, stomach***, parts of the ***small intestine, pancreas, kidneys, adrenal glands*** and the ***ovaries*** in females and ***testes*** in males also have endocrine functions. See figure 17.24.

The liver synthesizes the hormone ***angiotensin***. Angiotensin causes blood vessels to constrict, thereby elevating blood pressure. When the kidneys sense low blood pressure, they release the enzyme renin. Renin activates ***angiotensinogen***. Angiotensinogen raises blood pressure. The production of angiotensinogen ***by the liver is important during fetal development*** but in the adult, angiotensinogen production ***by the kidneys is more important***. The liver and the kidneys synthesize ***erythropoietin***. Erythropoietin is a hormone that stimulates the production of red blood cells in the bone marrow.

---

542   A narrow strip of land or some other material. In this case a narrow strip if thyroid tissue connecting both wings of the thyroid gland.

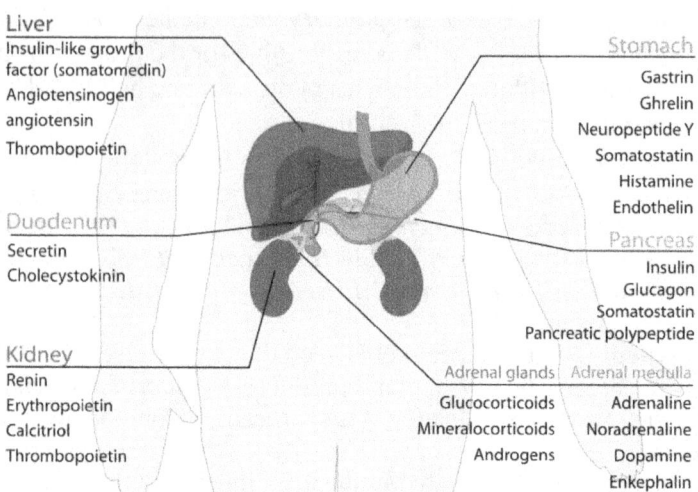

Liver
Insulin-like growth
factor (somatomedin)
Angiotensinogen
angiotensin
Thrombopoietin

Duodenum
Secretin
Cholecystokinin

Kidney
Renin
Erythropoietin
Calcitriol
Thrombopoietin

Stomach
Gastrin
Ghrelin
Neuropeptide Y
Somatostatin
Histamine
Endothelin

Pancreas
Insulin
Glucagon
Somatostatin
Pancreatic polypeptide

Adrenal glands
Glucocorticoids
Mineralocorticoids
Androgens

Adrenal medulla
Adrenaline
Noradrenaline
Dopamine
Enkephalin

Figure 17.24. Endocrine glands in the abdominal cavity.

The Stomach

The stomach, besides having a role in digestion, also has endocrine function. It secretes the hormones *ghrelin*, *somatostatin* and *endothelin*. Ghrelin regulates *hunger*. If the stomach is empty, ghrelin is secreted, but when the stomach is full or stretched, secretion of ghrelin stops. Ghrelin acts on the hypothalamus.

The stomach, intestine and delta cells of the pancreas produce somatostatin. Somatostatin *inhibits insulin* production and *glucagon* secretion. Endothelin is the body's most potent vasoconstrictor and plays a role in the regulation of blood flow and blood pressure.

The Duodenum

The duodenum is the first part of the small intestine. The structure is about 12 inches long (from the Latin, *duodecim* meaning twelve). Although the duodenum plays a major part in the digestion of food substances, it also has endocrine functions. It secretes *secretin* and *cholecystokinin*. Both hormones help regulate emptying of the stomach and activity of the pancreas.

The Pancreas

The *beta cells* of the pancreas produce the hormone *insulin*. Insulin regulates carbohydrate metabolism by facilitating the absorption of glucose into cells. If *insulin levels fall*, glucose will not be able to enter cells to be metabolized for energy. As a result, glucose levels in the blood rise. Consistently high levels of glucose in the blood may indicate diabetes. *Glucagon*, produced by the alpha cell of the pancreas, lowers glucose levels in the blood. If glucose levels in the blood fall, glucagon will stimulate the liver to convert stored glycogen into glucose.

The Kidney

The kidney secretes many hormones. Three of these are *erythropoietin*, *calcitriol* and *thrombopoietin*.

Erythropoietin is secreted in response to low oxygen levels in the tissues. Increased erythropoietin levels stimulate red blood cell production (*erythropoiesis*).

Calcitriol (1,25-dihydroxycholecalciferol or 1,25-dihydroxyvitamin D_3) allows for increased absorption of dietary calcium (Ca^{2+}) from the gastrointestinal tract and increased reabsorption of Ca^{2+} in the renal tubules. Calcitriol also increases the release of Ca^{2+} from bone.

Thrombopoietin is a hormone that regulates the production of blood platelets.

The Adrenal Glands (suprarenal glands)

The adrenal glands are located superior to each kidney. The adrenal glands release hormones that the body needs when placed in a *stress* situation. The adrenal gland produces *cortisol*, *aldosterone*, *androgens*, *adrenaline* and *norepinephrine*, also called noradrenaline

Cortisol is released by the adrenal glands in response to stress and a low blood sugar level. The hormone stimulates the production of glucose (gluconeogenesis) in the liver. The release of any substances that can cause inflammation is suppressed by cortisol. Aldosterone is an important regulator of blood pressure. A mineralocorticoid hormone increases the reabsorption of water and ions by the kidney; sodium is conserved, and potassium is secreted causing an increase in blood pressure.

Ovaries

The ovaries are located on each side of the uterus. See figure 17.25. Each ovary is attached to the uterus by the ovarian ligament.

Estrogen, testosterone and progesterone are three hormones secreted by the ovaries. *Estrogen* produces secondary sex characteristics at puberty. In females, breasts enlarge, voice remains high pitched and the pelvis widens. In males, enlargement of the testes and penis occur. The voice deepens, muscle and bone increase and facial hair appears.

Testosterone, an anabolic steroid, although produced by males and females, is the main male sex hormone. Testosterone is responsible for the development of male secondary sexual characteristics at puberty in males. The testes and penis enlarge, the voice deepens, muscle and bone mass increase and facial hair appears.

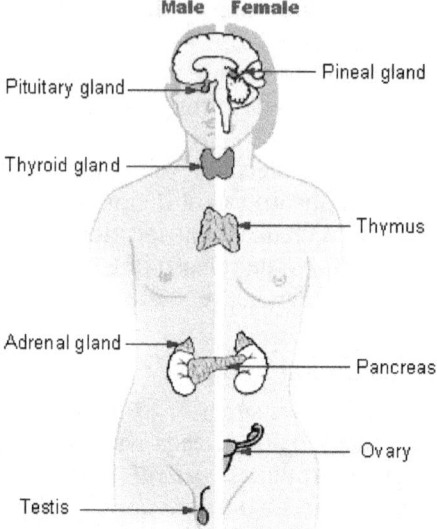

Major Endocrine Glands

Male   Female

Pituitary gland —

Pineal gland

Thyroid gland —

Thymus

Adrenal gland —

Pancreas

Ovary

Testis —

Figure 17.25. Location of the pancreas, kidneys, adrenal glands, ovaries and testis. Male structures are on the left and female structures are on the right.

Progesterone (pregn-4-ene-3,20-dione; abbreviated as **P4**) plays an important role in the female menstrual cycle, pregnancy and embryogenesis. See figure 16.26.

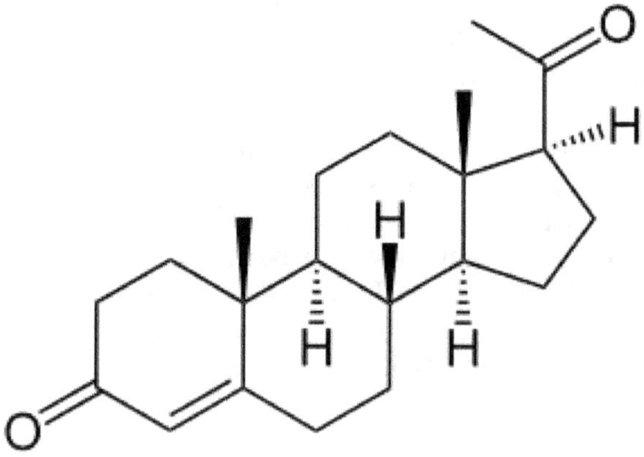

Figure 17.26. The structural formula for progesterone.

Table 17.2. Major hormones of the human body.

Gland and Secretion	Target Tissue	Response
Pituitary, anterior lobe		
Human growth hormone (HGH/somatotropin)	Ends of long bones and most tissues	Increases height in children and adolescents
Thyroid stimulating hormone (TSH)	Thyroid gland	Growth of thyroid, stimulates metabolism, produces thyroxine (T4), and then triiodothyronine (T3)
Adrenocorticotropic hormone (ACTH)	Adrenal glands (adrenal cortex)	Stimulates the production of glucocorticoids, mineralocorticoids, and androgens (male sex hormones)
Beta endorphins	Central and peripheral nervous system	Acts as a pain killer in the after traumatic injury
Prolactin	Ovaries and testes	Milk production in humans
Lactotrophins	Corpus luteum	Stimulates and maintains production of milk (lactation)
Gonadotropins: follicle-stimulating hormone (FSH), luteinizing hormone (LH)	Ovaries and testes	Stimulates testosterone production. FSH regulates woman's cycle and stimulates ovaries to produce eggs, LH triggers ovulation and development, growth, pubertal maturation. FSH and LH act synergistically.

Table 17.2. Major hormones of the human body.		
Pituitary, intermediate lobe		
Melanocyte-stimulating hormone (MSH)	Melanocytes of hair and skin	Responsible for hair and skin color
Pituitary, posterior lobe		
Stores the antidiuretic hormone vasopressin	Distal tubule and collecting duct of nephron	Reabsorption of water constricts blood vessels
Oxytocin	Uterus	Increases contractions at time of birth
Pineal gland		
Melatonin	Brain	Regulates circadian rhythm (day night cycle)
Hypothalamus		
Neurohormones (releasing hormones)	Anterior pituitary	Stimulates the pituitary to release or inhibit release of pituitary hormones
Thyroid gland		
Triiodothyronine (T3)	Almost all tissues of the body	Growth and development, metabolism, body temperature, and heart rate
Thyroxine (tetraiodothyronine -T4)	Most tissues of the body	Prohormone of (T3)
Parathyroid gland		
Parathyroid hormone	Bones, kidneys and small intestine	Decreases blood calcium levels

11. The Lymphatic System

The lymphatic system, part of the circulatory system, consists of lymphatic vessels, lymphatic organs (spleen, lymph nodes and tonsils) and a clear fluid called lymph. See figure 17.27. The lymphatic system returns lost plasma in the form of lymph back to the heart. The lymphatic system is an open system without a pump. The circulatory system is a closed system of vessels with the heart as a pump.

When blood reaches capillaries of the circulatory system, plasma moves out of the capillaries at the arterial end and into the spaces between the cells. The space between cells is the interstitial space and the fluid that fills this space is interstitial fluid. Much of the plasma that enters the interstitial space is reabsorbed back into the capillaries at the venous end. The interstitial fluid that is not reabsorbed, is picked up by lymphatic capillaries and returned to the blood.

The lymphatic system maintains blood volume by transporting lymph back to the heart. The lymph replaces the plasma lost to interstitial fluid. Like blood plasma, lymph contains white blood cells or leucocytes that have an immune function.

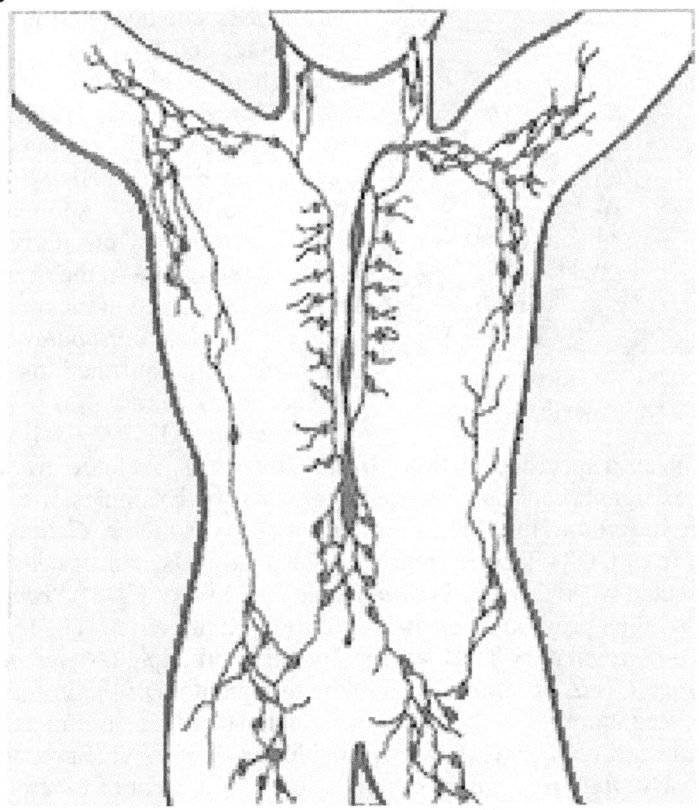

Figure 17.27. Generalized diagram of the lymphatic system.

12. The Immune System

The immune system protects the human body against infectious disease caused by many different kinds of *pathogenic organisms*. A healthy immune system has to be able to detect pathogens and distinguish between healthy body cells and foreign invaders. The immune system consists of the *innate immune system* (nonspecific system) and the *adaptive immune system* (cell-mediated and humoral-mediated immune system).

The innate immune system is made up of intact *skin* and *white blood cells* (leukocytes). The skin is a mechanical barrier to pathogens. If unbroken, it denies the passage of microbes into the body. White blood cells pass freely through the tissues and phagocytize cell debris or foreign microbes such as bacteria. Natural killer cells, mast cells, eosinophils, basophils and phagocytic cells (macrophages, neutrophils and dendritic cells) are all part of the nonspecific innate cellular response to foreign invaders.

The *adaptive immune system* (acquired immunity) consists of *B lymphocytes* and *T lymphocytes*. See figure 17.28. The designation of B lymphocytes and T lymphocytes reflect where each cell matures. *Both kinds of lymphocytes originate in the bone marrow*. B cells are bursa or bone marrow derived cells. B cells remain in bone marrow and mature there. T cells are thymus derived cells because they mature in the *thymus*. Some T cells mature in the tonsils.

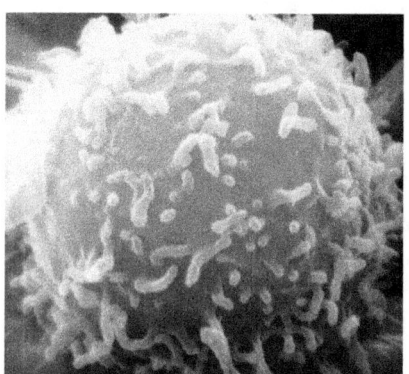

Figure 17.28. An electron micrograph of a lymphocyte.

B cells secrete *antibodies*. T cells have many different functions.

There are several kinds of T cells. See figure 17.29. Each subset of T cells has a specific function. *Helper T-cells* (T_H) release *cytokines*. Cytokines signal other and change their behavior. Examples of cytokines are interferon, interlukins and tumor necrosis factor. *Cytotoxic T cells* (Tc or CD8+ T cells) get rid of damaged cells, cancer cells and cells infected with viruses. *Memory T cells* (CD4+ or CD8+) *"remember"* an antigen previously encountered through infection. The memory exhibited by regulatory T cells is the operative principle of *vaccination*. *Regulatory T cells* (suppressor T cells) are regulators of the immune system. Regulatory T cells help protect against antigens produced by the organism itself (self-antigens). *Natural killer T cells* (NKT) recognize lipids of *Mycobacteria tuberculosis*, the causative agent of tuberculosis. When natural killer T cells go awry, autoimmune disease such as Lupus erythematosus may result.

Antigen-presenting cells (APCs), also called **Langerhans**[543] or dendritic cells, detect antigens and bring the antigen to helper T cells. Antigen presenting cells are present in all layers of the epidermis. Special surface receptors on helper T cells recognize antigens and activates helper T cell causing helper T cells to release *cytokines* and other *stimulatory* signals (green arrows). *Cytokines* and other *stimulatory* signals stimulate the activity of macrophages, killer T cells and B cells. B cells produce *antibodies*. See figure 17.29.

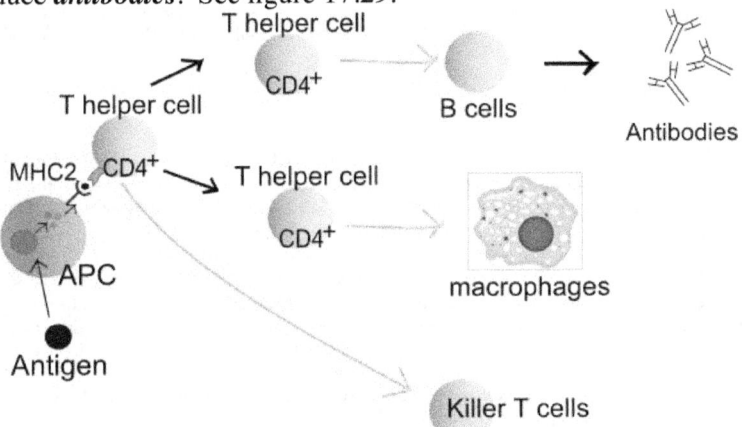

Figure 17.29. The cascade of immunological events that lead to antibody and macrophage production.

An *antibody* (Ab) is a protein that is responsible for destroying foreign invaders such as bacteria and viruses. Antibodies target specific parts of an invader. Antibodies can mark the invader for destruction by the immune system or the antibody can enhance phagocytosis, activate **complement**[544] and neutralize toxins. The antibody has an antigen binding site that may be part of a bacterial cell that is necessary for the bacterial cell to survive.

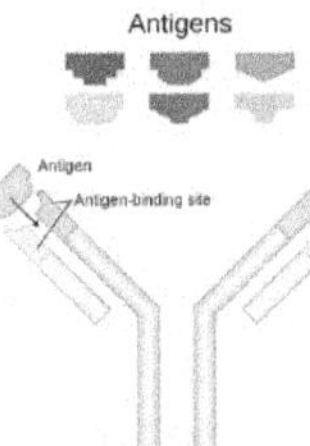

Antibodies are "Y" shaped glycoproteins made up of two chains, a light chain and a heavy chain. See figure 17.30. Gerald Edelman (1929 -2014) of Rockefeller Institute for Medical Research, now the Rockefeller University, discovered the structure of an antibody.

Figure 17.30. Antibody structure.

---

543   German pathologist that discovered antigen presenting cells as a medical student.

544   Complement is a series of molecules that magnify the immune response by enabling the release of cytokines such as interferon.

Credits for Figures

Figure 1.2. The work of art depicted in this image and the reproduction thereof are in the public domain worldwide. The reproduction is part of a collection of reproductions compiled by *The Yorck Project*. The compilation copyright is held by *Zenodot Verlagsgesellshaft mbH* and licensed under the *GNU Free Documentation License*.

Figure 2.1 a and b. Photos by M. Anzelone

Figure 3.4. September 2006. NASA Astrobiology Institute. Public Domain.

Figure 3.5. A small subunit of ribosomal RNA. Modified image taken from Rfam database. Figure is in the public domain.

Figure 3.6. Photograph of a tobacco leaf infected with TMV. *Courtesy CDC. This image is in the public domain.*

Figure 3.8. Electron micrograph of the Ebola virus. Courtesy CDC. This image is in the public domain. This image is a work of the Centers for Disease Control and Prevention, part of the United States Department of Health and Human Services, taken or made as part of an employee's official duties. As a work of the U.S. federal government, the image is in the public domain.

Figure 4.1. Negative staining of *B. burgdorferi*, Courtesy CDC. This image is in the public domain. This image is in the public domain and thus free of any copyright restrictions. As a matter of courtesy we request that the content provider be credited and notified in any public or private usage of this image.

Figure 4.2. The deer tick, *Ixodes scapularis*. Courtesy of the Agricultural Research Service of the United States Department of Agriculture. The image is in the public domain.

Figure 4.3. Electron micrograph of *Treponema palllidum pallidum. Courtesy CDC.* This image is a work of the Centers for Disease Control and Prevention, part of the United States Department of Health and Human Services, taken or made as part of an employee's official duties. As a work of the U.S. federal government, the image is in the public domain.

Figure 4.5. *Amoeba histolytica.* Wet mount. Courtesy CDC. This image is a work of the Centers for Disease Control and Prevention, part of the United States Department of Health and Human Services, taken or made as part of an employee's official duties. As a work of the U.S. federal government, the image is in the public domain.

Figure 4.6. *Giardia lamblia.* Courtesy CDC. This image is in the public domain. This image is a work of the Centers for Disease Control and Prevention, part of the United States Department of Health and Human Services, taken or made as part of an employee's official duties. As a work of the U.S. federal government, the image is in the public domain.

Figure 4.8. Life cycle of *Toxoplasma Gondii. Courtesy CDC.* This image is in the public domain and thus free of any copyright restrictions.

Figure 4.10. Generalized diagram of a mature flower.
Released into the public domain by its author, LadyofHats.

Figure 4.11. Generalized diagram of a dicot seed. Released into the public domain by its author, LadyofHats.

Figure 4.12. Leaves of the poison ivy plant. Courtesy of Robert H. Mohlenbrock, USDA, SCS. 1991. *Southern wetland flora: A guide to plant species. South National Technical Center, Fort Worth, TX. USDA NRCS Wetland Science Institute.*

Figure 4.14. Left: Egg of *T. trichiura* in an iodine-stained wet mount. Right: Egg of *T. trichiura* in an unstained wet mount. Center: Micrograph of an adult female Trichuris human whipworm that is approximately 4 cm long. *CDC's Division of Parasitic Diseases and Malaria (DPDM).* This image

is in the public domain.
Figure 4.15. Adult *Necator americanus*. Courtesy CDC. This image is in the public domain.

Figure 4.16. The parasitic roundworm *A. lumbricoides*. Courtesy CDC. Laboratory Identification of Parasitic Diseases of Public Health Concern.

Figure 4.17. *T. spiralis*. Courtesy CDC. This work is in the public domain in the United States because it is a work prepared by an officer or employee of the United States Government as part of that person's official duties.

Figure 4.18. *Taenia saginata,* the beef tapeworm.

Figure 4.19. A bed bug. CDC/ Harvard University, Dr. Gary Alpert; Dr. Harold Harlan; Richard Pollack. Photo Credit: Piotr Naskrecki. This image is in the public domain.

Figure 4.20. *Musca domestica*. Image is in the public domain.

Figure 4.21. *Aedes aegypti*. This image (or other media file) is in the public domain because its copyright has expired.

Figure 4.22. *Anopheles gambiae*. This image is a work of the Centers for Disease Control and Prevention, part of the United States Department of Health and Human Services, taken or made as part of an employee's official duties. As a work of the U.S. federal government, the image is in the public domain.

Figure 4.23. *Metanephrops japonicus*. Released into the public domain by its author, Dieno.

Figure 8.10. A typical compound light microscope. Photo by M. Anzelone.

Figure 8.15. Refraction of light waves reflected from a pencil in a glass of water. Photo by M. Anzelone.

Figure 8.17. An equal arm balance depicted on an Egyptian papyrus c. 2000 BC. It shows a heart being weighed against the feather of truth. This image (or other media file) is in the public domain because its copyright has expired. This applies to Australia, the European Union and those countries with a copyright term of life of the author plus 70 years.

Figure 8.18. A single pan balance. Photo by M. Anzelone.

Figure 9.1. A mixed culture of many different kinds of bacteria on a nutrient agar plate. Photo M. Anzelone

Figure 9.2. A streak plate showing a pure culture. Photo M. Anzelone.

Figure 9.10. Analogous structures. Comparison of bones of: 1. a pterosaur, 2. bat and 3. bird. This image is in the public domain because its copyright has expired.
Figure 9.13. A polymerase chain reaction or thermal cycler machine. Courtesy CDC.

Figure 9.14. Gel electrophoresis produces a DNA fingerprint. Courtesy CDC.

Figure 9.15. Normal x-ray of the human chest. This work is in the public domain in the United States because it is a work prepared by an officer or employee of the United States Government as part of that person's official duties. Courtesy CDC.

Figure 9.16. Computed tomography (CAT) scan of the human brain. The person who associated a work with this deed has dedicated the work to the public domain by waiving all of his or her rights to the work worldwide under copyright law, including all related and neighboring rights, to the extent allowed by law. You can copy, modify, distribute and perform the work, even for commercial purposes, all without asking permission.

Figure 9.18. A sonogram of a human bladder. Anterior view at left, lateral view at right. Image M. Anzelone.

Figure 9.19. PQRST tracing. Courtesy of Paul Maccaro, M.D.

Figure 9.20. Placement of electrodes on a patients chest. The person who associated a work with this deed has dedicated the work to the public domain by waiving all of his or her rights to the work worldwide under copyright law, including all related and neighboring rights, to the extent allowed by law.

Figure 12.4. Four major stages of mitosis. Faithfully reproduced from Gray's Anatomy, 17th edition, copyright has expired.

Figure 12.5. Stages of meiosis. This image is from the Science Primer a work of the National Center for Biotechnology Information, part of the National Institutes of Health. As a work of the U.S. federal government, the image is in the public domain.

Figure 12.6. Fertilization of an egg by a sperm cell. This image is in the public domain. A work of the Federal government.

Figure 13.8. The DNA double helix and base pairing. Work of the U.S. Government. In the public domain.

Figure 14.2. Lichen growing on the north side of a rock. Photo: M. Anzelone.

Figure 14.8. Adapted from National Oceanographic and Atmospheric Administration (NOAA) Courtesy National Oceanographic and Atmospheric Administration.

Figure 14.9. United States temperature ranges from 1895 through 2009. Courtesy National Oceanographic and Atmospheric Administration.

Figure 15.1 and figure 15.2. The anatomical position and surface gross anatomy. From the (medical) gallery of Mikael Häggström. The person who associated a work with this deed has dedicated the work to the public domain by waiving all of his or her rights to the work worldwide under copyright law, including all related and neighboring rights, to the extent allowed by law. You can copy, modify, distribute and perform the work, even for commercial purposes, all without asking permission.

Figure 15.3. Quadrants of the body. This work has been released into the public domain by its author, LadyofHats. This applies worldwide. LadyofHats grants anyone the right to use this work for any purpose, without any conditions, unless such conditions are required by law.

Figure 15.4. Body cavities. This work is in the public domain in the United States because it is a work prepared by an officer or employee of the United States Government as part of that person's official duties.

Figure 16.1. A generalized epithelial cell. Modified from A. N. Lin and D. Martin Carter. *Epidermolysis Bullosa*. 1992.

Figure 17.1. A cross-section of the human brain. A cross-section of the human brain. This media file is in the public domain in the United States. This applies to U.S. works where the copyright has expired, often because its first publication occurred prior to January 1, 1923.

Figure 17.3. The human nervous system. Permission is granted to copy, distribute and/or modify this document under the terms of the GNU Free Documentation License, Version 1.2 or any later version published by the Free Software Foundation; with no Invariant Sections, no Front-Cover Texts, and no Back-Cover Texts.

Figure 17.4. The axial skeleton. This work has been released into the public domain by its author, LadyofHats. This applies worldwide.

Figure 17.5. Major bones of the human skeleton. Grayed area denotes the axial skeleton. This work has been released into the public domain by its author, LadyofHats. This applies worldwide.

Figure 17.6. Curves and regions of the human spine. This work is in the public domain in the United States because it is a work prepared by an officer or employee of the United States Government as part of that person's official Duties.

Figure 17.7. The pulmonary circulation. This work is in the public domain in the United States because it is a work prepared by an officer or employee of the United States Government as part of that person's official duties.

Figure 17.8. Circulation of blood through the human heart. Figure Permission is granted to copy, distribute and/or modify this document under the terms of the GNU Free Documentation License, Version 1.2 or any later version published by the Free Software Foundation

Figure 19.9. This work has been released into the public domain by its author, LadyofHats. This applies worldwide. In some countries this may not be legally possible; if so: LadyofHats grants anyone the right to use this work for any purpose, without any conditions, unless such conditions are required by law.

Figure 17.10. The human respiratory system. This work has been released into the public domain by its author, LadyofHats. This applies worldwide. In some countries this may not be legally possible; if so: LadyofHats grants anyone the right to use this work for any purpose, without any conditions, unless such conditions are required by law.

Figure 17.11. The human digestive system. This work has been released into the public domain by its author, LadyofHats. This applies worldwide. In some countries this may not be legally possible; if so: LadyofHats grants anyone the right to use this work for any purpose, without any conditions, unless such conditions are required by law.

Figure 17.12. A cross section of human skin. This work has been released into the public domain by its author, LadyofHats. This applies worldwide. In some countries this may not be legally possible; if so: LadyofHats grants anyone the right to use this work for any purpose, without any conditions, unless such conditions are required by law.

Figure 17.13. The scapula. Lateral view. This image is in the public domain because its copyright has expired. it is from the 20th U.S. edition of Gray's Anatomy of the Human Body, originally published in 1918 and therefore lapsed into the public domain.

Figure 17.14. Point of insertion of biceps brachii at the radial tuberosity of the radius. This image is in the public domain because its copyright has expired. it is from the 20th U.S. edition of Gray's Anatomy of the Human Body, originally published in 1918 and therefore lapsed into the public domain.

Figure 17.15 and 17.16. Anterior view of the human muscular system. This faithful reproduction of a lithograph plate from Gray's Anatomy, it is from the 20th U.S. edition of Gray's Anatomy of the Human Body, originally published in 1918 and therefore lapsed into the public domain.

Figure 17.17. Diagram of the male reproductive system. I, the copyright holder of this work, release this work into the public domain. This applies worldwide. In some countries this may not be legally possible; if so: I grant anyone the right to use this work for any purpose, without any conditions, unless such conditions are required by law. This work is in the public domain in the United States because it is a work prepared by an officer or employee of the United States Government as part of that person's official duties.

Figure 17.18. Diagram of the female reproductive system. This image is a work of the Centers for Disease Control and Prevention, part of the United States Department of Health and Human Services, taken or made as part of an employee's official duties. As a work of the U.S. federal government, the image is in the public domain.

Figure 17.19. The human kidneys. Posterior view. This faithful reproduction of a lithograph plate from Gray's Anatomy, it is from the 20th U.S. edition of Gray's Anatomy of the Human Body, originally published in 1918 and therefore lapsed into the public domain.

Figure 17.20. Diagrammatic representation of a nephron. The person who associated a work with this deed has dedicated the work to the public domain by waiving all of his or her rights to the work worldwide under copyright law, including all related and neighboring rights, to the extent allowed by law. You can copy, modify, distribute and perform the work, even for commercial purposes, all without asking permission.

Figure 17.21. Endocrine glands of the head and neck. This work has been released into the public domain by its author, LadyofHats. This applies worldwide. In some countries this may not be legally possible; if so: LadyofHats grants anyone the right to use this work for any purpose, without any conditions, unless such conditions are required by law.

Figure 17.22. Diagrammatic representation showing the location of the pituitary gland in the sella turcica. This image is in the public domain because it contains materials that originally came from the National Institutes of Health

Figure 17.23. Endocrine glands in the abdominal cavity. This work has been released into the public domain by its author, LadyofHats. This applies worldwide. In some countries this may not be legally possible; if so: LadyofHats grants anyone the right to use this work for any purpose, without any conditions, unless such conditions are required by law.

Figure 17.24. Location of the pancreas, kidneys, adrenal glands, ovaries and testis. Male structures are on the left and female structures are on the right. This work is in the public domain in the United States because it is a work prepared by an officer or employee of the United States Government as part of that person's official duties

Figure 17.26. Generalized diagram of the lymphatic system. This work is in the public domain in the United States because it is a work prepared by an officer or employee of the United States Government as part of that person's official duties.

Figure 17.27. An electron micrograph of a lymphocyte. This image is in the public domain and may be used, linked, or reproduced without permission. If an image is used, credit should be given to the listed source and/or author. Author is unknown. Colorization of the original black and white image was done by Dr. Triche National Cancer Institute.

Figure 17.29. The cascade of immunological events that lead to antibody and macrophage production. I, Mikael Häggström, the copyright holder of this work, release this work into the public domain. This applies worldwide.

Figure 17.30. Antibody structure. I, Fvasconcellos, the copyright holder of this work, release this work into the public domain. This applies worldwide.

www.ingramcontent.com/pod-product-compliance
Lightning Source LLC
Chambersburg PA
CBHW070850180526
45168CB00005B/1766